T0335630

Rapid Detection of Food Adulterants and Contaminants

Rapid Detection of Food Adulterants and Contaminants
Theory and Practice

Shyam Narayan Jha

*Central Institute of Post-Harvest Engineering
and Technology, India*

ELSEVIER

AMSTERDAM • BOSTON • HEIDELBERG • LONDON
NEW YORK • OXFORD • PARIS • SAN DIEGO
SAN FRANCISCO • SINGAPORE • SYDNEY • TOKYO

Academic Press is an imprint of Elsevier

Academic Press is an imprint of Elsevier
125 London Wall, London EC2Y 5AS, UK
525 B Street, Suite 1800, San Diego, CA 92101-4495, USA
225 Wyman Street, Waltham, MA 02451, USA
The Boulevard, Langford Lane, Kidlington, Oxford OX5 1GB, UK

Copyright © 2016 Elsevier Inc. All rights reserved.

No part of this publication may be reproduced or transmitted in any form or by any means, electronic or
mechanical, including photocopying, recording, or any information storage and retrieval system, without
permission in writing from the publisher. Details on how to seek permission, further information about the
Publisher's permissions policies and our arrangements with organizations such as the Copyright Clearance
Center and the Copyright Licensing Agency, can be found at our website: www.elsevier.com/permissions.

This book and the individual contributions contained in it are protected under copyright by the Publisher
(other than as may be noted herein).

Notices
Knowledge and best practice in this field are constantly changing. As new research and experience broaden
our understanding, changes in research methods, professional practices, or medical treatment may become
necessary.

Practitioners and researchers must always rely on their own experience and knowledge in evaluating and using
any information, methods, compounds, or experiments described herein. In using such information or methods
they should be mindful of their own safety and the safety of others, including parties for whom they have a
professional responsibility.

To the fullest extent of the law, neither the Publisher nor the authors, contributors, or editors, assume any
liability for any injury and/or damage to persons or property as a matter of products liability, negligence or
otherwise, or from any use or operation of any methods, products, instructions, or ideas contained in the
material herein.

ISBN: 978-0-12-420084-5

British Library Cataloguing in Publication Data
A catalogue record for this book is available from the British Library

Library of Congress Cataloging-in-Publication Data
A catalog record for this book is available from the Library of Congress

For information on all Academic Press publications
visit our website at www.elsevier.com

  Working together
to grow libraries in
developing countries

www.elsevier.com • www.bookaid.org

Publisher: Nikky Levy
Acquisition Editor: Patricia Osborn
Editorial Project Manager: Karen R. Miller
Production Project Manager: Lisa Jones
Designer: Mark Rogers

Typeset by TNQ Books and Journals
www.tnq.co.in

Printed and bound in the United States of America

This book is **dedicated to my Grand Father (*baba*)** who inculcated high values and *sanskaras* in me. I always studied with his inspiration, keeping lantern near his head while sleeping in dark night of my rural village

Contents

Foreword

Consumers the world over are demanding quality and safe foods, be they processed or fresh. Factors such as increasing individual and family incomes, two-member working families, urbanization, and fast lifestyles have resulted in greater demands for processed foods such as ready-to-eat and ready-to-cook foods. Raw agricultural commodities need be handled, processed, transported, and kept safe until they reach consumers. Every single unit operation and every single player in the postharvest chain—and the system as a whole—have vital roles to play in maintaining the quality and safety of the foods we eat.

The unit operations selected for processing, the methods of processing, the equipment used for processing, the cleanliness of the plant and machinery, the selection of packaging materials, the modes of transport, the safety of stored raw and processed foods, and the cleanliness and hygienic conditions of individuals working in the system are all very important elements in delivering safe foods to consumers. Currently, only a handful of companies attempt to maintain strict regulations in maintaining the safety of foods, whereas numerous others pay little or no attention to safety issues.

Contamination of foods during processing and handling can and must be avoided. Properly adopting Hazard Analysis and Critical Control Points norms can help in reducing contamination levels. Adulteration, in some sense, can be considered to be "forced" contamination. Food regulatory authorities have vital roles to play in containing contamination and adulteration. Rapid and reliable methods of detecting contamination will help to identify unsafe foods and to implement their timely recall. Food companies, researchers, and regulatory authorities should have a thorough understanding of food adulteration and contamination, as well as methods of detecting and avoiding them. Comprehensive reading materials on these subjects are scarce and often unavailable.

I am glad Dr. S. N. Jha has come out with this book, titled *Rapid Detection of Food Adulterants and Contaminants: Theory and Practice*. This book comprehensively covers subjects encompassing food safety and quality, common adulterants and contaminants, standards and permissible limits, basic detection techniques, biosensors, spectroscopy and chemometrics, and imaging methods. These chapters discuss in detail the basics of each selected subject and the advances made through research and experience. The illustrations, supporting data, and the entire presentation of the book are meticulous.

This book is timely for the food industry and academia, and the food-processing sector. I appreciate Dr. Jha for his untiring efforts in bringing together such a wonderful book.

K. Alagusundaram
Deputy Director General (Engineering)
Indian Council of Agricultural Research
New Delhi, India
June 8, 2015

Preface

Nowadays food safety is of paramount importance for consumers, retailers, as well as regulators who are engaged in enacting food laws. Adulteration of high-value foods with low-value ones, color- and taste-enhancing ingredients, and mislabeling are rampant, particularly in processed food markets. Insecticide/pesticide residues, heavy metals, toxins beyond permissible limits, dangerous bacteria, and so on in fresh as well as processed foods are now found in various countries. These mostly are checked by collecting samples from markets and using traditional laboratory (wet chemistry) methods, which are time-consuming, laborious, and expensive. Test reports usually are made available several days or months after testing; by then many are already affected by eating those infected foods. Rapid testing methods that can detect possible adulterants, contaminants, and toxins in a very short time and preferably at the market site are urgently required and therefore are increasingly being tried the world over.

Numerous articles and research papers that deal with applications of various rapid and often nondestructive methods of food testing are being published every year. Food standards and permissible limits of various contaminants and toxins are, however, different in different countries. The accuracy of the method being used also varies, and it is difficult to have knowledge of all standards and permissible limits, though this knowledge changes time to time for everyone. This book was developed after receiving an overwhelming response to my earlier book, *Nondestructive Evaluation of Food Quality: Theory and Practice*, to fill the gap of food safety, including one chapter on each of the topic of food safety and quality, common adulterants and contaminants, standards and permissible limits, as well as other chapters on different techniques covering theory/basics in brief and their practical applications in terms of a few recent works reported in the literature.

The purpose of this book is to provide basic knowledge and information to postgraduate students and scientists interested in the upstream research on food safety aspects. It is particularly useful to beginners. It also may be useful to researchers directly engaged in developing fast, reliable, and nondestructive methods for the evaluation of food quality and safety aspects, as well as the food industries and regulators indirectly responsible to check food quality to minimize public hazards. The book may be used as a textbook; however, I am not responsible for any loss and/or damage while practicing or using any information and knowledge provided by this book.

Rapid Detection of Food Adulterants and Contaminants: Theory and Practice contains wide information including common adulterants and contaminants in various foods, standards and permissible limits of different food regulatory authorities, and related detection techniques; it would not have been possible to prepare a manuscript without the help of my colleagues, Dr. K. Narsaiah, Dr. Pranita Jaiswal, Dr. Yogesh Kumar, Dr. Rahul Kumar Anurag, as well as various research associates in my laboratory. I sincerely acknowledge the contribution of each of them.

I am indebted for childhood contributions and teaching of my mother, father, uncle (*Lal kaka*), and late elder brother in a rural village, which made me reach this status of writing a book of international standing. The sacrifices of my wife, Bandana, and daughters, Priya and Preeti, especially during my extramural projects for the National Agricultural Innovation Project and National Agricultural Science Fund, which gave me in-depth understanding of this subject, are greatly appreciated. I thank in advance the readers who send their suggestions, if any, for further improvements to future versions of this book.

S. N. Jha
Ludhiana
June 28, 2015

FOOD SAFETY AND QUALITY

Food is defined as a substance—whether processed, semiprocessed, or raw—that is intended for human consumption and is essential for nourishment and subsistence of life. Food safety and quality are therefore of paramount importance and have drawn attention of all stakeholders. With the increasing liberalization of the agro-industrial market and consumer awareness, the food supply chain is becoming integrated. Food safety and quality therefore have become major concerns. It is not enough to just produce food; the food must first be safe to consume and, second, be wholesome and nutritious throughout the supply chain, from production to consumption.

The media often reports epidemics of food poisoning, prompting the recall of an entire lot of that food from the marketplace, which causes enormous health hazards and economic loss. This situation arises as a result of foods being contaminated by harmful bacteria, adulterants, or toxins either during the production process or supply and storage. If foods are safe to consume, the question of their quality in terms of physical and nutritious parameters arises. Producers or marketers sometimes use additives and spike the product with similar-looking cheap material without declaring the same on packages to dupe consumers and get quick returns; this is not allowed by any food regulatory authority. Therefore, the competitiveness of food production nowadays is more dependent on foods' safety and quality than on quantity and price. In contrast to the quantity-oriented markets that are often subsidized and producers can always sell everything they produce, quality-oriented markets are market-driven, whereby commodities, production areas, production chains, and brands compete with each other. Therefore, ensuring safety and quality of food arriving from all corners of the world is essential for trade sustainability and to avoid risks to public health. This chapter deals with definitions of food safety and quality; types of physical, chemical, and microbiological adulterants, contaminants, residues, and toxins; as well as Hazard Analysis and Critical Control Points (HACCP), with an example of producing and supplying safe food.

1.1 FOOD SAFETY

Food safety is a scientific discipline describing handling, preparing, and storing food in ways that prevent food-borne illness. Food safety is, therefore, an increasingly important public health issue. Food can transmit disease from person to person, as well as serve as a growth medium for bacteria that can cause food poisoning. As the common saying goes, "Prevention is better than cure"; this assumes the utmost significance in the case of food products consumed by humans. In theory, food poisoning is 100% preventable. There are five key principles of food hygiene, according to the World Health Organization (2010):

1. Prevent contaminating food with pathogens spreading from people, pets, and pests.
2. Separate raw and cooked foods to prevent contaminating the cooked foods.

3. Cook foods for the appropriate amount of time and at the appropriate temperature to kill pathogens.
4. Store food at the proper temperature.
5. Do use safe water and cooked materials.

Food safety is related to microbiological, chemical, and physical hazards resulting in short-term (e.g., infectious disease, kidney problems) or long-term (chronic) health problems (e.g., cancer, reduced life expectancy). A food safety hazard is an agent or condition that could potentially cause an adverse human health effect. Furthermore, the condition of the food itself can also be hazardous. Food-borne illness is defined as diseases, usually either infectious or toxic, caused by agents that enter the body through the ingestion of food. Food-borne diseases are a widespread and growing public health problem, both in developed and developing countries. While less well documented, developing countries bear the brunt of the problem because of the presence of a wide range of food-borne diseases, including those caused by parasites. The high prevalence of diarrheal diseases in many developing countries suggests major underlying food safety problems. Governments all over the world are therefore intensifying their efforts to improve food safety. Broadly, then, *food safety* may be defined as a scientific discipline describing the handling, preparation, and storage of food in ways that prevent food-borne illness. Most of the time, food-borne hazards cannot be seen, which may be the result of adulteration and contamination by physical, chemical, or biological objects in food or drink that can cause injury or illness (Washington State Department of Health, 2013). Food adulteration is the act of intentionally debasing the quality of food offered for sale either by the admixture or substitution of inferior substances or by the removal of some valuable ingredient. Food is declared adulterated if a substance is added that depreciates or injuriously affects it; cheaper or inferior substances are substituted wholly or in part, or any valuable or necessary constituent has been wholly or in part abstracted; it is an imitation; it is colored or otherwise treated to improve its appearance; or if it contains any added substance that is injurious to health. An adulterant, therefore, is a substance found within other substances (in this case food and beverages), although it is not allowed for legal or other reasons. An adulterant is distinct from permitted food additives. There is little difference between an adulterant and an additive. Chicory, for example, may be added to coffee to reduce the cost. This would be considered adulteration if it is not declared, but it may be stated on the label. The term *contamination* is usually used for the inclusion of unwanted substances by accident or as a result of negligence rather than intent. Adulterants added to reduce the amount of expensive product in illicit drugs are called "cutting agents," whereas the deliberate addition of toxic adulterants to food or other products for human consumption is "poisoning." Microbiological pathogens and toxins, introduced either by contamination or by the condition of the food itself, often become threats to life. These adulterants and contaminants are described in the following sections.

1.1.1 PHYSICAL ADULTERANTS AND CONTAMINANTS

Physical adulterants are normally cheaper substances added in larger amounts to food products to get more profit or gain. Physical adulterants usually do not react with the original food. They increase the mass or volume, or improve the food's color and appearance. Types of adulterants are broadly grouped into three categories (Table 1.1). A few common physical adulterants and their health effects are listed in Table 1.2. Physical contamination of food refers to an event where a foreign object falls into prepared food. This may happen as a result of external exposure to poisonous or polluting substances and the

Table 1.1 Type of Adulterants and Contaminants

S. No.	Type	Substances Added
1	Intentional adulterants	Sand, marble, chips, stone, mud, other filth, talc, chalk powder, water, mineral oil, and harmful color
2	Incidental adulterants	Pesticide residues, droppings of rodents, larvae in foods, or similar unwanted materials
3	Metallic contaminants	Arsenic from pesticides, lead from water, effluent from chemical industries, tin from cans, or any similar materials

Table 1.2 Physical Adulterants and Their Effects on Health

S. No.	Physical Adulterants	Foods Commonly Involved	Health Effects
1	Argemone seeds Argemone oil	Mustard seeds Edible oils and fats	Epidemic dropsy, glaucoma, cardiac arrest
2	Artificially colored foreign seeds	As a substitute for cumin seed, poppy seed, black pepper	Injurious to health
3	Foreign leaves or exhausted tea leaves, artificially colored sawdust	Tea	Injurious to health, cancer
4	Tricresyl phosphate	Oils	Paralysis
5	Rancid oil	Oils	Destroys vitamins A and E
6	Sand, marble chips, stones, filth	Food grains, pulses, etc.	Damages digestive tract
7	*Lathyrus sativus*	*Khesari dal* alone or mixed in other pulses	Lathyrism (crippling spastic paraplegia)
8	Nonfood grade or contaminated packing materials	A food	Blood clot, angiosarcoma, cancer, etc.

environments where food is being prepared. Common physical contaminants in food include hair, dirt, insects, and even glass. Food can be contaminated by a variety of different methods. If you are in a fast food chain or a restaurant, the most common method of physical contamination of food is through dirty hands. If hands remain unwashed, it is possible that workers could contaminate a food supply. Food can also be contaminated in factories. Poor sanitary conditions can taint food easily and cause food poisoning. Some possible common physical contaminants in food are listed in Table 1.3.

1.1.2 CHEMICAL ADULTERANTS AND RESIDUES

Chemical adulterants are chemicals and residues present where the food is grown, harvested, processed, packaged, stored, transported, marketed, and consumed. Food gets adulterated or contaminated by these harmful chemicals beyond the permissible limits fixed by regulators. Chemicals sometimes

Table 1.3 Common Physical Adulterants/Contaminants in Different Foods

S. No.	Food Articles	Adulterants
1	Milk	Water, starch, urea, detergent Vegetable oil, synthetic milk, formalin
2	Ghee, cottage cheese, condensed milk, *Khoa*, milk powder, etc.	Coal tar, dyes
3	Sweet curd	Vanaspati
4	Rabdi	Blotting paper
5	Khoa and its product	Starch
6	*Chhana* or paneer	Starch
7	Ghee	Vanaspati or margarine
8	Butter	Vanaspati or margarine
9	Edible oil	Prohibited color
10	Coconut oil	Any other oil
11	Sugar	Chalk powder, urea, color
12	Honey	Sugar solution
13	Jaggery	Washing soda, chalk powder, sugar solution, metanil (yellow color)
14	Sweetmeats, ice cream, and beverages	Metanil yellow (a nonpermitted coal tar color), saccharin
15	Wheat, rice, maize, jawar, bajra, chana, barley, etc.	Dust, pebble, stone, straw, weed seeds, damaged grain, "weevilled" grain, insects, rodent hair and excreta
16	Maida	Resultant atta or cheap flour
17	Maida/rice	Boric acid
18	Wheat, bajra, and other grains	Ergot (a fungus containing a poisonous substance), dhatura kanel bunt
19	Parboiled rice	Metanil yellow (a nonpermitted coal tar color), turmeric (for golden appearance)
20	Parched rice	Urea
21	Wheat flour	Chalk powder
22	Dal, whole and split	Khesari dal, clay, stone, gravels, webs, insects, rodent hair and excreta, metanil yellow (a nonpermitted coal tar color)
23	Atta, maida, suji (rawa)	Sand, soil, insect, webs, lumps, rodent hair and excreta, iron filings
24	Bajra	Ergot-infested bajra
25	Pulses	Lead chromate
26	Mustard seed	Argemone seed
27	Powdered spices	Added starch, common salt
28	Turmeric powder	Colored sawdust
29	Turmeric whole	Lead chromate, chalk powder, or yellow soapstone powder
30	Chili powder	Brick powder, salt powder or talc powder, artificial color, water-soluble coal tar color
31	Common salt	White powder
32	Tea leaves	Exhausted tea

are also added intensely to increase foods' shelf life, taste, and appearance. They also sometimes come into food in smaller amounts from the air, water, and soil. Some examples are arsenic, mercury, and nitrates. Chemical residues in foods are common today because of the rampant use of agrochemicals in agricultural practices and animal husbandry with the intent to increase crops and reduce costs. Such agents include pesticides (e.g., insecticides, herbicides, rodenticides); plant growth regulators; veterinary drugs (e.g., nitrofuran, fluoroquinolones, malachite green, chloramphenicol); and bovine somatotropin. There is a separate issue of genetically modified food, or the presence in foods of ingredients from genetically modified organisms. The impact of chemical contaminants on consumer health and well-being is often apparent only after many years of prolonged exposure at low levels (e.g., cancer). Chemical contaminants can be classified (Tables 1.4 and 1.5) according to the source of contamination and the mechanism by which they enter the food product.

Table 1.4 Chemical Contaminants in Foods and Their Ill Effects on Health

S. No.	Contaminants	Foods Commonly Involved	Health Effects
1	Mineral oil (white oil, petroleum fractions)	Edible oils and fats, black pepper	Cancer
2	Lead chromate	Turmeric (whole and powdered), mixed spices	Anemia, abortion, paralysis, brain damage
3	Methanol	Alcoholic liquors	Blurred vision, blindness, death
4	Diethylstilbestrol (additive in animal feed)	Meat	Impotence, fibroid tumors, etc.
5	3,4-Benzopyrene	Smoked food	Cancer
6	Excessive solvent residue	Solvent extracted oil, oil cake, etc.	Possibility of numbness in feet and hand
7	Nonpermitted color or permitted food color beyond safe limit	Colored food	Mental retardation, cancer, other toxic effects
8	Butylated hydroxyanisol and butylated hydroxytoluene beyond safe limit	Oils and fats	Allergy, liver damage, increase in serum cholesterol, etc.
9	Monosodium glutamate (flour) (beyond safe limit)	Chinese food, meat and meat products	Brain damage, mental retardation in infants
10	Coumarin and dihydrocoumarin	Flavored food	Blood anticoagulation
11	Food flavors beyond safe limit	Flavored food	Chances of liver cancer
12	Brominated vegetable oils	Cold drinks	Anemia, enlargement of heart
13	Sulfur dioxide and sulfite beyond safe limit	In a variety of foods as preservatives	Acute irritation of the gastrointestinal tract, etc.
14	Artificial sweeteners beyond safe limit	Sweet foods	Chance of cancer
15	Arsenic	Fruits such as apples sprayed with lead arsenate	Dizziness, chills, cramps, paralysis, death
16	Barium	Foods contaminated by rat poisons (barium carbonate)	Violent peristalsis, arterial hypertension, muscular twitching, convulsions, cardiac disturbances

Continued

Table 1.4 Chemical Contaminants in Foods and Their Ill Effects on Health—cont'd

S. No.	Contaminants	Foods Commonly Involved	Health Effects
17	Cadmium	Fruit juices, soft drinks, etc., in contact with cadmium-plated vessels or equipment; cadmium-contaminated water and shellfish	Increased salivation, acute gastritis, liver and kidney damage, prostate cancer
18	Cobalt	Water, liquors	Cardiac insufficiency and myocardial failure
19	Lead	Water, natural and processed foods	Lead poisoning (foot-drop, insomnia, anemia, constipation, mental retardation, brain damage)
20	Copper	Any Food	Vomiting, diarrhea
21	Tin	Any Food	Colic, vomiting
22	Zinc	Any Food	Colic, vomiting
23	Mercury	Mercury fungicide-treated seed grains or mercury-contaminated fish	Brain damage, paralysis, death

1.1.3 MICROBIOLOGICAL AND PATHOGENIC CONTAMINANTS

Food-borne illness caused by microorganisms is a large and growing public health problem. Most countries with systems for reporting cases of food-borne illness have documented significant increase since the 1970s in the incidence of diseases caused by microorganisms in food, including pathogens such as *Salmonella*, *Campylobacter jejuni*, and enterohemorrhagic *Escherichia coli*, and parasites such as *Cryptosporidium*, *Cryptospora*, and trematodes. Some important bacterial and fungal contaminants, and pathogens, in commonly available foods and their ill effects on health are listed in Tables 1.6–1.8, respectively.

1.1.4 TOXINS

Toxins are substances created by plants and animals that are poisonous to humans. Toxins also include medications that are helpful in small doses but poisonous when used in a large amount. Toxins can be small molecules, peptides, or proteins that are capable of causing disease on contact with or upon absorption by body tissues, interacting with biological macromolecules such as enzymes or cellular receptors.

Natural toxins are chemicals that are naturally produced by living organisms. These toxins are not harmful to the organisms themselves but they may be toxic to other creatures, including humans, when eaten. Some plants have the capacity to naturally produce compounds that are toxic to humans when consumed (Health Canada, 2011).

Mycotoxins are another group of natural toxins. The word *mycotoxin* is derived from the Greek word for fungus, *mykes*, and the Latin word *toxicum*, meaning "poison." Mycotoxins are toxic chemical products formed by fungi that can grow on crops in the field or after harvest. Foods that can

Table 1.5 Natural Chemical Contaminants and Their Ill Effects on Health

S. No.	Natural Contaminants	Foods Commonly Involved	Health Effects
1	Fluoride	Drinking water, seafoods, tea, etc.	Excess fluoride causes fluorosis (mottling of teeth, skeletal and neurological disorders)
2	Oxalic acid	Spinach, amaranth, etc.	Renal calculi, cramps, failure of blood to clot
3	Gossypol	Cottonseed flour and cake	Cancer
4	Cyanogenetic compounds	Bitter almonds, apple seeds, cassava, some beans, etc.	Gastrointestinal disturbances
5	Polycyclic aromatic hydrocarbons	Smoked fish, meat, mineral oil-contaminated water, oils, fats, and fish, especially shellfish	Cancer
6	Phalloidin (alkaloid)	Toxic mushrooms	Mushroom poisoning (hypoglycemia, convulsions, profuse watery stools, severe necrosis of liver leading to hepatic failure and death)
7	Solanine	Potatoes	Solanine poisoning (vomiting, abdominal pain, diarrhea)
8	Nitrates and nitrites	Drinking water, spinach rhubarb, asparagus, etc., and meat products	Methemoglobinemia, especially in infants; cancer and tumors in the liver, kidney, trachea, esophagus, and lungs. The liver is the initial site but afterward tumors appear in other organs.
9	Asbestos (may be present in talc, kaolin, etc., and in processed foods)	Polished rice, pulses, processed foods containing anticaking agents, etc.	Absorption in particulate form by the body may produce cancer
10	Pesticide residues (beyond safe limit)	All types of food	Acute or chronic poisoning with damage to nerves and vital organs such as the liver and kidney
11	Antibiotics (beyond safe limit)	Meats from antibiotic-fed animals	Multiple-drug resistance, hardening of arteries, heart disease

be affected include cereals, nuts, fruits and dried fruits, coffee, cocoa, spices, oilseeds, and milk (Health Canada, 2011). The toxin produced by the bacteria *Clostridium botulinum* is one of the most potent toxins known. It is very common in soil, survives indefinitely, and grows when conditions are favorable (Bagley et al., 1997).

Nitrate toxicity can be caused by cropped plants such as oats, corn, millet, sorghum, and Sudangrass, which may contain high nitrate levels under heavy fertilization, frost, or drought conditions. Weeds that are most likely to cause nitrate toxicity are kochia, pigweed, and lambsquarter (Bagley et al., 1997).

Aflatoxins include aflatoxin (AF) B1, AFB2, AFG1, and AFG2. In addition, AFM1 has been identified in the milk of dairy cows consuming AFB1-contaminated feed. The aflatoxigenic

Table 1.6 Common Bacterial Contaminants and Their Ill Effects on Health

S. No.	Bacterial Contaminants	Foods Involved	Health Effects
1	*Bacillus cereus*	Cereal products, custards, puddings, sauces	Food infection (nausea, vomiting, abdominal pain, diarrhea)
2	*Salmonella* spp.	Meat and meat products, raw vegetables, salads, shellfish, eggs and egg products, warmed-up leftovers	Salmonellosis (food infection usually with fever and chills)
3	*Shigella sonnei*	Milk, potatoes, beans, poultry, tuna, shrimp, moist mixed foods	Shigellosis (bacillary dysentery)
4	*Staphylococcus aureus* entero-toxins A, B, C, D, or E	Dairy products, baked foods (especially custard or cream-filled foods), meat and meat products, low-acid frozen foods, salads, cream sauces, etc.	Increased salivation, vomiting, abdominal cramp, diarrhea, severe thirst, cold sweats, prostration
5	*Clostridium botulinus* toxins A, B, E, or F	Defectively canned low- or medium-acid foods; meats, sausages, smoked vacuum-packed fish, fermented foods, etc.	Botulism (double vision, muscular paralysis, death due to respiratory failure)
6	*Clostridium perfringens* (Welchii) type A	Milk improperly processed or canned meats, fish, and gravy stocks	Nausea, abdominal pains, diarrhea, gas formation

Aspergillus spp. are generally regarded as storage fungi, proliferating under conditions of relatively high moisture/humidity and temperature. Aflatoxin contamination is, therefore, almost exclusively confined to tropical feeds such as oilseed by-products derived from groundnuts, cottonseeds, and palm kernels. In addition, some plant toxins naturally available in different crops (Table 1.9) and some other contaminants with their health effects are listed in Table 1.10.

1.1.5 PREVENTION OF ADULTERATION AND CONTAMINATION

"Prevention is better than cure" is a well-established idiom. One should always ensure that food should not, knowingly or unknowingly, be adulterated or contaminated. To prevent food adulteration and contamination, one should adhere to the following five key elements.

1.1.5.1 Keep Clean

- Wash hands before handling food and often during food preparation.
- Wash hands after going to the toilet.
- Wash and sanitize all surfaces and equipment used for food preparation.
- Protect kitchen areas and food from insects, pests, and other animals.
- Use proper cap, gloves, and dress to avoid contamination from hair and other contaminants during food preparation.

Table 1.7 Common Fungal Contaminants and Their Ill Effects on Health

S. No.	Fungal Contaminants	Foods Involved	Health Effects
1	Aflatoxins	*Aspergillus flavus*-Contaminated foods such as groundnuts, cottonseeds, etc.	Liver damage and cancer
2	Ergot alkaloids from *Claviceps purpurea*, toxic alkaloids, ergotamine, ergotoxin, and ergometrine groups	Ergot-infested bajra, rye meal or bread	Ergotism (St. Anthony's fire-burning sensation in extremities, itching of skin, peripheral gangrene)
3	Toxins from *Fusarium sporotrichioides*	Grains (millet, wheat, oats, rye, etc.)	Alimentary toxic aleukia (epidemic panmyelotoxicosis)
4	Toxins from *Fusarium sporotrichiella*	Moist grains	Urov disease (Kaschin-Beck disease)
5	Toxins from *Penicillium inslandicum*, *Penicillium atricum*, *Penicillium citreovirede*, *Fusarium*, *Rhizopus*, *Aspergillus*	Yellow rice	Toxic moldy rice disease
6	Sterigmatocystin from *Aspergillus versicolour*, *Aspergillus nidulans*, and *Bipolaris*	Food grains	Hepatitis
7	*Ascaris lumbricoides*	Any raw food or water contaminated by human feces containing eggs of the parasite	Ascariasis
8	*Entamoeba histolytica* (viral)	Raw vegetables and fruits	Amebic dysentery
9	Virus of infectious hepatitis (virus A)	Shellfish, milk, unheated foods contaminated with feces, urine, and/or blood of infected human	Infectious hepatitis
10	Machupo virus	Foods contaminated with rodent urine, such as cereals	Bolivian hemorrhagic fever

1.1.5.2 Separate Raw and Cooked Materials
- Keep raw meat, poultry, and seafood, vegetables, and so on separate from other foods.
- Use separate equipment and utensils, such as knives and cutting boards, for handling raw foods.
- Store food in containers to avoid contact between raw and prepared foods.

1.1.5.3 Cook Thoroughly
- Be sure that meat and poultry juices are clear and not discolored.
- Cook food thoroughly, especially meat, poultry, eggs, and seafood.
- Bring foods like soups and stews to boiling to make sure that they have reached 70 °C.
- Reheat cooked food properly before eating.

Table 1.8 Common Food-Borne Pathogens

Pathogen	Symptoms	Food Items
Bacillus cereus	Diarrhea, abdominal cramps, nausea, and vomiting (emetic type)	Meats, milk, vegetables, fish, rice, potatoes, pasta, and cheese
Campylobacter jejuni	Nausea, abdominal cramps, diarrhea, headache (varying in severity)	Raw milk, eggs, poultry, raw beef, cake icing, water
Clostridium botulinum	Nausea, vomiting, diarrhea, fatigue, headache, dry mouth, double vision, muscle paralysis, respiratory failure	Low-acid canned foods, meats, sausage, fish
Clostridium perfringens	Abdominal cramps and diarrhea; some include dehydration	Meats and gravies
Cryptosporidium parvum	Watery diarrhea accompanied by mild stomach cramping, nausea, loss of appetite; symptoms may last 10–15 days	Contaminated water or milk
Escherichia coli 0157:H7	Hemorrhagic colitis, possibly hemolytic uremic syndrome	Ground beef, raw milk
Giardia lamblia	Infection of the small intestine, diarrhea, loose or watery stool, stomach cramps, and lactose intolerance	Food and water
Hepatitis A	Fever, malaise, nausea, abdominal discomfort	Water, fruits, vegetables, iced drinks, shellfish, and salads
Listeria monocytogenes	Meningitis, septicemia, miscarriage	Vegetables, milk, cheese, meat, seafood
Salmonella	Nausea, diarrhea, abdominal pain, fever, headache, chills, prostration	Meat, poultry, egg or milk products
Staphylococcus	Severe vomiting, diarrhea, abdominal cramping	Custard- or cream-filled baked goods, ham, tongue, poultry, dressing, gravy, eggs, potato salad, cream sauces, sandwich fillings
Shigella	Abdominal pain, cramps, diarrhea, fever, vomiting, blood, and pus	Salads, raw vegetables, dairy products, and poultry
Vibrio	Diarrhea, abdominal cramps, nausea, vomiting, headache, fever	Fish and shellfish
Yersiniosis	Enterocolitis (may mimic acute appendicitis)	Raw milk, chocolate milk, water, pork, other raw meats

1.1.5.4 Use Safe Water and Raw Materials

- Use safe water or treat it to make it safe.
- Select fresh and wholesome foods.
- Choose foods processed for safety, such as pasteurized milk.
- Wash fruits and vegetables, especially if eaten raw.
- Do not use food beyond its expiration date.

Table 1.9 Plant Toxins and Their Typical Concentrations in Different Food Crops

S. No.	Toxins	Principal Sources	Typical Concentrations
1	Lectins	Jack beans	73 units/mg protein
		Winged beans	40–320 units/mg
		Lima beans	59 units/mg protein
2	Trypsin inhibitors	Soybeans	88 units/mg
3	Antigenic proteins	Soybeans	–
4	Cyanogens	Cassava root	186 mg HCN/kg
5	Condensed tannins	*Acacia* spp.	65 g/kg
		Lotus spp.	30–40 g/kg
6	Quinolizidine alkaloids	Lupin	10–20 g/kg
7	Glucosinolates	Rapeseeds	100 mmol/kg

1.1.5.5 Keep Food at Safe Temperatures

Keep your food at the right temperature to avoid the temperature danger zone—the temperature in which bacteria can widely spread. To avoid such danger, use aseptic techniques, if possible, and follow the following precautions:

- Do not leave cooked food at room temperature for more than 2 h.
- Refrigerate promptly all cooked and perishable foods, preferably below 5 °C.
- Keep cooked food hot (more than 60 °C) before serving.
- Do not store food too long, even in the refrigerator.
- Do not thaw frozen food at room temperature.

Chemical contaminants present in foods are often unaffected by thermal processing (unlike most microbiological agents). Rampant use of insecticides and pesticides during production, storage, and transport of agricultural produce, particularly fruits and vegetables, makes them unsafe. To avoid and or minimize the residues of insecticides and pesticides in such food, regulatory authorities should enforce good agricultural, storage, and transport practices, and educate the stakeholders about not using excess chemicals and pesticides, as well as their legal and negative effects on health.

Food regulatory authorities should frequently analyze risk and have a raw material/processed food tracing, surveillance, inspection, and compliance system in place to prevent food hazards from occurring. It is necessary to impose food facility registration for the purpose of inspection. There should be prior notice of imported foods. Adequate infrastructure for a testing laboratory for rapid detection of food adulteration is essential.

1.2 HAZARD ANALYSIS AND CRITICAL CONTROL POINTS

Food safety assurance is an exercise that covers a whole gamut of interactions among raw materials, food ingredients, processing methods, manufacturing environments, and the other critical areas of the manufacturing process that may affect foods' microbiological quality and safety. It is nearly impossible to monitor every aspect of a large manufacturing process/food manufacturing plants and

Table 1.10 Other Food Contaminants and Their Health Effects

Categories	Contaminants	Potential Health Concerns
Microorganisms		
	Cryptosporidium	• Gastrointestinal illness (such as diarrhea, vomiting, and cramps)
	Giardia lamblia	• Gastrointestinal illness (such as diarrhea, vomiting, and cramps)
	Heterotrophic plate count	• Heterotrophic plate count has no health effects; it is an analytic method used to measure the variety of bacteria that are common in water. The lower the concentration of bacteria in drinking water, the better maintained the water system is.
	Legionella	• Legionnaire's disease, a type of pneumonia
	Total *coliforms* (including fecal coliform and *Escherichia coli*)	• Not a health threat in itself; it is used to indicate whether other potentially harmful bacteria may be present.
	Viruses (enteric)	• Gastrointestinal illness (such as diarrhea, vomiting, and cramps)
	Brucella	• According to pathogenicity of individual microorganism
	E. coli	• According to pathogenicity of individual microorganism
	Staphylococcus sp.	• According to pathogenicity of individual microorganism
	Streptococcus sp.	• According to pathogenicity of individual microorganism
	Listeria sp.	• According to pathogenicity of individual microorganism
	Mycobacterium tuberculosis	• According to pathogenicity of individual microorganism
	Mycobacterium bovis	
	Salmonella	• According to pathogenicity of individual microorganism
	Clostridium	• According to pathogenicity of individual microorganism
	Yeast and fungus	• According to pathogenicity of individual microorganism
Disinfection by-products		
	Bromate	• Increased risk of cancer
	Chlorite	• Anemia; nervous system effects in infants and young children
	Haloacetic acids (HAA5)	• Increased risk of cancer
	Total trihalomethanes	• Liver, kidney, or central nervous system problems; increased risk of cancer

Table 1.10 Other Food Contaminants and Their Health Effects—cont'd

Categories	Contaminants	Potential Health Concerns
Disinfectants		
	Chloramines (as Cl_2)	• Eye/nose irritation, stomach discomfort, anemia
	Chlorine (as Cl_2)	• Eye/nose irritation, stomach discomfort
	Chlorine dioxide (as ClO_2)	• Anemia; nervous system effects in infants and young children
Inorganic chemicals		
	Antimony	• Increase in blood cholesterol; decrease in blood glucose
	Barium	• Increase in blood pressure
	Copper	• Short-term exposure causes gastrointestinal distress
		• Long-term exposure causes liver or kidney damage
		• People with Wilson's disease should consult their personal doctor if the amount of copper in their water exceeds the action level.
	Cyanide (as free cyanide)	• Nerve damage or thyroid problems
	Fluoride	• Bone disease (pain and tenderness of the bones); children may get mottled teeth
	Mercury (inorganic)	• Kidney damage
	Nitrates (measured as nitrogen)	• Infants younger than the age of 6 months who drink water containing nitrates in excess of the MCL could become seriously ill and, if untreated, may die. Symptoms include shortness of breath and blue-baby syndrome.
	Nitrites (measured as nitrogen)	• Infants younger than the age of 6 months who drink water containing nitrites in excess of the MCL could become seriously ill and, if untreated, may die. Symptoms include shortness of breath and blue-baby syndrome.
	Selenium	• Hair or fingernail loss; numbness in fingers or toes; circulatory problems
	Thallium	• Hair loss; changes in blood; kidney, intestine, or liver problems
Inorganic chemicals (pesticides/insecticides/ fungicides/environmental contaminants/processing contaminants etc.)	Acrylamide	• Nervous system or blood problems; carcinogenic
	Benzene	• Anemia; decrease in blood platelets; carcinogenic
	Benzopyrene (polycyclic aromatic hydrocarbons)	• Reproductive difficulties; carcinogenic effects
	Carbofuran	• Problems with blood, nervous system, or reproductive system
	Chlordane	• Liver or nervous system problems; carcinogenic effects

Continued

Table 1.10 Other Food Contaminants and Their Health Effects—cont'd

Categories	Contaminants	Potential Health Concerns
	Chlorobenzene	• Liver or kidney problems
	2,4-D	• Kidney, liver, or adrenal gland problems
	Dalapon	• Minor kidney changes
	Dioxin (2,3,7,8-TCDD)	• Reproductive difficulties; carcinogenic effects
	Diquat	• Cataracts
	Endothall	• Stomach and intestinal problems
	Endrin	• Liver problems
	Glyphosate	• Kidney problems; reproductive difficulties
	Heptachlor	• Liver damage; carcinogenic effects
	Lindane	• Liver or kidney problems
	Methoxychlor	• Reproductive difficulties
	Polychlorinated biphenyls	• Skin changes; thymus gland problems; immune deficiencies; reproductive or nervous system difficulties; carcinogenic
	Hexachlorobenzene	• Damage to reproductive and endocrine systems
	Aldrin-dieldrin	• Neurotoxic, teratogenic, carcinogenic, immunotoxic and other effects
	Dieldrin	• Adverse health effects
	Hexachlorocyclohexane	
	Dichloro diphenyl trichloroethane	
	Monochloro propanediol	• Carcinogenic, mutagenic effects
	Glycidyl esters	• Carcinogenic, mutagenic effects
	Phthalates	• Toxic effects
	Tetraconazole	• Is moderately toxic to freshwater and estuarine/marine fish and freshwater invertebrates, but it is highly toxic to estuarine/marine invertebrates; tetraconazole is slightly toxic to mammals and moderately toxic to birds (reproductive chronic effects).
	Tebuconazole	• Causes significant reproductive toxicity on laboratory mice; a possible human carcinogen
	Hydroxymethylfurfuryl (HMF)	• Products containing high levels of HMF may also contain a lot of acrylamide. In humans, HMF is metabolized to 5-hydroxymethyl-2-furoic acid, which is excreted in urine, and 5-sulfoxymethylfurfural (SMF). SMF can form adducts with DNA or proteins, and rodent toxicology studies have indicated potential toxicity and carcinogenicity.
	Furan	• Causes liver cancer in rodents, possibly carcinogenic to humans

Table 1.10 Other Food Contaminants and Their Health Effects—cont'd

Categories	Contaminants	Potential Health Concerns
Veterinary drug residues		
Antibiotics	Benzylpenicillin Tetracycline Oxytetracycline Chlortetracycline Trimethoprim Ceftiofur Streptomycin Oxfendazole Sulfonamides	• Antibiotic resistance, nausea, vomiting, anaphylaxis, and allergic shock toxicity (like aplasia of the bone marrow); carcinogenic effects
Parasiticide drugs	Albendazole	• Carcinogenic effects
Hormones	Progesterone	• Physiological problems
	Estrone	
	Testosterone	
	Bovine growth hormone or Barium strontium titanate	
Toxins		
	Aflatoxin	• Aflatoxin M1 in milk is a carcinogenic metabolite of aflatoxin B1, which induces liver cancer (hepatocellular carcinoma)
	Fumonisins	• Toxicity
	Trichothecenes	
	Zearalenone	
	Citrinine	
	Ochratoxin	
	Patulin	• Induces acute symptoms including agitation, convulsions, edema, ulceration, intestinal inflammation and vomiting, and could cause chronic neurotoxic, immunotoxic, genotoxic, and teratogenic effects in rodents
Heavy metals		
	Arsenic	• Skin damage or problems with circulatory systems, and may have increased risk of getting cancer
	Chromium	• Allergic dermatitis
	Lead	• Infants and children: delays in physical or mental development; children could show slight deficits in attention span and learning abilities; Adults: kidney problems; high blood pressure
	Cadmium	• Kidney damage
Adulterants		
	Mineral oil	• Health-related problems
	Weeds/inferior oils	• Toxic effects

facilities. It is, however, possible to monitor certain key areas—or, the "critical points"—where safety measures can reasonably be monitored and assured. This is the basis for the HACCP system, which is a systematic approach for identifying, assessing, and controlling hazards. The system offers a rational approach to the control of potential hazards in foods, which avoids many weaknesses inherent in the inspection process or approach and circumvents the shortcomings of reliance on end-product testing. Focusing attention on the factors that directly affect the safety of a food eliminates wasteful use of resources on extraneous considerations, while ensuring that the desired levels of safety and quality are maintained.

1.2.1 HACCP PRINCIPLES

Seven basic principles summarize the theme of HACCP from an international perspective:

Principle 1: Identify the potential hazard(s) associated with food production at all stages, from growth, processing, manufacture, and distribution to the point of consumption. Assess the likelihood of occurrence of the hazard(s) and identify the preventive measures for control.

Principle 2: Determine the point/procedures/operational steps that can be controlled to eliminate the hazard(s) or minimize its likelihood of occurrence (critical control point (CCP)).

Principle 3: Fix target level(s) and tolerances that must be met to ensure the CCP is under control.

Principle 4: Establish a monitoring system to ensure control of CCPs by scheduled testing or observation.

Principle 5: Establish a mechanism for the corrective action to be taken when monitoring indicates that a particular CCP is not under control.

Principle 6: Document (keep records) concerning all procedures and records appropriate to these principles and their application.

Principle 7: Establish verification procedures, which include appropriate supplementary tests and procedures to confirm that HACCP is working effectively.

1.2.2 IMPLEMENTATION OF HACCP SYSTEMS

HACCP implementation usually involves two separate stages: preliminary preparation and application of HACCP principles.

1.2.2.1 Stage 1: Preliminary Preparation

1. *Create the HACCP team*: The multidisciplinary HACCP team should comprise a quality assurance specialist who understands the biological, chemical, or physical hazards connected with a particular product group; a production specialist who has the responsibility of looking after the whole process for manufacturing the product; an engineer who has a good knowledge of hygiene, and the design and operation of a plant and equipment; a packaging specialist who has a thorough knowledge of the effect and nature of packaging material for the desired product; a distribution specialist who has expertise in the area of handling, storage, and transportation from production to consumer; a hygiene specialist who has the responsibility of looking at the process from a hygiene and sanitation point of view, with a proactive approach; and a microbiologist who can identify the "gray areas" of microbial contamination, enumerate microorganisms when required, and suggest safety measures (Codex Alimentarius Commission, 1995a,b).

2. *Describe the food product*: The main purpose of this section is to provide as much information as possible to the HACCP team for proper evaluation. The description must include the following items: composition of the product (e.g., list of ingredients, including description or specifications of the raw materials); characteristics of the product (e.g., solid, liquid, emulsion, pH, Brix); processing methods (heating, smoking, cutting/slicing, freezing, thawing); packaging methods/ system (vacuum, modified atmosphere, controlled and/or shrink packaging); storage and distribution conditions; expected shelf life; instructions for use.
3. *Identify intended use*: State the intended use of the product by the consumer and the consumer target group, for example, the general public, institutional caterers, infants.
4. *Construct a flow diagram*: The purpose of this step is to provide a clear, simple picture of all steps involved in producing the product. The flow diagram must cover all steps in the process that are under the direct control of the manufacturing unit, from receipt of raw materials through distribution of the finished product. (A process flow chart for manufacturing tomato puree is presented as a guideline in Figure 1.1).
5. *Verify the flow diagram on site*: It is important for the HACCP team to verify the flow diagram on site during operating hours. Any deviation must result in an amendment of the original flow diagram. If the analyses are applied to a proposed line, preproduction runs must be observed carefully.

An effective HACCP program works only on a specific product and process and must take into account the actual procedure that is in use. For a HACCP program to be useful, data generated from an initial HACCP study need to be constantly updated and implemented to ensure maximum product safety.

1.2.2.2 Stage 2: Application of HACCP Principles

Principle 1: Identify the potential hazard(s) associated with food production at all stages, from growth, processing, manufacture, and distribution to the point of consumption.
Assess the likelihood of occurrence of the hazard(s) and identify the preventive measures for control.

On the basis of the flow diagram generated, the HACCP team should be able to identify all the potential hazards that are expected to occur at each step. The nature of the hazards must be such that their elimination or reduction to acceptable levels is essential to the production of safe food. Once all potential hazards have been identified, the HACCP team may then consider and describe the control measures to be adopted. More than one control measure may be required to control one hazard; similarly many potential hazards may be controlled by a single control measure.

Principle 2: Determine the point/procedures/operational steps that can be controlled to eliminate the hazard(s) or minimize its likelihood of occurrence (CCP).

The identification of CCPs requires a logical approach such as the CCP decision tree (Figure 1.2). The sequence given in the flow diagram must be strictly followed. At each step, the decision tree must be applied to each hazard whose occurrence is probable and each control measure identified. The CCP is specific for a specific product; every product, by and large, requires a different manufacturing process and must not have unwanted or unnecessary critical points.

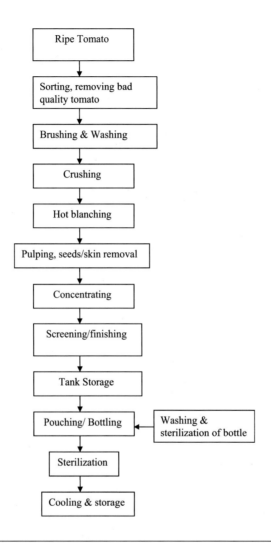

FIGURE 1.1

Tomato puree manufacturing process flow chart.

Principle 3: Fix target level(s) and tolerances that must be met to ensure that the CCP is under control.

The critical limits for each CCP or control measure should represent some quantitative (measurable) parameters that can be measured relatively quickly and easily, for example, temperature, time, pH, preservative level, firmness, texture, and appearance. Those levels should be commensurate with those required by food standards fixed by the related regulatory authority in marketing areas.

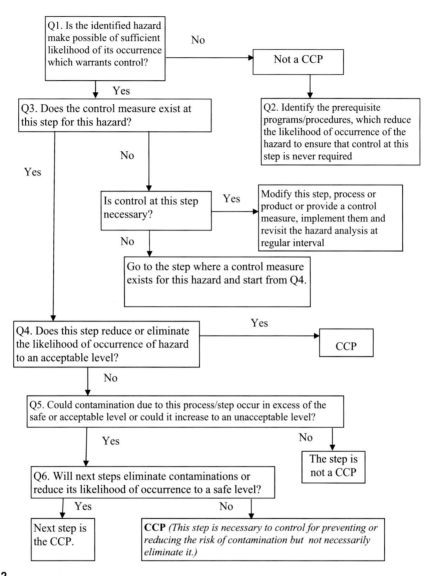

FIGURE 1.2

A typical example of a critical control point (CCP) decision tree for the identification of CCPs in a food processing plant.

Principle 4: Establish a monitoring system to ensure control of the CCP by scheduled testing or observation.

The program should describe the procedure, frequency, and personnel responsible for carrying out the measurements or observations. The monitoring system could be online (flow rate, temperature) or off-line (measurement of total solids, carbon dioxide concentrations, etc.). Online systems give an immediate indication of performance, so it is desirable to have online continuous monitoring systems for each CCP; practically, however, this is not always possible. It is therefore important for the HACCP team to ensure that the results obtained are directly relevant to the CCP and limitations, if any, are fully understood.

Principle 5: Establish a mechanism for the corrective action to be taken when monitoring indicates that a particular CCP is not under control.

Since the HACCP team is multidisciplinary, it should be able to specify the action once the monitoring results show a deviation in CCP. Facilities and planning should be available for immediate action when the CCP goes out of the specified limits.

Principle 6: Document (keep records) concerning all procedures and records appropriate to these principles and their application.

A comprehensive record-keeping system for ingredients, processes, and product controls should be established to facilitate tracing and recall of the product when necessary. In addition, this also helps to find and correct deviations in CCPs. HACCP records must include the following:

- Product description and intended use
- Complete flow diagram of the process, including the CCPs
- Hazards, control limits, monitoring, and corrective action for each CCP
- Verification procedures and data

Principle 7: Establish verification procedures, which include appropriate supplementary tests and procedures, to confirm that HACCP is working effectively.

Verification is necessary to ensure that the HACCP system is working correctly. *Verification* is defined as those activities, other than monitoring, that determine the validity of the HACCP plan and that the system is operating according to the plan. The major science in an HACCP system centers on proper identification of the hazards, CCPs, critical limits, and proper verification procedures in place. These processes should take place during the development and implementation of the HACCP plans and maintenance of the HACCP system.

One aspect of verification is evaluating whether the HACCP system is functioning according to the HACCP plan. An effective HACCP system requires little end-product testing, since sufficient validated safeguards are built in early in the process. Therefore, rather than relying on end-product testing, firms should rely on frequent reviews of their HACCP plan, verification that the HACCP plan is being correctly followed, and review of CCP monitoring and corrective action records.

Another important aspect of verification is the initial validation of the HACCP plan to determine that the plan is scientifically and technically sound, that all hazards have been identified, and that these hazards will be effectively controlled when the HACCP plan is properly implemented (Sperber, 1991). Information needed to validate the HACCP plan often includes (1) expert advice and scientific studies, and (2) in-plant observations, measurements, and evaluations. For example, validation of the manufacturing and packaging process for tomato puree should include the scientific justification for the heating times and temperatures needed to appropriately destroy pathogenic microorganisms and studies to confirm that the conditions will deliver the required time and temperature to each pack of juice.

Subsequent validations are performed and documented by an HACCP team or an independent expert, as needed. For example, validations are conducted when there is an unexplained system failure; when a significant product, process, or packaging change occurs; or when new hazards are recognized. In addition, an unbiased, independent authority should conduct a periodic comprehensive verification of the HACCP system. Such authorities can be internal or external to the food operation. This independent verification should include a technical evaluation of the hazard analysis and each element of the HACCP plan, as well as an on-site review of all flow diagrams and appropriate records of operation of the plan. A comprehensive verification is independent of other verification procedures and must be performed to ensure that the HACCP plan is controlling the hazards. If the results of the comprehensive verification identify deficiencies, the HACCP team must modify the HACCP plan as necessary. Individuals within a company, third-party experts, and regulatory agencies carry out verification activities. It is important that an individual who is doing verification has appropriate technical expertise to perform this function.

1.3 FOOD QUALITY
1.3.1 DEFINITION

The term *quality* refers to the rated ability of a product, whether it is a food or fruits and vegetables, to perform its functions. *Quality* implies the degree of excellence of a product or its suitability for a particular use. In other words, quality can be viewed as an absence of defects in a product. However, *quality* means different things to different handlers within a distribution chain. According to the International Organization for Standardization, the quality of a food can be defined as the totality of the features and characteristics of a product that bear its ability to satisfy stated or implied needs. *Food quality* therefore embraces both sensory attributes that are readily perceived by the human senses and hidden attributes, such as safety and nutrition, that require sophisticated instrumentation to measure. Quality is thus a human construct for a product comprising many desired properties and characteristics. Quality produce encompasses sensory properties (appearance, texture, taste, and aroma), nutritive values, chemical constituents, mechanical properties, functional properties, and defects. Quality is not a single, well-defined attribute but comprises many properties and characteristics.

In the ISO 9000 standard (developed by the International Organization for Standardization) *quality* is defined as "the totality of the features and characteristics of product or service that bear on its ability to satisfy stated or implied needs." Kader (1997) defines it, in relation to fruits and vegetables, as "the combination of attributes or properties that give them value in terms of human food." Quality may be equated to meeting the standards required by a selective customer. In this context the customer is the person or organization receiving the product at each point in the production chain. This is important because quality is perceived differently depending on the needs and intentions of the particular customer. If something is not a quality product, this implies that the product does not meet a certain standard that has been adopted by the customer. In this case the market price can be adversely affected. Conversely, if a product is perceived to be a quality product, it can be sold at a better price.

1.3.2 **PRODUCT-ORIENTED QUALITY**

The quality of a food product changes as the product proceeds from processors to handlers after harvest. The relative importance of different quality attributes changes from handling to purchase to consumption. Shewfelt (1999) points out that quality is often defined from either a product orientation or a consumer orientation. An understanding of the different perspectives of different participants in postharvest distribution is essential in any attempt to improve the quality of a fresh fruit or vegetable for the consumer.

A product orientation views quality as a bundle of attributes that are inherent in a product and can be readily be quantified throughout handling and distribution. For example, Jha et al. (2006) have tried to determine the maturity of mangoes. A consumer orientation defines quality in terms of consumer satisfaction, a much less tangible and less quantifiable concept. Both orientations have strengths and limitations in the delivery of fresh items from harvest to the consumer. Selection of measurement techniques and development of product standards depend on the orientation. Failure to appreciate the differences in perspective results in barriers to improvements in fresh fruit and vegetable quality.

Most postharvest research (physiological as well as technological) assumes a product orientation to quality. *Quality* is defined as a series of attributes selected on the basis of accuracy and precision of measurement. These attributes are in turn used to evaluate the effect of a breeding line or transgenic product, chemical or quarantine treatment, handling technique or system, set of storage conditions, or other postharvest variables. Product-oriented quality is readily defined and clearly understood. Quality changes can be plotted as a function of time and directly related to changes that occurred, such as an increase in free fatty acid in oil and the rancidity of rice bran during handling and storage. These data can be used to develop a mechanistic understanding of effects on loss of quality. Product-oriented quality is usually measured with analytical instruments, and the data can readily be analyzed and results reproduced. The accuracy and precision of measurement provides "internal validity" to any scientific study. A product orientation provides a clear assessment of which treatment(s) is superior or inferior within the context of study objectives.

Product-oriented quality has its limitations, however. Measurements that are less readily quantified carry less weight than those that are readily quantified. Such biases tend to favor handling and storage treatments that maintain appearance (a readily quantifiable attribute) over texture (less quantifiable) over flavor (very difficult to quantify). Likewise, sugar and acid measurement (readily quantified) are likely to predominate over volatile compound analysis. Instrumental methods are preferred to sensory evaluation, which is preferred over consumer testing. While the generation of large data sets provides a wide range of attributes to separate effects of postharvest treatments, the results lack "external validity," or "the extent to which the test results can be generalized to market behavior." Thus, it is not possible to determine whether the significant differences in appearance after treatment are even detectable by many consumers, much less lead to a change in purchase behavior. Likewise, it is not possible to determine whether large differences in predominant volatile compounds affect flavor perception any more than small differences in compounds present in trace amounts. In addition, the product-oriented approach is unable to keep pace with changes in consumer desires and expectations.

A product orientation to quality is the best for assessing the effectiveness of change in a handling system like cultivar selection, harvest technique, or postharvest treatment. It can be adjusted to become more responsive to the marketplace if the quality attributes important to consumers are known, and if accurate and precise measurements can be obtained.

1.3.3 CONSUMER-ORIENTED QUALITY

A consumer orientation to quality requires an understanding of consumer behavior and is focused on predicting product performance in the marketplace. When performed well, consumer-oriented studies provide external validity, thus giving a better appreciation of a product's potential performance in the marketplace. Such studies focus more on how to measure human perception and behavior than on how to measure specific quality attributes. Measurement of consumer attitudes can be simplified to determine either acceptability (superior, acceptable, or unacceptable) or willingness to purchase. Qualitative consumer studies can be used to identify quality attributes that drive acceptability at the points of purchase and consumption. Judicious coupling of quantitative consumer panels with sensory descriptive analysis can either verify or refute the accuracy of consumer statements about critical quality attributes.

A consumer-oriented approach to quality has its own limitations. A consumer is frequently viewed as a monolith with consistent preferences. Realistically, however, consumer preferences vary widely from one cultural or demographic perspective to another, from one consumer to another within a cultural or demographic group, and even within the same consumer depending on many factors, including current mood and intended use of the product. Critical quality attributes that drive product acceptability can be more easily identified using a consumer-oriented approach, but these attributes may be difficult to measure accurately and precisely. While consumers represent only valid sources of preference or acceptability, they are not good at expressing the rationale for these preferences. Furthermore, it may be difficult to quantify these attributes during handling and storage.

A consumer orientation to quality is best at identifying consumer needs and expectations. It can be made more useful to physiologists if the consuming population can be segmented into distinct user groups, based on quality preference rather than demographic groupings, and expressed as a percentage distribution of the population. It is an exact tool in assessing a particular experimental treatment within the distribution chain. Consumer satisfaction from both food quality and safety point of views is necessary for human health and sustainable marketing.

REFERENCES

Bagley, C.V., DVM and Extension Veterinarian Utah State University, 1997. Toxic Contaminants in Harvested Forages. Utah State University Extension, Electronic Publishing.

Codex Alimentarius Commission, 1995a. Recommended international code of practice - general principles of food hygiene (CAC/RCP 1-1969, Rev. 2(1985)). In: Codex Alimentarius. General Requirements (Food Hygiene), vol. 1 B, pp. 1—20.

Codex Alimentarius Commission, 1995b. Guidelines for the application of the hazard analysis critical control point (HACCP) system (CAC/GL 18-1993). In: Codex Alimentarius. General Requirements (Food Hygiene), vol. 1 B, pp. 21—30.

Health Canada, 2011. Food and Nutrition.

Jha, S.N., Kingsly, A.R.P., Chopra, S., 2006. Physical and mechanical properties of mango during growth and storage for determination of maturity. J. Food Eng. 72 (1), 73—76.

Kader, A.A., 1997. A summary of CA recommendations for fruits other than apples and pears. In: 7th Intl. Contr. Atmos. Res. Conf. Univ. of California, Davis, pp. 1—34.

Shewfelt, R.L., 1999. What is quality? Postharvest Biol. Technol. 15, 197—200.

Sperber, W.H., 1991. The modern HACCP system. Food Technol. 45, 116—120.

Washington State Department of Health, 2013. Food Safety Is Everybody's Business. Washington State Food & Beverage Workers' Manual.

World Health Organization, 2010. Prevention of Food Borne Disease: Five Keys to Safer Food.

COMMON ADULTERANTS AND CONTAMINANTS

Food adulterants and contaminants, and their sources, are unlimited in nature. Chapter 1 dealt with common types of physical, chemical, microbiological, and toxicological adulterants and contaminants and their ill effects on health. There are also possibilities of food conversion of various contaminants to harmful toxins during food processing and value addition. This chapter deals with such common contaminants, toxins, and heavy metals, and their sources.

2.1 CHEMICAL CONTAMINANTS

2.1.1 ACRYLAMIDE

Acrylamide (C_3H_5NO; 2-propenamide, propenamide, acrylic amide, prop-2-enamide, ethylene carboxamide, acrylic acid amide) is a chemical that can form in some foods during high-temperature cooking, such as frying, roasting, and baking of foods that are rich in carbohydrates (Stadler et al., 2002; Tareke et al., 2002). Acrylamide forms in foods as a result of the Maillard reaction between natural sugars and the amino acid asparagine (Mottram et al., 2002; Stadler et al., 2002; Zyzak et al., 2003; Hamlet et al., 2008). It does not come from food packaging or the environment. It has been found primarily in food made from plants, such as potatoes, grain products, coffee, peanuts, lentils, and asparagus, and typically it is not associated with meat, dairy, or seafood products (Friedman and Levin, 2008; Forstova et al., 2014; EFSA, 2011). Acrylamide exposure through various means has been associated with neurological and carcinogenic effects in human and animals (Calleman et al., 1994; Friedman et al., 1995; Johnson et al., 1986). Given the prevalence of acrylamide in the human diet and its toxicological effects, the risk of dietary intake of acrylamide is a general public health concern.

Comparing frying, roasting, and baking potatoes, frying causes the formation of the largest amount of acrylamide (FDE, 2011). Roasting potato pieces causes less acrylamide to form, followed by baking whole potatoes. Boiling potatoes and microwaving whole potatoes with skin on to make "microwaved baked potatoes" does not produce acrylamide. Soaking raw potato slices in water for 15—30 min before frying or roasting helps reduce acrylamide formation during cooking. Storing potatoes in the refrigerator can result in increased acrylamide during cooking. Generally, more acrylamide accumulates when cooking is done for longer periods or at higher temperatures. Cooking cut potato products, such as frozen french fries or potato slices, to a golden yellow color rather than a brown color helps reduce acrylamide formation (FDE, 2011). Toasting bread to a light brown color, rather than a dark brown color, reduces the amount of acrylamide. Acrylamide forms in coffee when coffee beans are roasted, not when coffee is brewed at home or in a restaurant. So far, scientists have not found good ways to reduce acrylamide formation in coffee.

Copyright © 2016 Elsevier Inc. All rights reserved.

Acrylamide is carcinogenic, teratogenic, or mutagenic; it is also neurotoxic to the central nervous system (Baum et al., 2005; Hashimoto and Aldridge, 1970; Dearfield et al., 1988). Many authorities classify acrylamide as a probable human carcinogen (IARC, 1994; EU, 2000) based on its carcinogenicity and toxicity in some animal studies. Glycidamide and the α,β-unsaturated amide system of acrylamide may bind with biological nucleophilic groups of proteins and DNA. This reaction plays a significant role in the underlying biological events resulting in toxicity (Shipp et al., 2006). Acrylamide also resulted in abnormalities such as decreased motor strength, abnormal gait, and skin abnormalities among highly exposed workers. It seems likely that the mode of action for acrylamide may be a combination of DNA damage, interference with mitosis, meiosis, and oxidative stress (Banerjee and Segal, 1986; Park et al., 2002; Watzek et al., 2012). A recent study demonstrates structural and ultrastructural evidence of neurotoxic effects of fried potato chips on rat postnatal development (El-Sayyad et al., 2011). The authors reveal for the first time that rat fetal exposure to acrylamide, as a pure compound or from a maternal diet of fried potato chips, causes cerebellar cortical defects and myodegeneration of the gastrocnemius muscle during the postnatal development of pups.

Several competing theories have been proposed to explain acrylamide's mechanism of action to decrease the release of neurotransmitters, including biochemical changes, such as the alteration in ion levels or the inhibition of glycolytic enzymes by acrylamide adduct formation (Friedman, 2003). It is known that the interference of acrylamide and its metabolite glycidamide with kinesin motor proteins in neurofilaments causes failure of the transport of nerve signals between axons, and this may be one of the mechanisms involved in its neurotoxicity (JECFA, 2002). Acrylamide, and not glycidamide, is primarily responsible for the induction of neurotoxicity. Morphological changes also were demonstrated, such as changes in nerve structures or their degeneration, which could be visualized with microscopic examinations. It has recently been postulated that neurotoxicity of acrylamide might be cumulative. The same neurotoxic effects can be seen at low and high doses of acrylamide, with the low doses just requiring longer exposure; thus dietary exposure might not be negligible (LoPachin, 2004).

Acrylamide administered to male and female rodents via drinking water caused reproductive abnormalities and neurotoxicological problems at various doses, such as a significant decrease in number of live fetuses per litter, a negative effect on copulatory behavior as well as on sperm motility and morphology, degeneration of the epithelial cells of the seminiferous tubules, and decreased fertility rates and retarded development of pups (Shipp et al., 2006; Tyl et al., 2000; Tyl and Friedman, 2003). These toxic effects may be attributed to the interfering effect of acrylamide on the kinesin motor proteins, which also exist in the flagella of sperm, resulting in reduced sperm motility and fewer fertilization events.

Acrylamide also produces oxidative stress in some animals. After administration of acrylamide in drinking water at doses as low as 0.025 mg/kg/day for 10 weeks, oxidative stress was observed more in male rat liver, lung, and testes, as well as in cultured HepG2 cells (Yousef and El-Demerdash, 2006; Jiang et al., 2007). Ingestion of acrylamide-containing products contributes to the creation of oxidative stress and induces a proinflammatory state, a risk factor for progression of atherosclerosis (Naruszewicz et al., 2009).

Thus, acrylamide induces (1) formation of covalent adducts with DNA in vivo, (2) formation of covalent adducts with hemoglobin, (3) induction of gene mutations and chromosomal aberrations in germ cells, (4) induction of chromosomal aberrations in somatic cells in vivo, (5) induction of gene mutations and chromosomal aberrations in cultured cells in vitro, and (6) induction of cell transformation in cell lines. Clear evidence of carcinogenic activity among laboratory animals is demonstrated by studies that are interpreted as showing a dose-related (1) increase of malignant neoplasms, (2) increase of a combination of malignant and benign neoplasms, or (3) marked increase of benign neoplasms if there is an indication from this or other studies of the ability of such tumors to progress to malignancy.

2.1.2 BENZENE

Benzene (C_6H_6; benzol, benzole, cyclohexatriene, pyrobenzole, benzine, phenyl hydride, pyrobenzol, phene, mineral naphtha) is a chemical that can form at very low concentrations in some beverages that contain both benzoate salts and ascorbic acid (vitamin C) or erythorbic acid (Gardner and Lawrence, 1993). Accumulation in foods also arises from other causes, such as transfer from packaging materials, preservation environments, degradation of preservatives, heat, cooking processes, and irradiation techniques used for sterilization. For instance, exposure to transition-metal catalysts (e.g., Cu(II) or Fe(III) ions), heat, and light can stimulate the formation of benzene in some beverages that contain benzoate salts and ascorbic acid (Gardner and Lawrence, 1993; Lachenmeier et al., 2008; Barshick et al., 1995). Sodium or potassium benzoate may be added to beverages to inhibit the growth of bacteria, yeasts, and molds. Benzoate salts also are naturally present in some fruits and their juices, such as cranberries. Vitamin C may be present naturally in beverages or added during processing to prevent spoilage or to provide additional nutrients.

The occurrence of benzene in food and beverages has been reported during earlier investigations. Several factors either related to food processing conditions or to bad manufacturing practices are potential sources of benzene in foodstuffs. Food processing conditions such as irradiation treatments and high-temperature processes (e.g., roasting) have been linked to benzene formation (Barshick et al., 1995). Under such conditions, the decomposition of certain amino acids (Kjallstrand and Petersson, 2001; Zhu et al., 2004) and the decarboxylation of benzoate may take place, whereby benzene formation occurs. Further research regarding benzene formation from the oxidative decarboxylation of benzoate in the absence of irradiation treatments and the presence of hydroxyl radicals in food have been reported in the literature (Lachenmeier et al., 2008). Accordingly, hydroxyl radical formation is promoted by the presence of ascorbic acid and transition metal ions (e.g., Cu^{2+} or Fe^{3+} ions). Other variables such as temperature and pH also influence benzene formation from benzoate.

Benzene can also be present in food through migration from various packaging materials or from contamination of the environment and water supply. Contaminated carbon dioxide was reported as a benzene source in beer and in water (Wu et al., 2006). The use of hexane in solvent extraction of vegetable oils may also contribute to benzene occurrence in food (Masohan et al., 2000). Finally, flavors incorporated into foodstuffs through smoking (with wood or charcoal) or adding liquid smoke are another source of benzene in foods (Wittkowski et al., 1990). Based on results from a Center for Food Safety and Applied Nutrition survey, of almost 200 samples of soft drinks and other beverages tested for benzene from 2005 through May 2007, a small number of products sampled contained more than 5 ppb of benzene.

Benzene is a volatile organic compound known to be carcinogenic to humans and is classified as a group 1 carcinogen (IARC, 1987). It is generally accepted that benzene is a risk factor for acute myeloid leukemia in humans (Descatha et al., 2005). Benzene is an immunotoxicant (Dean et al., 1979; Snyder, 1987; McMurry et al., 1991) that causes changes in circulating leukocytes, including lymphocytes (Aksoy et al., 1971, 1987). It also has a more depressive effect on T lymphocytes than on

B cells, and it affects both humoral and cellular acquired immunity in animals (Cronkite et al., 1982). The metabolites of benzene affect the immune system of animals: inhibition of interleukin 2 production and inhibition of maturation and proliferation of B lymphocytes in the marrow, spleen, and thymus. Another study found decreased white blood cell count, platelet count, and other hematological values in persons exposed to 1 ppm benzene (Lan et al., 2004). Although the mechanisms of benzene hematoxicity and carcinogenicity are not completely understood, it is accepted that one or more of the reactive metabolites are involved, including benzene oxide (BO), 1,2- and 1,4-benzoquinone (BQ), and muconaldehyde (Snyder, 2000, 2002). BO first spontaneously rearranges to phenol, or undergoes further metabolism to catechol via dihydrodiol dehydrogenases. Two other minor BO metabolites are S-phenylmercapturic acid, which is derived from the conjugation of glutathione with BO (Henderson et al., 2005), and muconaldehyde, which is produced as a result of a second CYP oxidation of oxepin that ultimately gives rise to t,t-muconic acid. Phenol could be further converted by cytochrome P-450 2E1 (CYP2E1) to produce hydroquinone.

Subsequent oxidation of hydroquinone and catechol, either spontaneously or via peroxidases, produces 1,4-BQ and 1,2-BQ, respectively. These reactive electrophiles, including BO, 1,2-BQ, and 1,4-BQ, can ultimately form covalent bonds with a wide variety of macromolecules, including DNA and proteins, and presumably are involved in the cancer process by inducing cell toxicity in bone marrow cells (Snyder, 2000, 2002).

2.1.3 DIOXINS AND POLYCHLORINATED BIPHENYLS

Polychlorinated dibenzo-*p*-dioxins (PCDDs), polychlorinated dibenzofurans (PCDFs), and polychlorinated biphenyls (PCBs) are persistent toxic lipophilic substances, which are widespread throughout the environment and accumulate in the food chain (Wania and MacKay, 1996; Erickson and Kaley, 2011). PCDDs and PCDFs are commonly referred to as dioxins, which comprise a group of 210 congeners, of which 17 are considered highly toxic. The 209 different PCB congeners are classified as dioxin-like (dl-PCBs) or non-dioxin-like (ndl-PCBs), according to their toxicological properties. The 17 highly toxic dioxins and the 13 dl-PCBs, together referred to as dl-compounds, exert their effect via activation of the aryl hydrocarbon receptor, whereas the ndl-PCBs work through different mechanisms (Safe and Hutzinger, 1984).

The primary route of human exposure to PCDD/PCDFs and PCBs is through consumption of contaminated food (ATSDR, 2000). PCDD/PCDF and PCB concentrations in foods have been widely reported in the literature to evaluate the potential health risk posed by dietary exposure worldwide (Tilson et al., 1990; Olanca et al., 2014; Baars et al., 2004; Senthil Kumar et al., 2001). The intake of PCDD/PCDFs and PCBs by humans is estimated to account for greater than 90% through diet (van Leeuwen et al., 2000). It has been reported that meat, dairy products, and fish make up more than 90% of the intake of PCDD/PCDFs and PCBs among the general population (Bocio and Domingo, 2005; Charnley and Doull, 2005; Huwe and Larsen, 2005). The highest concentrations of dioxins and PCBs are found in animals that are high in the food chain, such as large fish and seagulls in the marine food chain. High concentrations can also be found in products from animals in the terrestrial food chain, for example, in liver or eggs.

In 1968, thousands of people were accidentally orally exposed to very high concentrations of PCBs and PCDFs in Fukuoka, Japan. This exposure was caused by PCB contamination of rice oil, which was subsequently used for cooking. The disease described by these symptoms was called

Yusho rice oil disease. The second major human exposure occurred in 1978 in Taiwan. High concentrations of PCBs were ingested by humans through use of contaminated rice oil. From these two specific incidents, and other less noted PCB exposure situations, the effects of PCBs on humans have been collected together.

Toxicological effects of PCBs and dioxins include carcinogenicity, immunotoxicity, and reproductive and developmental toxicity (Van den Berg et al., 2006; Lundqvist et al., 2006). Oxidation of PCBs may produce their oxides, which potentially could act as tumor promoters. Other symptoms of PCB ingestion are weakness, nausea, headache, impotence, insomnia, loss of appetite, loss of weight, and abdominal pain. Muscle spasms and muscular pain also may be related to PCB poisoning. PCBs are capable of causing inflammation and burning, edema of the eyelids, and cysts of the tarsal glands and conjunctiva. The sebaceous gland of the eyelids, the meibomian gland, may also become hypertrophic and result in cheese-like discharges from the eyes. Autopsies of humans exposed to PCBs in the Yusho rice oil poisoning revealed typical chloracne and pigmentation of cutaneous tissue. In addition, follicular hyperkeratosis, dilation of hair follicles, and melanin increase (pigmentation) in the epidermis were observed in histological preparations of skin. Other PCB effects on the skin include xerosis, nail deformity, hair loss, and hyperhidrosis. Organ system impairment such as fibrosis, hepatocellular necrosis, enlargement of the liver, and reduction of air capacity in the lungs has been observed. Prolonged exposure to PCBs seems to cause liver damage. In addition, hypertrophia, hyperplastic gastritis, and ulcer formation have been observed in the gastrointestinal tract.

2.1.4 ETHYL CARBAMATE

Ethyl carbamate ($C_3H_7NO_2$; urethane, ethyl urethane, leucethane, pracarbamin, pracarbamine) is a compound that can occur naturally in fermented foods and beverages, such as spirits, wine, beer, bread, soy sauce, and yogurt (Dennis et al., 1989; Battaglia et al., 1990; Le Kim et al., 2000). Therefore, the major source of dietary exposure to ethyl carbamate in the human population is through the consumption of fermented foods and beverages. In addition to the fermentation process, it is also produced in these foods during storage. A number of precursors in food and beverages can form ethyl carbamate, including hydrocyanic acid, urea, citrulline, cyanogenic glycosides, and other N-carbamyl compounds. Cyanate is probably the ultimate precursor in most cases, reacting with ethanol to form the carbamate ester.

The reaction mechanisms that give rise to ethyl carbamate vary significantly among different kinds of foods. In the case of wine, the use of urea as a yeast nutrient can result in an elevated concentration of ethyl carbamate in the finished product, whereas light has little influence. Ethyl carbamate formation increases over time, with the reaction rate being exponentially accelerated at higher temperatures. Urea is formed when the wine yeast metabolizes arginine, a major α-amino acid in grape juice that is available to yeast. This reaction is dependent on the yeast strain. Yeasts differ in their ability to produce urea and to reuse urea secreted into the wine. Lactic acid bacteria also metabolize arginine and release citrulline, an amino acid, which then reacts with ethanol to form ethyl carbamate. Over-fertilized vineyards, in general, yield wines with higher urea and thus high ethyl carbamate concentrations. Much higher ethyl carbamate concentrations were detected in spirits derived from stone fruit such as cherries, plums, mirabelles, and apricots. The formation of cyanogenic glycosides such as amygdalin in stone fruit by enzymatic action leads to the generation of cyanide, which is the most important precursor of ethyl carbamate in these spirits. As stated earlier, this cyanide is oxidized to cyanate, which reacts with ethanol to form ethyl carbamate. In stone fruit distillation, when exposed to light ethyl carbamate forms from the natural precursors of fruit mash and ethyl alcohol (Lachenmeier et al., 2005; Schehl et al., 2007).

The International Agency for Research on Cancer (IARC) classified ethyl carbamate as "possibly carcinogenic to humans" (group 2B) in 1974 (IARC, 1974), and updated the classification to "probably carcinogenic to humans" (group 2A) in 2007 (IARC, 2007). The toxicology of ethyl carbamate is well documented. It is absorbed rapidly and nearly completely through the gastrointestinal tract and the skin. It is evenly distributed in the body, followed by fast elimination; more than 90% is eliminated as carbon dioxide within 6 h in mice. Ethyl carbamate metabolism involves three main pathways: hydrolysis, N-hydroxylation or C-hydroxylation, and side-chain oxidation. Hydrolysis is mediated by esterases and leads to the production of ethanol, carbon dioxide, and ammonia. The N-hydroxylation, C-hydroxylation, and side-chain oxidation are mediated by CYP2E1 to form N-hydroxycarbamate, α-hydroxy ethyl carbamate, and vinyl carbamate, respectively. Hydroxycarbamate is conjugated and excreted in the urine, α-hydroxy ethyl carbamate is metabolized to ammonia and carbon dioxide, and vinyl carbamate is oxidized to vinyl carbamate epoxide. The vinyl carbamate epoxide is further metabolized via glutathione conjugation to carbon dioxide and ammonia (Guengerich et al., 1991; Hoffler et al., 2003), and has been recognized as the main metabolite responsible for the carcinogenicity of ethyl carbamate since it binds covalently to nucleic acids (DNA, RNA) and proteins (Park et al., 1993; FAO/WHO, 2006). Ethyl carbamate exposure resulted in dose-dependent increased incidences of alveolar and bronchiolar, hepatocellular, and Harderian gland adenoma or carcinoma, hepatic hemangiosarcoma, and mammary gland adenoacanthoma or adenocarcinoma (in females only). Statistically significant but smaller increases in incidence of hemangiosarcoma of the heart (males only) and spleen (females only), squamous cell papilloma or carcinoma of the forestomach and skin (males only), and benign or malignant ovarian granulosa cell tumors, with a dose-related increase in nonneoplastic lesions affecting liver, heart, and uterine blood, together with hepatic eosinophilic foci, were also observed. Those sites for which a significant increase in tumors was observed at the lowest dose tested were the lung and the Harderian gland (FAO/WHO, 2006; IARC, 1974; IARC, 2007).

2.1.5 MELAMINE

Melamine ($C_3N_6H_6$; 2,4,6-triamino-1,3,5-triazine) is a synthetic triazine compound and an organic base that is quite stable during heat processing but dissociates at a low pH. It is an industrial chemical and has been used as a component in various products. The best known use of melamine is in combination with formaldehyde to produce melamine resin, a very hard-wearing thermosetting plastic. The compound is a component in many plastics, adhesives, glues, and laminated products such as plywood, cement, cleansers, fire-retardant paint, fertilizers, and pigments. Melamine in plasticware may be leached from the product by acid and thus can migrate into food (Lu et al., 2009; Lund and Petersen, 2006). It was also intentionally added to certain food products such as milk, infant formula, frozen yogurt, pet food, biscuits, and coffee drinks (WHO, 2011). Melamine also has analogs that are produced by successive deamination reactions, such as ammeline, ammelide, and cyanuric acid. It is now generally accepted that melamine in food, especially in combination with an analog, can have potentially serious health consequences for animals and humans.

Melamine contamination in food became a food safety issue when this was detected in pet foods and caused kidney failure in thousands of dogs and cats in North America in 2007 (Tyan et al., 2009).

Since then, however, it has become apparent that a similar incident affecting an estimated 6000 dogs in Asia in 2004—and first attributed to mycotoxin contamination—was also likely to have been caused by melamine. An investigation of the 2007 incident found that melamine and its analog cyanuric acid were present in wheat gluten and rice protein concentrate used as thickening and binding ingredients (Le et al., 2013). It has also been found in beverages, including coffee and orange juice, at levels of up to 2 mg/kg, but this is thought to be as a result of migration from plastic cups at high temperatures. Very low concentrations of melamine are thought to be occasionally present in some processed foods as a result of migration from packaging or processing equipment. It is also possible that very low concentrations of melamine could be generated as a by-product of processing.

Melamine is not approved for addition to foods or feeds, nor is it permitted for use as a fertilizer anywhere in the world. Nitrogen content has long been used as a surrogate for assessing the protein content of foods, and melamine contains a substantial amount of nitrogen (66% by mass); thus, its addition to a food product would give a falsely high result in tests designed to determine protein content and cause the material to be assigned a higher quality rating and commercial value (Chan et al., 2008). It has been estimated that the addition of 1 g of melamine to 1 L of milk would increase the apparent protein content by approximately 0.4%. If this is indeed the case, then melamine is an adulterant and has been deliberately added to milk, wheat gluten, and rice protein concentrate in a fraudulent attempt to increase profits and disguise poor-quality products. In animals it is also produced as a metabolite of the insecticide cyromazine, which is widely used to prevent insect damage to fruit and vegetables.

In 2008, melamine was found in dairy products, especially powdered milk used to make an infant formula that was associated with widespread kidney disease in babies. Samples of dairy products (including infant formula) were reported to contain melamine at concentrations between 0.09 and 6196 mg/kg. Melamine has also been found in liquid milk (highest concentration, 8.6 mg/kg) and a wide variety of other products made using contaminated milk powder. These include chocolate and milk-based confectionery, biscuits and other bakery products, coffee and tea whiteners, and milk-based beverages. It was also detected in Chinese fresh eggs at concentrations of 3.1—4.7 mg/kg. Contaminated foods have been found all over the world, not particularly in Asian countries, but also in the European Union, the United States, Canada, and Australia. Reported contamination levels were variable, ranging from 0.38 to 945 mg/kg in dairy products and from 0.6 to 6694 mg/kg in processed foods and food ingredients.

There is little information on the likely dietary exposure that would result from such concentrations in processed foods, but the European Food Safety Authority has estimated that chocolate with large amounts of contaminated milk powder could result in an exposure of 1.35 mg/kg body weight/day, more than six times the current World Health Organization (WHO) tolerable daily intake (TDI) of 0.2 mg/kg body weight for children and adults. WHO has also recommended a TDI for cyanuric acid of 1.5 mg/kg body weight.

Melamine and its congeners are toxic to humans and animals, though the amount leading to adverse effects depends on the rate of exposure. Ingestion of melamine may lead to reproductive damage, bladder stones, or kidney stones, which in turn may lead to bladder carcinogenesis (ECD, 2008a,b). Another study suggests that the effects observed in animals were caused by the formation of crystals in the urine, leading to kidney and bladder calculi, blocking the renal tubules in severe cases and potentially causing fatal kidney failure. The crystals were composed of a stable, insoluble melamine—cyanuric acid complex, which formed a lattice-like structure held together by hydrogen bonds (Kim et al., 2010). It has been reported that the two compounds are absorbed separately in different regions of the gut because melamine is a base and has a much lower pKa value than cyanuric acid. The insoluble complex is thought to then form and produce the crystals within the kidneys (Tsujihata, 2008). The other opinion about the mechanism responsible for the kidney problems is somewhat different. Cyanuric acid was not present in significant quantities in the urinary tract calculi obtained from affected children; instead, the stones were found to consist of melamine and uric acid (Lawley et al., 2012). It is thought that these stones formed in infants rather than adults because they typically have higher concentrations of uric acid in their urine. There is still some uncertainty about the exact toxicological mechanisms involved.

2.2 MYCOTOXINS

Mycotoxins are metabolites produced by fungi when they grow on a suitable matrix, and they cause some toxicological problems when ingested through contaminated feed or food by animals or humans. Mycotoxins are secondary metabolites, with no apparent function in the normal metabolism of fungi. Their structures vary from simple heterocyclic rings with molecular weights of up to 50 Da to groups with six to eight irregularly arranged heterocyclic rings with a total molecular weight of >500 Da, and they do not show immunogenicity. More than 400 different types of mycotoxins have already been identified. The major mycotoxin-producing fungi are species of *Aspergillus*, *Fusarium*, and *Penicillium*. Most important groups of mycotoxins are aflatoxins, ochratoxin A (OTA), trichothecenes (deoxynivalenol, nivalenol), zearalenone, and fumonisins. The occurrence of mycotoxins in foods and derivatives is not only a problem in developing countries; in fact, mycotoxins affect agribusiness in many countries, influencing or even impeding export and reducing livestock and crop production in addition to affecting human health. Major food commodities affected are cereals, nuts, dried fruit, coffee, cocoa, spices, oil seeds, dried peas, beans, and fruit (Turner et al., 2009).

Mycotoxins usually enter the body the via oral route (ingestion of contaminated foods), but the nasal route (inhalation of toxicogenic spores) and direct dermal route (contact) are also important. Mycotoxins can enter the human and animal food chains through direct or indirect contamination. The indirect contamination of foodstuffs and animal feed occurs when any ingredient has been previously contaminated by a toxigenic fungus; even though the fungus is eliminated during processing, the mycotoxins remain in the final product. Direct contamination, on the other hand, occurs when the product, food, or feed becomes infected by a toxigenic fungus, with the subsequent formation of mycotoxins. It is known that the majority of food and feed products can allow the growth and development of toxigenic fungi during their production, processing, transport, and storage. The ingestion of mycotoxins by humans occurs mainly through eating contaminated plant products, as well as through products derived from foods such as milk, cheese, meat, and other animal products.

Some of these toxins are heat stable (Zinedine and Manes, 2009) and thus constitute a potential risk for human and animal health because these cannot be destroyed or inactivated during the pasteurization process. Mycotoxins are well known for their toxic effects: carcinogenicity, genotoxicity, teratogenicity, nephrotoxicity, hepatotoxicity, and immunotoxicity (CAST, 2003). Naturally occurring mixtures of aflatoxins are classified as carcinogenic to humans (group 1), OTA and fumonisins as possible carcinogens to humans (group 2B), and zearalenone as a carcinogen (group 3) (IARC, 1993a, 1999, 2002).

2.2.1 AFLATOXINS

The main known aflatoxins are called B1, B2, G1, and G2, based on their fluorescence under ultraviolet light (B refers to blue and G to green) and their mobility during thin-layer chromatography. They are mainly produced by *Aspergillus flavus* and *Aspergillus parasiticus*. However, the species *Aspergillus nomius*, *Aspergillus bombycis*, *Aspergillus pseudotamari*, and *Aspergillus ochraceoroseus* have also been shown more recently to be aflatoxigenic but are encountered less frequently (Klich et al., 2000; Peterson et al., 2001; Bennett and Klich, 2003). In terms of mycology, there are enormous qualitative and quantitative differences in the aflatoxigenic capacity of isolates of *A. flavus*. It is known that only 50% of the strains of these species produce aflatoxins (Klich and Pitt, 1988), and that some of the aflatoxigenic isolates produce up to 106 μg/kg of aflatoxins (Cotty et al., 1994). Aflatoxin B1 is the most potent natural carcinogen known (Squire, 1981) and is usually the major aflatoxin produced by toxigenic strains.

Some substrates are extremely good for the growth of aflatoxigenic fungi and the formation of aflatoxins. Natural contamination is a common occurrence worldwide. Both the genetic ability for the formation of aflatoxins and the capacity to contaminate foods with these toxins are highly variable among fungi. Aflatoxins have been found in peanuts, peanut derivatives, animal feed, milk, nuts, and various grains (Van Egmond, 1989; Wild et al., 1987; Battacone et al., 2009; D' Mello, 2003; El Khoury et al., 2008; Georgiadou et al., 2012; Pietri et al., 2012).

Aflatoxin is associated with both toxicity and carcinogenicity in human and animal populations (Newberne and Butler, 1969; Peers and Linsell, 1973; Shank et al., 1972). The diseases caused by aflatoxin consumption are loosely called aflatoxicoses. Acute aflatoxicosis results in death; chronic aflatoxicosis results in cancer, immune suppression, and other sluggish pathological conditions. The liver is the primary target organ, with liver damage occurring when poultry, fish, rodents, and nonhuman primates are fed aflatoxin B1. There are substantial differences in species susceptibility. Moreover, within a given species, the magnitude of the response is influenced by age, sex, weight, diet, exposure to infectious agents, and the presence of other mycotoxins and pharmacologically active substances. Because of their capacity to bind with the DNA of cells, aflatoxins affect protein synthesis; they also contribute to the occurrence of thymic aplasia (congenital absence of thymus and the parathyroid glands, with a consequent deficiency in cell immunity; also known as DiGeorge syndrome). Aflatoxins also have immunosuppressive properties, inducing infections in people contaminated with these substances.

2.2.2 FUMONISINS

Fumonisins are fungal toxins produced by several species of the genus *Fusarium*, especially *Fusarium verticillioides*, *Fusarium proliferatum*, and *Fusarium nygamai*, in addition to *Alternaria alternate* f.sp. *lycopersici* (Marasas et al., 2001; Rheeder et al., 2002). Other species, such as *Fusarium anthophilum*, *Fusarium dlamini*, *Fusarium napiforme*, *Fusarium subglutinans*, *Fusarium polyphialidicum*, and *Fusarium oxysporum*, have also been included in the group of producers of these mycotoxins. Fumonisins constitute a group of 16 substances referred to as B1 (FB1, FB2, FB3, and FB4), A1, A2, A3, AK1, C1, C3, C4, P1, P2, P3, PH1a, and PH1b (Seo and Lee, 1999; Musser and Plattner, 1997). Only FB1, FB2, and FB3 are present in naturally contaminated foods (Soriano and Dragacci, 2004).

The IARC has evaluated the cancer risk of *fumonisins* to humans and classified them as group 2B (probably carcinogenic) (Rheeder et al., 2002). The carcinogenic character of *fumonisins* does not seem to involve interaction with DNA. On the other hand, its similarity with sphingosine suggests a probable intervention in the biosynthesis of sphingolipids. The inhibition of sphingolipid biosynthesis leads to serious problems related to cell activity, since these substances are essential for membrane composition, cell communication, intracellular as well as extracellular interactions, and growth factors.

The natural occurrence of *fumonisins* in home-grown corn was statistically associated with the high rate of human esophageal cancer in Africa (Rheeder et al., 1992; Sydenham et al., 1990; Thiel et al., 1992), northern Italy (Franceschi et al., 1990), Iran (Shephard et al., 2000), and the southeast United States (Gelderblom et al., 1992; Rheeder et al., 1992), and with the promotion of primary liver cancer in certain endemic areas of the People's Republic of China (Li et al., 2001). *Fumonisins* are also responsible for leukoencephalomalacia in equine species and rabbits (Marasas et al., 1988; Bucci et al., 1996); pulmonary edema and hydrothorax in pigs (Harrison et al., 1990); hepatotoxic, carcinogenic, and apoptotic (programmed cell death) effects in rat livers (Gelderblom et al., 1991, 1996; Pozzi et al., 2001); atherogenic effects in vervet monkey (Fincham et al., 1992); medial hypertrophy of pulmonary arteries in swine (Casteel et al., 1994); immunosuppression in poultry (Li et al., 1999); and brain hemorrhage in rabbits (Bucci et al., 1996).

A possible case of acute exposure to *fumonisin* B1 involved 27 villages in India, where consumption of unleavened bread made from moldy sorghum or corn caused transient abdominal pain, borborygmus, and diarrhea (Bhat et al., 1997). Finally, *fumonisins* can cause neural tube defects in experimental animals and thus may also have a role in human cases. It has been hypothesized that a cluster of anencephaly and spina bifida cases in southern Texas may have been related to *fumonisins* in corn products (Hendricks, 1999).

2.2.3 TRICHOTHECENES

Trichothecenes constitute a group of more than 180 metabolites produced by fungi of the genera *Fusarium*, *Myrothecium*, *Phomopsis*, *Stachybotrys*, *Trichoderma*, *Trichotecium*, *Verticimonosporium*, and possibly others. The term *trichothecene* is derived from trichothecin, the first member of the family to be identified. The trichothecenes are a family of related cyclic sesquiterpenoids, which are divided into four groups (types A–D) according to their characteristic functional groups. Type A and B trichothecenes are the most common. Type A is represented by neosolaniol, HT-2 toxin, T-2 toxin, T-2 triol, T-2 tetraol, and diacetoxyscirpenol (DAS). Type B is most frequently represented by deoxynivalenol (DON), 3-acetyl-DON, 15-acetyl-DON, nivalenol, and fusarenon X (Krska et al., 2007). All of the trichothecenes are characterized by the possession of a tetracyclic 12,13-epoxy trichothene skeleton. Despite the large number of molecules identified, few of them occur naturally. The most important trichothecenes are DON, nivalenol, toxin T-2, toxin HT-2, and DAS. Other trichothecenes are widely produced by the fungi *Myrothecium*, *Stachybotrys*, and *Trichothecium*. Notable among these are atranone, roridin, satratoxin, and verrucarin. Recent studies concluded that these mycotoxins are chemically stable during heating and can survive food processing (Bullerman and Bianchini, 2007; Cetin and Bullerman, 2006). Different trichothecenes have been detected in various grains, animal feed, animal products, beverages, cereals, and bakery products (Aniolowska and Steininger, 2014; Montes et al., 2012; Placinta et al., 1999;

Eriksen and Pettersson, 2004; Schothorst and Jekel, 2003; Juan et al., 2014). DON is the mycotoxin most commonly found in grains.

Human and animals are exposed simultaneously to several trichothecenes because (1) most fusariums are able to produce a number of mycotoxins simultaneously, (2) food commodities can be contaminated by several fungi simultaneously or in quick succession, and (3) a complete diet is made up of various different commodities. Humans may also be exposed to multiple trichothecenes via products from animals that have eaten contaminated feed. Oral exposure to these mycotoxins can result in alimentary hemorrhage and vomiting; direct contact causes dermatitis.

The trichothecenes are known for their strong capacity to inhibit eukaryotic protein synthesis, interfering in the initiation, elongation, and termination steps of protein synthesis (Bennett and Klich, 2003). The trichothecenes were compounds proven to be involved in the inhibition of peptidyl transferase activity (Shifrin and Anderson, 1999). At the molecular level, trichothecenes inhibit the peptidyl transferase reaction by binding to the 60S ribosomal subunit, suggesting that one of the cytotoxic mechanisms is translational inhibition. It was later shown that trichothecenes induce mitogen-activated protein kinases that induce the production of proinflammatory cytokines, implying the immune system as the most important target of trichothecenes (Shifrin and Anderson, 1999; Pestka et al., 2004). The exact mechanisms of trichothecenes' action are not fully understood, but when they are ingested in high doses by animals they cause nausea, vomiting, and diarrhea (Pestka, 2007). When ingested by pigs and other animals in small doses trichothecenes can cause weight loss and the refusal to eat (vomitoxin). Chronic intake of small amounts of trichothecenes leads to an increased susceptibility to infectious diseases as a result of the suppression of the immune system (Schlatter, 2004; Pestka, 2007), as well as other symptoms such as anorexia, anemia, neuroendocrine changes, and immunologic effects (Pestka and Smolinski, 2005; Pestka, 2007). It has been hypothesized that toxin T-2 and DAS are associated with alimentary toxic aleukia disease, which affected thousands of people in Orenburg, a region in the former Soviet Union, during World War II. Those affected had eaten grain infected with *Fusarium sporotrichioides* and *Fusarium poae*. The symptoms of the disease include inflammation of the skin, vomiting, and hepatic tissue damage.

2.2.4 ZEARALENONE

Zearalenone is a nonsteroidal β-resorcyclic acid lactone with the systematic name 3,4,5,6,9,10-hexahydro-14,16-dihydroxy-3-methyl-[S-(E)]-1H-2-benzoxacycl-otetradecin-1,7(8H)-dione. This is a secondary metabolite produced mainly by *Fusarium graminearum*, although other species, such as *Fusarium culmorum*, *Fusarium equisetii*, and *Fusarium crookwellense* also produce these substances and other analogs. These fungal species are widely found as contaminants in many countries. The classification of zearalenone as a toxin is considered inappropriate because, although it is biologically potent, it is rarely toxic. Its structure, in fact, resembles 7-β-estradiol, the main hormone produced in the human female ovary. Zearalenone would better fit the classification of a nonsteroidal estrogen or a mycoestrogen.

It is one of the most significant mycotoxins worldwide in maize, sorghum, wheat, and their derived foodstuffs and feed (Liu et al., 2012; Wu et al., 2011); zearalenone has been shown to competitively bind to the estrogen receptor and cause impaired fertility and abnormal fetal development in farm animals. In addition to its adverse hormonal effects, hepatic and renal lesions in rodents and the reduction of milk production in cows have also been observed (Baldwin et al., 1983;

Maaroufi et al., 1996; Dong et al., 2010). It also induces genotoxic effects by inducing DNA adducts, DNA fragmentation, apoptosis, micronuclei, and chromosome aberrations (Abid-Essefi et al., 2003). Hassen et al. (2005) demonstrated inhibition of cell proliferation and macronucleus synthesis in different cell lines. The association between the consumption of moldy grains and hyperestrogenism in pigs has been observed since 1920. High concentrations of zearalenone in pig feed may cause disturbances related to conception, abortion, and other problems. Reproductive problems have also been observed in cows and ovine species. Studies on experimental animals have not yet verified the carcinogenic capacity of zearalenone.

2.2.5 CITRININ

Citrinin was first isolated from secondary metabolites of *Penicillium citrinum*. Other species of *Penicillium* (*Penicillium expansum* and *Penicillium viridicatum*), and even of *Aspergillus* (*Aspergillus niveus* and *Aspergillus terreus*), also were subsequently confirmed to produce these substances. Certain isolates of *Penicillium camemberti*, used in cheese production, and *Aspergillus oryzae*, used in the production of Asiatic foods such as sake, miso, and soy sauce, can also produce citrinin. More recently, citrinin was isolated from the metabolites of the fungi *Monoascus ruber* and *Monoascus purpureus*, species that are used industrially in the production of red pigments.

Citrinin has been known to be nephrotoxic, hepatotoxic, and carcinogenic to humans and animals. Citrinin, like OTA, has been reported to be a potential risk factor for human Balkan endemic nephropathy, originally described as a chronic tubulointerstitial kidney disease in southeastern Europe (Bamias and Boletis, 2008). Citrinin was associated with yellow rice syndrome in Japan in 1971 because of the regular presence of *P. citrinum* in this food product. It has also been considered responsible for nephropathy in pigs and other animals, although its acute toxicity varies depending on the animal species. Oat, rye, barley, corn, and wheat grains are excellent substrates for the formation of citrinin (Li et al., 2012; Dietrich et al., 2001). This mycotoxin, which is present in the structure of polyketide, has also been found in products naturally colored with pigments of *Monoascus*, as well as sausages naturally fermented in Italy.

2.2.6 PATULIN

Patulin (4-hydroxy-4H-furo[3,2-c]pyran-2(6H)-one) is a metabolite that was first isolated as a substance with antimicrobial properties from the fungus *Penicillium patulum*, later called *Penicillium urticae* and currently known as *Penicillium griseofulvum*. Patulin was later isolated from other fungal species, too, and received different names, such as clavacin, claviformin, expansin, micoine C, and penicidin. It was used as a nose and throat spray in the treatment of the common cold and as an ointment for the treatment of skin infections. During the 1960s, however, it became clear that although it did have antibacterial, antiviral, and antiprotozoan activity, patulin was toxic to animals and plants. After this revelation it was reclassified as a true mycotoxin. The disease known as blue mold, which is common in apple, pear, cherry, and other fruits, is caused by the fungus *P. expansum*, currently considered the most efficient producer of patulin in nature. Patulin is commonly found in nonfermented juices and other food products (Pique et al., 2013; Zaied et al., 2013), but it is not resistant to fermentation in products derived from cider, where it is efficiently metabolized by yeasts.

Patulin caused gastrointestinal effects such as distension, ulceration, and hemorrhage in acute and short-term in vivo studies. Recent studies have also demonstrated that patulin alters the intestinal barrier's function. In chronic studies of rats, patulin causes neurotoxicity, immunotoxicity, and genotoxicity. Reproductive and teratogenicity in vivo studies showed that patulin is embryotoxic. Regarding its potential as a human carcinogen, it has been classified as group 3 (not classifiable as to its carcinogenicity to humans) by the IARC (Moake et al., 2005). Patulin has electrophilic properties and high reactivity to cellular nucleophiles. At a cellular level it can cause enzyme inhibition and chromosomal damage. Patulin causes cytotoxic and chromosome-damaging effects, mainly by forming covalent adducts with essential cellular thiols (Fliege and Metzler, 2000; Glaser and Stopper, 2012).

2.2.7 OCHRATOXIN A

OTA was discovered in 1965 as a metabolite of *Aspergillus ochraceus*. It has a chemical structure similar to that of aflatoxins, being represented by an isocoumarin substitute bound to an L-phenyl-alanine group. Not all of the isolates of *A. ochraceus* are capable of producing OTA. *Aspergillus alliaceus*, *Aspergillus auricomus*, *Aspergillus carbonarius*, *Aspergillus glaucus*, *Aspergillus meleus*, and *Aspergillus niger*, as well as *Penicillium nordicum* and *Penicillium verrucosum*, are also producers of OTA. Because *A. niger* is a species widely used in industry for the production of enzymes and citric acid for human consumption, it is important to certify that industrial isolates are not producers of OTA. This toxin can contaminate a wide variety of foods as a result of fungal infection in crops, in fields during growth, at harvest, or during storage and shipment. In addition to cereals and cereal products, it is also found in a range of other food commodities, including coffee, cocoa, wine, beer, pulses, spices, dried fruits, grape juice, pig kidney, and other meat and meat products of nonruminant animals exposed to feed contaminated with this mycotoxin (Petzinger and Weidenbach, 2002; Ozden et al., 2012; Bircan, 2009; Shundo et al., 2009; Amezqueta et al., 2005; Brera et al., 2011; Gopinandhan et al., 2008; Coronel et al., 2011; Battilani et al., 2006; Kabak, 2009).

This toxin has been shown to be nephrotoxic, hepatotoxic, teratogenic, and immunotoxic. The IARC classified OTA in group 2B (possibly carcinogenic). It has been found in the blood and other tissues of animals and in milk, including human milk, as well as in pork meat intended for human consumption. OTA has been found to be responsible for pig nephropathy, which has been widely studied in Scandinavian countries. The disease is endemic among pigs in Denmark, where it is also associated with bird deaths. Studies have revealed that although small quantities of OTA can survive the processing and metabolism of pigs and birds, it is improbable that this could be detected in milk or bovine meat.

2.3 HEAVY METALS

The main threats to human health from heavy metals are associated with exposure to chromium, lead, cadmium, mercury, and arsenic. These metals have been extensively studied and their effects on human health regularly reviewed by international bodies such as WHO. Heavy metals have been used by humans for thousands of years. Although several adverse health effects of heavy metals have been known for a long time, exposure to heavy metals continues, and is even increasing in some parts of the world, particularly in less developed countries. In general, lethal heavy metals can interact with DNA and proteins, causing oxidative degradation of biological macromolecules. Thus the process of a breakdown of metal−ion homeostasis has been implicated in a plethora of diseases. Metals are known to modulate gene expression by interfering with signal transduction pathways that play important roles in cellular growth and development. Deviation in the growth and differentiation of cell is a typical characteristic of the cancer phenotype. Metals can activate various factors that ultimately control cell cycle progression and apoptosis. There are two types of metal: one is redox-active metals such as iron, copper, chromium, and cobalt, which may undergo cyclic reactions participating in the transfer of electrons between metals and substrates; therefore they may play an important role in the maintenance of redox homeostasis, a phenomenon tightly linked with metal homeostasis. However, an abnormal change in metal homeostasis may lead to the formation of deleterious free radicals participating in modifications to DNA bases, enhanced lipid peroxidation, and altered calcium and sulfydryl homeostasis. The second type of metals are redox-inert elements such as cadmium and arsenic, which have no known biological function and are even known to be toxic at low concentrations. In general, the primary route for their toxicity and carcinogenicity is depletion of glutathione, bonding to sulfydryl groups of proteins and other mechanisms of action. Humans may be exposed to both types of these elements from a variety of natural sources, including contaminated food.

2.3.1 CHROMIUM

Chromium (Cr), one of the most common elements, exists in several oxidation states (Manova et al., 2007). The most important stable states are 0 (elemental metal), +III, and +VI. The health effects and toxicity/carcinogenicity of Cr are primarily related to the oxidation state of the metal at the time of exposure. Trivalent (Cr[III]) and hexavalent (Cr[VI]) compounds are thought to be the most biologically significant.

The intake of Cr by humans occurs via contamination from food and beverage processing and packaging (Berg et al., 2000). Aquatic species can be affected by Cr concentrations from sediments and waters. Crab and fish are marine species frequently consumed by humans and considered to be potential transporters of Cr through the food web. Acids contained in fruits (malic and citric) can remove Cr from stainless steel cooling vats during the process of packing canned fruit. Ascorbic acid and other reducing agents convert Cr(VI) to Cr(III) inside cells, and Cr(III) is strongly bound to macromolecules such as cysteine or glutathione in ternary complexes. Trivalent Cr can also be co-ordinated with two guanine bases from DNA, causing DNA alterations and mutations.

Cr(VI) at high doses is considered to cause the greatest health risk (Carlosena et al., 1997). Its toxic effect on biological systems is attributed to the ability of Cr(VI) to migrate across the cell membrane, thus enhancing the intracellular Cr concentration (Bobrowski et al., 2009). Adverse health effects occurring after oral exposure include gastrointestinal symptoms, hypotension, and hepatic and renal failure. An increase in the rate of stomach tumors was observed in humans and animals exposed to Cr(VI) in drinking water. Sperm damage and damage to the male reproductive system have also occurred in laboratory animals exposed to Cr(VI). Recent studies using cell cultures revealed a much greater potential for Cr(VI) to cause chromosomal damage and mutations than was previously expected.

2.3.2 **CADMIUM**

Food is the main source of cadmium (Cd) among the nonsmoking population. Fertilizers produced from phosphate ores constitute a major source of spread of Cd pollution. In addition, the inappropriate disposal of Cd-containing wastes has increased its emission in populated areas around the world (Jarup, 2003). It has been found in various food types (Bjerregaard et al., 2005). Cd is an element that represents serious hazards because it can be absorbed via the alimentary tract, penetrates the placenta during pregnancy, and damages membranes and DNA. Furthermore, Cd is the metal of most concern because it is the "only metal that might pose human or animal health risks at plant tissue concentrations that are not generally phytotoxic (Peijnenburg et al., 2000)." The preliminary results of a number of studies of total dietary Cd currently in progress indicate that dietary intake probably varies according to country, from ≤ 50 to 150 µg/day. According to Sridhara Chary et al. (2008) populations that restrict their diet to locally grown produce, such as subsistence farmers, are particularly at risk from soil contamination because the Cd in their diet is not diluted by food from other noncontaminated areas, as it is in the majority of the developed world. The IARC (1993b) classified Cd as a human carcinogen. The values for provisional tolerable dietary intake set by the Food and Agriculture Organization of the WHO is 1 g Cd/kg body weight.

Different sources of Cd, such as dairy products and vegetables, must be considered as risk factors. The highest Cd concentrations are found in rice, wheat, oyster, mussels, and the kidney cortex of animals (Oymak et al., 2009). In several places around the world, cropland is irrigated with wastewater, and studies of metal transfer from crops to humans are just starting to be performed in these countries. Among crop plants, rice has a special place because of its capacity to absorb Cd. Chaney et al. (2004) reported that rice has the "ability to accumulate soil Cd in grains, excluding Fe, Zn and Ca (even though the soil contains 100 times more Zn than Cd)." This poses a real threat for farmers consuming polished rice that is deficient in iron, zinc, and calcium. The daily intake of Cd from crop plants normally cultivated in wastewater-irrigated land seems to be too low, based on health risk indices, to pose a threat for the human population (Khan et al., 2008). A study that included data collected for 16 years, however, found that in the Jinzu River basin, the increased total Cd intake by humans seems to be related to an adverse effect of this element on life span (Kobayashi et al., 2002). These researchers found that the mortality rate among people ingesting more than 2.0 mg Cd from rice cultivated in a Cd-polluted area was higher compared with people ingesting <2.0 mg of Cd.

In mammals, Cd is transported as a Cd−protein complex, especially a Cd−metallothionein complex (Wong and Rainbow, 1986). This complex is distributed to various tissues and organs and is ultimately reabsorbed in kidney tubuli. There is no mechanism for the excretion of Cd in humans; thus it accumulates in tissues. The half-life of Cd in the kidney cortex is 20−35 years. In humans, the largest amount of Cd is deposited in the kidneys, liver, pancreas, and lungs (Davis et al., 2006). Cd itself is unable to generate free radicals directly; however, indirect formation of reactive oxygen species (ROS) and reactive nitrogen species (RNS) involving superoxide radicals, hydroxyl radicals, and nitric oxide has been reported. Some experiments also confirmed the generation of nonradical hydrogen peroxide, which itself may in turn be a significant source of radicals via Fenton chemistry. Cd can activate cellular protein kinases (protein kinase C), which results in enhanced phosphorylation of various transcription factors, which in turn leads to activation of target gene expression. An interesting mechanism explaining the indirect role of Cd in free radical generation was presented, in which it was proposed that Cd can replace iron and copper in various cytoplasmic and membrane proteins (e.g., ferritin, apoferritin), thus increasing the amount of unbound free or poorly chelated copper and iron ions participating in oxidative stress via Fenton reactions (Boffetta, 2014). Displacement of copper and iron by Cd can explain the enhanced Cd-induced toxicity: copper, displaced from its binding site, is able to catalyze the breakdown of hydrogen peroxide via the Fenton reaction. The toxic mechanisms of Cd are not well understood, but it is known to act intracellularly, mainly via free radical–induced damage, particularly to the lungs, kidneys, bone, central nervous system, reproductive organs, and heart. The testis is a good marker of Cd exposure. Cd-induced testicular damage and testicular necrosis have been documented by many researchers. Various studies of Cd-induced testicular toxicity in rat models have been carried out. Cd is a potent human carcinogen that causes preferentially prostate, lung, and gastrointestinal (kidney and pancreas) cancers. New findings in the explanation of Cd-induced carcinogenicity with respect to cell adhesion have recently been published. E-cadherin, a transmembrane Ca(II)-binding glycoprotein playing an important role in cell–cell adhesion, can bind Cd to Ca(II) binding regions, changing the glycoprotein conformation. Thus the disruption of cell–cell adhesion induced by Cd could play an important role in tumor induction and promotion.

2.3.3 ARSENIC

Most of arsenic (As) compounds are colorless and do not have any smell. The presence of As in food, water, or air is a serious human health risk. It can get into the human body in several ways, including water and food (Abernathy et al., 2001; EFSA, 2009). Inorganic As (iAs) (arsenite or As(III) and arsenate or As(V)) is considered the most dangerous form because of its biological availability, as well as its physiological and toxicological effects (iAs is classified as a nonthreshold, class 1 human carcinogen) (ATSDR, 2007). Because of the toxicity of As, the WHO has set a value of 2 g As/kg body weight/day as the tolerable daily intake. In countries where the basic diet consists mostly of vegetables, however, the stated value is not observed when cropland or irrigation water contains high concentrations of As. Worldwide, rice is one of the basic crops known for its high potential to accumulate As (Meharg et al., 2008; Signes-Pastor et al., 2009). This is because it is often watered with groundwater containing elevated As concentrations. In some countries, the As concentrations in rice are 10-fold higher when compared with that of other cereal crops. The situation is worse in countries where food is cooked with water containing As. iAs represents the highest risk for human dietary intake since it can bind to important molecules such as proteins or DNA. Trophic transfer of As occurs mainly in countries like India and Bangladesh, where the health effects of As in people are well documented.

In addition to As intake from cereals and vegetables, consumption of certain seafood can also be a potential source of As. The As concentration in marine animals can be double or triple compared with As in terrestrial foods. The dietary intake of As via seafood depends on the type of animal tissue consumed by humans. Cooking processes such as boiling, baking, and stewing, among others, can increase organic As and iAs in bivalves and squids.

The exact molecular mechanism of As toxicity and carcinogenesis is still not known. Current views of molecular mechanisms of As toxicity involve genetic changes, the involvement of increased oxidative stress, enhanced cell proliferation, and altered gene expression. As is a well-documented carcinogen in a number of studies. Chronic exposure to iAs from contaminated water is responsible for various adverse health effects such as tumors of the lung, skin, liver, bladder, and kidney. Skin lesions, peripheral neuropathy, and anemia are hallmarks of chronic As exposure. As is also a potential risk factor for atherosclerosis.

2.3.4 **LEAD**

Lead (Pb) is a persistent heavy metal, and because of its unusual physicochemical properties, it is used in various industrial applications. Pb is a toxic metal to humans and animals, and its persistence causes prolonged occurrence in the environment (in water, soil, dust) and in manufactured products containing Pb (Aboufazeli et al., 2013). Gastrointestinal absorption of Pb is higher among children (40–50%) than adults (3–10%). Pb toxicity is most commonly diagnosed through elevated blood concentrations. Blood concentrations of ≥ 10 g/dL (equivalent to 0.48 mol/L) are considered toxic and result in neurological disorders, cognitive impairments, hypertension, and other disorders.

The main way of exposure of humans and mammals to Pb is via the food chain (Oymak et al., 2009; Gama et al., 2006). Researchers have reported that in humans, two binding polypeptides are responsible for Pb binding in the kidneys: thymosin and acyl-coA binding protein. These researchers reported that these Pb-binding proteins have a kilo dalton (kD) of ≈ 14 nM and account for more than 35% of the Pb in kidney cortex tissue. In humans there is a correlation between Pb exposure and hearing loss. In addition to ROS, RNS have also been shown to play a significant role in the incidence of hypertension following Pb exposure in humans. Nitric oxide is known as an endothelium-derived relaxing factor. ROS formed as a consequence of Pb exposure may oxidize nitric oxide in vascular endothelial cells by forming peroxynitrite ($ONOO^-$), which is a highly reactive ROS capable of damaging DNA and lipids. Depleted NO• following Pb exposure causes hypertension in animal models.

2.3.5 **MERCURY**

Plants represent the main entry pathway for potentially health-threatening toxic metals into human and animal food. Agencies (both regulatory and health) overseeing food and health sectors have placed high priorities on monitoring, evaluating, and reducing risk to humans and wildlife from exposure to mercury (Hg). Hg ranks third on the Comprehensive Environmental Response, Compensation, and Liability Actlist of substances in terms of the risk that it poses to human morbidity and mortality. Exposure to Hg, both directly and through the food chain, is of significant concern.

Mercury contamination through organomercurials is more harmful than Hg itself. For example, organomercurials are readily absorbed into fish, where it travels through the food chain. Levels of mercurials result in contaminant concentrations that vary from highly toxic concentrations from an accidental spill to barely detectable concentrations that, after long-term exposure, can be detrimental to human health. The impact of ingested Hg on human populations depends on the form of Hg ingested. Elemental Hg (Hg(0)), when acutely or chronically ingested, results in a broad series of deleterious symptoms. Furthermore, oxidation of lipophilic Hg(0) results in its accumulation in the brain and liver. Organic Hg in the form of methyl and ethyl mercury makes its way up the food chain through fish. The major toxic effects of methyl-mercury (MeHg) occur in the central nervous system. In documented cases of Hg poisoning in Japan and Iraq, there were many resulting deaths. In addition, the surviving victims and their offspring displayed mental retardation, cerebral palsy, and muteness. This was particularly observed in children exposed while in the fetal stage. It has been demonstrated that $\sim 95\%$ of MeHg ingested by fish is absorbed in the gastrointestinal tract. Furthermore, MeHg crosses the placental barrier, and concentrations in fetal brains are five to seven times higher than in maternal blood. MeHg enters tissue via modification to sulfur adducts. Hg cysteine complexes are able to enter the endothelial cells of the blood–brain barrier. Although inorganic mercurials also pose a hazard to human health, they tend to induce fatigue, insomnia, weight loss, hypersalivation, renal dysfunction, and neurological disorders. Table 2.1 summarizes common contaminants and their sources among different foods.

Table 2.1 Common Contaminants and Their Sources in Different Food Items

Food Item	Category	Contaminants	Source(s)
A. Liquid food Drinking water			
	Microorganisms	*Cryptosporidium*	• Human and animal fecal waste
		Giardia lamblia	• Human and animal fecal waste
		Heterotrophic bacteria	• Heterotrophic plate count measures a range of bacteria that are naturally present in the environment
		Legionella	• Found naturally in water; multiplies in heating systems
		Total coliforms (including fecal coliform and *Escherichia Coli*)	• Coliforms are naturally present in the environment as well as feces; fecal coliforms and *E. coli* only come from human and animal fecal waste, soil runoff
		Viruses (enteric)	• Human and animal fecal waste
	Disinfection by-products	Bromate	• By-product of drinking water disinfection
		Chlorite	• By-product of drinking water
		Haloacetic acids	• By-product of drinking water
		Total trihalomethanes	• By-product of drinking water
	Disinfectants	Chloramines (as Cl_2)	• Water additive used to control microbes
		Chlorine (as Cl_2)	• Water additive used to control microbes
		Chlorine dioxide (as ClO_2)	• Water additive used to control microbes
	Inorganic chemicals	Antimony	• Discharge from petroleum refineries; fire retardants; ceramics; electronics; solder
		Arsenic	• Erosion of natural deposits; runoff from orchards, runoff from glass and electronics production waste
		Asbestos (fiber $>10\ \mu m$)	• Decay of asbestos cement in water mains; erosion of natural deposits
		Cadmium	• Corrosion of galvanized pipes; erosion of natural deposits; discharge from metal refineries; runoff from waste batteries and paints
		Chromium (total)	• Discharge from steel and pulp mills; erosion of natural deposits
		Copper	• Corrosion of household plumbing systems; erosion of natural deposits
		Cyanide (as free cyanide)	• Discharge from steel/metal factories; discharge from plastic and fertilizer factories

Table 2.1 Common Contaminants and Their Sources in Different Food Items—cont'd

Food Item	Category	Contaminants	Source(s)
		Fluoride	• Water additive that promotes strong teeth; erosion of natural deposits; discharge from fertilizer and aluminum factories
		Lead	• Corrosion of household plumbing systems; erosion of natural deposits
		Mercury (inorganic)	• Erosion of natural deposits; discharge from refineries and factories; runoff from landfills and croplands
		Nitrates (measured as nitrogen)	• Runoff from fertilizer use; leakage from septic tanks, sewage; erosion of natural deposits
		Nitrites (measured as nitrogen)	• Runoff from fertilizer use; leakage from septic tanks, sewage; erosion of natural deposits
		Selenium	• Discharge from petroleum refineries; erosion of natural deposits; discharge from mines
		Thallium	• Leaching from ore-processing sites; discharge from electronics, glass, and drug factories
	Organic chemicals	Acrylamide	• Added to water during sewage/wastewater treatment
		Benzene	• Discharge from factories; leaching from gas storage tanks and landfills
		Benzopyrene (PAHs)	• Leaching from linings of water storage tanks and distribution lines
		Carbofuran	• Leaching of soil fumigant used on rice and alfalfa
		Chlordane	• Residue of banned termiticide
		Chlorobenzene	• Discharge from chemical and agricultural chemical factories
		2,4-Dichlorophenoxyacetic acid	• Runoff from herbicide used on row crops
		Dalapon	• Runoff from herbicide used on rights of way
		Dioxin (2,3,7,8-tetrachlorodibenzodioxin)	• Emissions from waste incineration and other combustion; discharge from chemical factories
		Diquat	• Runoff from herbicide use
		Endothall	• do-
		Endrin	• Residue of banned insecticide
		Glyphosate	• Runoff from herbicide use
		Heptachlor	• Residue of banned termiticide

Continued

Table 2.1 Common Contaminants and Their Sources in Different Food Items—cont'd

Food Item	Category	Contaminants	Source(s)
		Lindane	• Runoff/leaching from insecticide used on cattle, lumber, gardens
		Methoxychlor	• Runoff/leaching from insecticide used on fruits, vegetables, alfalfa, livestock
		Polychlorinated biphenyls	• Runoff from landfills; discharge of waste chemicals
	Miscellaneous Plasticizer	Bisphenol A	• Plastic manufacturing; combustion of domestic waste; natural breakdown of plastics in environment; landfill leachates; linings for food and beverage packaging (can migrate from containers into a variety of foods and beverages)
	Perfluorinated compounds	Perfluorooctane sulfonate	• Manufacturing wastes; via volatilization, oxidation, and precipitation from pesticides, phencyclidine, flame retardants, coatings, food-packaging, etc.; landfill leachates
	Surfactants	4-Nonylphenol monoethoxylate (NP_1EO)	• Application of raw sludge to soils followed by landfill leachates; WWTP effluents
	Antibiotics	Sulfamethoxazole	• Manufacturing wastes; domestic and farm disposal of unused, expired antibiotics; land use of animal manure and sewage sludge; WWTP effluents
	Pharmaceuticals	Acetaminophen	• Manufacturing wastes; unused and expired drugs disposed by households; landfill leachates; WWTP effluents
	Hormones	Estrone (E1)	• Hormones injected to livestock and fish released from animal farms and aquacultures, respectively; human and animal excreta via WWTP effluents
	Artificial sweeteners	Sucralose	• Direct release from food industries, households, animal farming; WWTP effluents as excretion after consumption without undergoing change within the human body
	Antimicrobial preservatives	Benzylparaben	• Manufacturing wastes; runoff through use in cosmetics, toiletries, pharmaceuticals, food; WWTP effluents
	Toxins	Algal toxins	• Microcystin, many species of cyanobacteria produce toxins in water impacted by algal blooms

Table 2.1 Common Contaminants and Their Sources in Different Food Items—cont'd

Food Item	Category	Contaminants	Source(s)
Milk			
	Veterinary drug residue		
	Antibiotics	Benzylpenicillin	• Treatment of various systemic diseases
		Tetracycline	
		Oxytetracycline	
		Chlortetracycline	
		Trimethoprim	
		Ceftiofur	
		Streptomycin	
		Oxfendazole	
		Sulfonamides	
	Parasiticide drugs	Albendazole	• Anthelmentic drug
	Hormones	Progesterone	
		Estrone	
		Testosterone	
		BGH or BST (a protein hormone that increases milk production in cows)	
	Toxins	Aflatoxin	• Through ingestion of infected feed
		Fumonisins	
		Trichothecenes	
		Zearalenone	
		Citrinine	
		Ochratoxin	
	Heavy metals	Arsenic	• Through ingestion of contaminated feed and water
		Chromium	
		Lead	
		Cadmium	
	Microorganisms	*Brucella*	• Through infected animal, environment, fecal contamination, milking man and bad managemental practices
		E. coli	
		Staphylococcus sp.	
		Streptococcus sp.	
		Listeria sp.	
		Mycobacterium tuberculosis, Mycobacterium bovis	
		Salmonella	
		Clostridium	
		Yeast and fungus	
	Pesticides/insecticides	HCB	• Through environment (during pest/insect control)
		Aldrin-dieldrin	

Continued

Table 2.1 Common Contaminants and Their Sources in Different Food Items—cont'd

Food Item	Category	Contaminants	Source(s)
Edible oils		Dieldrin HCH DDT	
	Processing contaminants	MCPD	• Formed during the industrial processing of food oils during deodorization, processing contaminants
		Glycidyl esters	• Formed during the industrial processing of food oils during deodorization, processing contaminants
		Phthalates	• Formed during processing and storage by reaction with plastics, processing contaminants
	Mineral oil	Various types	• Man-made adulteration
	Weeds/other oils	Various types	• Due to contamination of source with weeds/by-products
	Fungal toxins	As described earlier in milk	• Due to faulty storage conditions (growth of toxin-producing fungal species/release of toxins before processing of oilseeds)
	Heavy metals	As described earlier in milk	• Environmental contaminants
	Pesticides	As described earlier in milk	• Used for pest control
	Insecticides	As described earlier in milk	• Used for insect control
Fruit juices/beverages	Heavy metals	As described earlier in milk	• Environmental contaminants
	Pesticides	As described earlier in milk	• Used for pest control
	Insecticides	As described earlier in milk	• Used for insect control
	Fungicides	Tetraconazole	• Is a broad-spectrum systemic fungicide with protective, curative, and eradicant properties; expected to be persistent and moderately to slightly mobile in soil, so it tends to accumulate in soil and has the potential to reach surface water via runoff and spray drift.

Table 2.1 Common Contaminants and Their Sources in Different Food Items—cont'd

Food Item	Category	Contaminants	Source(s)
		Tebuconazole	• Basic triazole fungicide widely used to control rusts, powdery mildews, and scabs. It has also been detected in many foods including fruits, vegetables, beverages, and wheat products.
	Microorganisms	Many	• Through infected fruits, environment, fecal contamination, bad management practices
	Yeast	*Saccharomyces cerevisiae* *Candida lambica* *Candida sake* *Rhodotorula rubra* *Geotrichum* spp. *Penicillium* *Fusarium* spp. Mold (*Aspergillus niger*)	• Contamination from various sources
	Fungal toxin	Patulin, others, as described earlier in milk	• Toxic metabolite produced by certain species of *Aspergillus*, *Penicillium*, and *Byssochlamys*; mainly found as a naturally occurring contaminant in apples and apple products, and occasionally found in other fruits such as pears, apricots, peaches, and grapes.
	Processing contaminants	HMF	• A common process contaminant that widely occurs in a range of processed foods; formed during heating and/or storage of foods as a result of the Maillard reaction and/or sugar caramelization. Honey, jams, juice concentrates, roasted coffee, caramels, balsamic vinegar, and dried fruits are among the foods containing significant amounts of HMF. Increasing temperature also increases the rate of HMF formation in foods during processing and storage. In addition, acidic conditions strongly accelerate HMF formation. The level of HMF is usually considered to be an index of product quality in processed foods.
	Benzene		• In infant nutrition, specifically in infant carrot juices. Many of these juices contain higher concentrations of benzene than are found in any other

Continued

Table 2.1 Common Contaminants and Their Sources in Different Food Items—cont'd

Food Item	Category	Contaminants	Source(s)
			beverage group, with an average content above the EU drinking water limit of 1 µg/L
B. Solid foods Food grains and oilseeds			
	Fungal toxin	As described earlier in milk	• Same source as described earlier
	Microorganisms	Various types	• Same source as described earlier
	Heavy metals	As described earlier in milk	• Same source as described earlier
	Yeast and molds	Various types	• Same source as described earlier
	Weeds	Various types	• Same source as described earlier
	Pesticides	As described earlier in milk	• Same source as described earlier
	Insecticides	As described earlier in milk	• Same source as described earlier
	Fungicides	As described earlier in fruit juices	• Same source as described earlier
Baked and other prepared food products			
	Processing contaminants	Acrylamide	• Via the Maillard reaction, which comprises a series of nonenzymic reactions between sugars and amino groups, principally those of amino acids. It takes place only at high temperatures and occurs mainly in cooked foods prepared by frying, baking, and roasting.
		Furans	• Furans form through several pathways, one of which is the thermal degradation of PUFAs. PUFAs are prone to oxidation during cooking and high-temperature processing, giving rise to lipid peroxides. One of these, 4-hydroxy-2-butenal, can form furan through cyclization and dehydration; another route is within the Maillard reaction, whereby the thermal degradation of free amino acids in the presence of reducing sugars gives rise to glycolaldehyde and acetaldehyde. These intermediates undergo aldol addition

Table 2.1 Common Contaminants and Their Sources in Different Food Items—cont'd

Food Item	Category	Contaminants	Source(s)
			to 2-deoxyaldotetrose, which further reacts to form furan. Furans can also form directly from sugar breakdown, via the intermediates 1-deoxyosone and 3-deoxyosone, and aldotetrose and its derivatives, 2-deoxyaldotetrose and 2-deoxy-3-ketoaldotetrose. Degradation of ascorbic acid and dehydroascorbic acid also gives rise to furan via aldotetrose and its derivatives.
		HMF	• HMF arises via the dehydration of fructose, used as an indicator of excessive heat treatment in biscuit manufacturing, partly because products containing high concentrations of HMF may also contain a lot of acrylamide.
	Fungal toxins	As described earlier in milk	• Same source as described earlier; heat-labile toxins are destroyed during heat processing and not found in heat-treated products.
	Microorganisms	Various types	• Same source as described earlier; heat-labile microbes are destroyed during heat processing and not found in heat-treated products; however, postprocessing cross-contamination may be a source of their presence in final products.
	Heavy metals	As described earlier in milk	• Same source as described earlier
	Yeast and molds	Various types	• Same source as described earlier; heat-labile organisms are destroyed during heat processing and not found in heat-treated products; however, postprocessing cross-contamination may be a source of their presence in final products; also formed during bad storage practices.
	Weeds	As described earlier	• Same source as described earlier
	Pesticides	As described earlier in milk	• Same source as described earlier
	Insecticides	As described earlier in milk	• Same source as described earlier
	Fungicides	As described earlier in milk	• Same source as described earlier

Continued

Table 2.1 Common Contaminants and Their Sources in Different Food Items—cont'd

Food Item	Category	Contaminants	Source(s)
Fruits and vegetables			
	Fungal toxins	As described earlier in milk	• Same source as described earlier
	Microorganisms	Various types	• Same source as described earlier
	Heavy metals	As described earlier in milk	• Same source as described earlier
	Yeast and molds	As described earlier in milk	• Same source as described earlier
	Pesticides	As described earlier in milk	• Same source as described earlier
	Insecticides	As described earlier in milk	• Same source as described earlier
	Fungicides	As described earlier	• Same source as described earlier
Meat and meat products			
	Microorganisms	Various types	• Same source as described earlier, heat-labile microbes are destroyed during heat processing and not found in heat-treated products; however, postprocessing cross-contamination may be a source of their presence in final products.
	Heavy metals	As described earlier in milk	• Same source as described earlier; due to the accumulation of some heavy metals in some organs, like the liver and kidney
	Yeast and molds	Various types	• Same source as described earlier; heat-labile organisms are destroyed during heat processing and not found in heat-treated products; however, postprocessing cross-contamination may be a source of their presence in final products; also formed during bad storage practices.
	Pesticides	As described earlier in milk	• Same source as described earlier; due to the accumulation in some organs, like the liver and kidney; some metabolites may accumulate in muscles
	Insecticides	As described earlier in milk	• Same source as described earlier; due to the accumulation in some organs, like the liver and kidney, some

Table 2.1 Common Contaminants and Their Sources in Different Food Items—cont'd

Food Item	Category	Contaminants	Source(s)
			metabolites may accumulate in muscles
	Fungicides	As described earlier	• Same source as described earlier; due to the accumulation in some organs, like the liver and kidney, some metabolites may accumulate in muscles
Fish and seafoods	Microorganisms	Various types	• Same source as described earlier; heat-labile microbes are destroyed during heat processing and not found in heat-treated products; however, postprocessing cross-contamination may be a source of their presence in final products.
	Heavy metals	As described earlier in milk	• Through water; due to the accumulation of some heavy metals in some organs, like the liver and kidney
	Yeast and molds	Various types	• During storage, heat-labile organisms are destroyed during heat processing and not found in heat-treated products; however, postprocessing cross-contamination may be a source of their presence in final products; also formed during bad storage practices
	Pesticides	As described earlier in milk	• Feed and water; due to the accumulation in some organs, like the liver and kidney; some metabolites may accumulate in muscles
	Insecticides	As described earlier in milk	• Feed and water; due to the accumulation in some organs, like the liver and kidney; some metabolites may accumulate in muscles
	Fungicides	As described earlier	• Feed and water, due to the accumulation in some organs, like the liver and kidney; some metabolites may accumulate in muscles
Animal feed and concentrates	Microorganisms	Various types	• Same source as described earlier
	Heavy metals	As described earlier in milk	• Same source as described earlier
	Yeast and molds	Various types	• Same source as described earlier

Continued

Table 2.1 Common Contaminants and Their Sources in Different Food Items—cont'd

Food Item	Category	Contaminants	Source(s)
	Pesticides	As described earlier in milk	• Same source as described earlier
	Insecticides	As described earlier in milk	• Same source as described earlier
	Fungicides	As described earlier	• Same source as described earlier
Tomato puree and pastes			• Same as described for fruits and vegetables
Fruit jams and jellies			• Same as described for fruits and vegetables
Pickles, curries, and other ethnic food products			• Same as described for fruits and vegetables
Honey and syrups			• Same as described for fruits and vegetables, acaricides, bee repellents

BGH, bovine growth hormone; BST, barium strontium titanate; DDT, dichloro diphenyl trichloroethane; HCB, hexachlorobenzene; HCH, hexachlorocyclohexane; HMF, hydroxymethylfurfural; MCPD, monochloro propanediol; PAH, polycyclic aromatic hydrocarbons; PUFA, polyunsaturated fatty acid; WWTP, wastewater treatment plant.

REFERENCES

Abernathy, C., Chakraborti, D., Edmonds, J.S., Gibb, H., Hoet, P., Hopenhayn-Rich, C., 2001. Environmental health criteria for arsenic and arsenic compounds. Environ. Health Criter. 1–521.

Abid-Essefi, S., Baudrimont, I., Hassen, W., Ouanes, Z., Mobio, T.A., Anane, R., Creppy, E.E., Bacha, H., 2003. DNA fragmentation, apoptosis and cell cycle arrest induced by zearalenone in cultured DOK, Vero and Caco-2 cells: prevention by Vitamin E. Toxicology 192 (2–3), 237–248.

Aboufazeli, F., Lotfi Zadeh Zhad, H.R., Sadeghi, O., Karimi, M., Najafi, E., 2013. Novel ion imprinted polymer magnetic mesoporous silica nano-particles for selective separation and determination of lead ions in food samples. Food Chem. 141 (4), 3459–3465.

Aksoy, M., Dincol, K., Akgun, T., Erdem, S., Dincol, G., 1971. Haematological effects of chronic benzene poisoning in 217 workers. Br. J. Ind. Med. 28 (3), 296–302.

Aksoy, M., Ozeris, S., Sabuncu, H., Inanici, Y., Yanardag, R., 1987. Exposure to benzene in Turkey between 1983 and 1985: a haematological study on 231 workers. Br. J. Ind. Med. 44 (11), 785–787.

Amezqueta, S., Gonzalez-Penas, E., Murillo, M., Lopez de Cerain, A., 2005. Occurrence of ochratoxin A in cocoa beans: effect of shelling. Food Addit. Contam. 22 (6), 590–596.

Aniolowska, M., Steininger, M., 2014. Determination of trichothecenes and zearalenone in different corn (*Zea mays*) cultivars for human consumption in Poland. J. Food Compos. Anal. 33 (1), 14–19.

ATSDR, 2000. Agency for Toxic Substances and Disease Registry, Toxicological Profile for Polychlorinated Biphenyls (PCBs). Toxicological Profiles ATSDR, Atlanta, GA. http://www.atsdr.cdc.gov/ToxProfiles/tp17.pdf (accessed 27.03.11).

ATSDR, 2007. Agency for Toxic Substances and Disease Registry, Toxicological Profile for Arsenic. Department of Health and Human Services, Public Health Service, United States.

Baars, A.J., Bakker, M.I., Baumann, R.A., Boon, P.E., Freijer, J.I., Hoogenboom, L.A.P., Hoogerbrugge, R., van Klaveren, J.D., Liem, A.K.D., Traag, W.A., de Vries, J., 2004. Dioxins, dioxin-like PCBs and non-dioxin-like PCBs in foodstuffs: occurrence and dietary intake in The Netherlands. Toxicol. Lett. 151 (1), 51−61.

Baldwin, R.S., Williams, R.D., Terry, M.K., 1983. Zeranol: a review of the metabolism, toxicology, and analytical methods for detection of tissue residues. Regul. Toxicol. Pharmacol. 3 (1), 9−25.

Bamias, G., Boletis, J., 2008. Balkan nephropathy: evolution of our knowledge. Am. J. Kidney Dis. 52 (3), 606−616.

Banerjee, S., Segal, A., 1986. In vitro transformation of C3H/10T1/2 and NIH/3T3 cells by acrylonitrile and acrylamide. Cancer Lett. 32 (3), 293−304.

Barshick, S.-A., Smith, S.M., Buchanan, M.V., Guerin, M.R., 1995. Determination of benzene content in food using a novel blender purge and trap GC/MS method. J. Food Compos. Anal. 8 (3), 244−257.

Battacone, G., Nudda, A., Palomba, M., Mazzette, A., Pulina, G., 2009. The transfer of aflatoxin M1 in milk of ewes fed diet naturally contaminated by aflatoxins and effect of inclusion of dried yeast culture in the diet. J. Dairy Sci. 92 (10), 4997−5004.

Battaglia, R., Conacher, H.B.S., Page, B.D., 1990. Ethyl carbamate (urethane) in alcoholic beverages and foods: a review. Food Addit. Contam. 7 (4), 477−496.

Battilani, P., Magan, N., Logrieco, A., 2006. European research on ochratoxin A in grapes and wine. Int. J. Food Microbiol. 111 (Suppl. 1), S2−S4.

Baum, M., Fauth, E., Fritzen, S., Herrmann, A., Mertes, P., Merz, K., Rudolphi, M., Zankl, H., Eisenbrand, G., 2005. Acrylamide and glycidamide: genotoxic effects in V79-cells and human blood. Mutat. Res. 580 (1−2), 61−69.

Bennett, J.W., Klich, M., 2003. Mycotoxins. Clin. Microbiol. Rev. 16 (3), 497−516.

Berg, T., Petersen, A., Pedersen, G.A., Petersen, J., Madsen, C., 2000. The release of nickel and other trace elements from electric kettles and coffee machines. Food Addit. Contam. 17 (3), 189−196.

Bhat, R.V., Shetty, P.H., Amruth, R.P., Sudershan, R.V., 1997. A foodborne disease outbreak due to the consumption of moldy sorghum and maize containing fumonisin mycotoxins. J. Toxicol. Clin. Toxicol. 35 (3), 249−255.

Bircan, C., 2009. Incidence of ochratoxin A in dried fruits and co-occurrence with aflatoxins in dried figs. Food Chem. Toxicol. 47 (8), 1996−2001.

Bjerregaard, P., Bjørn, L., Nørum, U., Pedersen, K.L., 2005. Cadmium in the shore crab Carcinus maenas: seasonal variation in cadmium content and uptake and elimination of cadmium after administration via food. Aquat. Toxicol. 72 (1−2), 5−15.

Bobrowski, A., Królicka, A., Zarębski, J., 2009. Characteristics of voltammetric determination and speciation of chromium − a review. Electroanalysis 21 (13), 1449−1458.

Bocio, A., Domingo, J.L., 2005. Daily intake of polychlorinated dibenzo-p-dioxins/polychlorinated dibenzofurans (PCDD/PCDFs) in foodstuffs consumed in Tarragona, Spain: a review of recent studies (2001−2003) on human PCDD/PCDF exposure through the diet. Envir. Res. 97 (1), 1−9.

Boffetta, S.A.P., 2014. Occupational Cancers. Springer, p. 221.

Brera, C., Debegnach, F., De Santis, B., Iafrate, E., Pannunzi, E., Berdini, C., Prantera, E., Gregori, E., Miraglia, M., 2011. Ochratoxin A in cocoa and chocolate products from the Italian market: occurrence and exposure assessment. Food Control 22 (10), 1663−1667.

Bucci, T.J., Hansen, D.K., Laborde, J.B., 1996. Leukoencephalomalacia and hemorrhage in the brain of rabbits gavaged with mycotoxin fumonisin B1. Nat. Toxins 4 (1), 51−52.

Bullerman, L.B., Bianchini, A., 2007. Stability of mycotoxins during food processing. Int. J. Food Microbiol. 119 (1−2), 140−146.

Calleman, C.J., Wu, Y., He, F., Tian, G., Bergmark, E., Zhang, S., Deng, H., Wang, Y., Crofton, K.M., Fennell, T., Costa, L.G., 1994. Relationships between biomarkers of exposure and neurological effects in a group of workers exposed to acrylamide. Toxicol. Appl. Pharmacol. 126 (2), 361−371.

Carlosena, A., Gallego, M., Valcarcel, M., 1997. Evaluation of various sample preparation procedures for the determination of chromium, cobalt and nickel in vegetables. J. Anal. At. Spectrom. 12 (4), 479−486.

CAST, 2003. Council for Agricultural Science and Technology. Mycotoxins: Risks in Plant, Animal, and Human Systems. Ames, IA, Task Force Report 139.

Casteel, S.W., Turk, J.R., Rottinghaus, G.E., 1994. Chronic effects of dietary fumonisin on the heart and pulmonary vasculature of swine. Fundam. Appl. Toxicol. 23 (4), 518–524.

Cetin, Y., Bullerman, L.B., 2006. Confirmation of reduced toxicity of deoxynivalenol in extrusion-processed corn grits by the MTT bioassay. J. Agric. Food Chem. 54 (5), 1949–1955.

Chan, E.Y.Y., Griffiths, S.M., Chan, C.W., 2008. Public-health risks of melamine in milk products. Lancet 372 (9648), 1444–1445.

Chaney, R., Reeves, P., Ryan, J., Simmons, R., Welch, R., Scott Angle, J., 2004. An improved understanding of soil Cd risk to humans and low cost methods to phytoextract Cd from contaminated soils to prevent soil Cd risks. Biometals 17 (5), 549–553.

Charnley, G., Doull, J., 2005. Human exposure to dioxins from food, 1999–2002. Food Chem. Toxicol. 43 (5), 671–679.

Coronel, M.B., Marin, S., Cano, G., Ramos, A.J., Sanchis, V., 2011. Ochratoxin A in Spanish retail ground roasted coffee: occurrence and assessment of the exposure in Catalonia. Food Control 22 (3–4), 414–419.

Cotty, P.J., Bayman, P., Egel, D.S., Elias, K.S., 1994. Agriculture, Aflatoxins and Aspergillus. In: Powell, K.A., Renwick, A., Peberdy, J.F. (Eds.), The Genus Aspergillus. Plenum Press, New York, NY, pp. 1–27.

Cronkite, E.P., Inoue, T., Carsten, A.L., Miller, M.E., Bullis, J.E., Drew, R.T., 1982. Effects of benzene inhalation on murine pluripotent stem cells. J. Toxicol. Environ. Health 9 (3), 411–421.

D' Mello, J.P.F., 2003. Food Safety Contaminants and Toxins. CABI Publishing, 65.

Davis, A.C., Wu, P., Zhang, X., Hou, X., Jones, B.T., 2006. Determination of cadmium in biological samples. Appl. Spectrosc. Rev. 41 (1), 35–75.

Dean, J.H., Padarathsingh, M.L., Jerrells, T.R., Keys, L., Northing, J.W., 1979. Assessment of immunobiological effects induced by chemicals, drugs or food additives. II. Studies with cyclophosphamide. Drug Chem. Toxicol. 2 (1–2), 133–153.

Dearfield, K.L., Abernathy, C.O., Ottley, M.S., Brantner, J.H., Hayes, P.F., 1988. Acrylamide: its metabolism, developmental and reproductive effects, genotoxicity, and carcinogenicity. Mutat. Res. 195 (1), 45–77.

Dennis, M.J., Howarth, N., Key, P.E., Pointer, M., Massey, R.C., 1989. Investigation of ethyl carbamate levels in some fermented foods and alcoholic beverages. Food Addit. Contam. 6 (3), 383–389.

Descatha, A., Jenabian, A., Fo, C., Ameille, J., 2005. Occupational exposures and haematological malignancies: overview on human recent data. Cancer Causes Control 16 (8), 939–953.

Dietrich, R., Schmid, A., Martlbauer, E., 2001. Citrinin in fruit juices. Mycotoxin Res. 17 (2), 156–159.

Dong, M., He, X.J., Tulayakul, P., Li, J.Y., Dong, K.S., Manabe, N., Nakayama, H., Kumagai, S., 2010. The toxic effects and fate of intravenously administered zearalenone in goats. Toxicon 55 (2–3), 523–530.

ECD, 2008a. European Commission Decision 2008/798/EC. Imposing Special Conditions Governing the Import of Products Containing Milk or Milk Products Originating in or Consigned from China, and Repealing Commission Decision 2008/757/EC.

ECD, 2008b. European Commission Decision 2008/921/EC. Amending Decision 2008/798/EC.

EFSA, 2009. European Food Safety Authority. Scientific Opinion on Arsenic in Food. EFSA CONTAM. 7 (10), 1351.

EFSA, 2011. Results on Acrylamide Levels in Food from Monitoring Years 2007–2009 and Exposure Assessment (Scientific report of EFSA. Parma, Italy).

El Khoury, A., Rizk, T., Lteif, R., Azouri, H., Delia, M.-L., Lebrihi, A., 2008. Fungal contamination and Aflatoxin B1 and Ochratoxin A in Lebanese wine–grapes and musts. Food Chem. Toxicol. 46 (6), 2244–2250.

El-Sayyad, H.I., El-Gammal, H.L., Habak, L.A., Abdel-Galil, H.M., Fernando, A., Gaur, R.L., Ouhtit, A., 2011. Structural and ultrastructural evidence of neurotoxic effects of fried potato chips on rat postnatal development. Nutrition 27 (10), 1066–1075 (Burbank, Los Angeles County, Calif.).

Erickson, M., Kaley II, R., 2011. Applications of polychlorinated biphenyls. Environ. Sci. Pollut. Res. 18 (2), 135−151.

Eriksen, G.S., Pettersson, H., 2004. Toxicological evaluation of trichothecenes in animal feed. Anim. Feed Sci. Technol. 114 (1−4), 205−239.

EU, 2000. Existing Chemicals Branch, Risk Assessment of Acrylamide. Office for Official Publication of the European Communities, Luxembourg.

FAO/WHO, 2006. Food and Agriculture Organisation of the United Nations/World Health Organisation. Safety evaluation of certain contaminants in food. Prepared by the Sixty-fourth meeting of the Joint FAO/WHO Expert Committee on Food Additives (JECFA). FAO Food Nutr. Pap. 82, 1−778.

FDE, 2011. Food Drink Europe Acrylamide Toolbox. http://www.fooddrinkeurope.eu/uploads/publications_documents/Toolboxfinal260911.pdf21/3/2013.

Fincham, J.E., Marasas, W.F.O., Taljaard, J.J.F., Kriek, N.P.J., Badenhorst, C.J., Gelderblom, W.C.A., Seier, J.V., Smuts, C.M., Faber, M., Weight, M.J., Slazus, W., Woodroof, C.W., van Wyk, M.J., Kruger, M., Thiel, P.G., 1992. Atherogenic effects in a non-human primate of Fusarium moniliforme cultures added to a carbohydrate diet. Atherosclerosis 94 (1), 13−25.

Fliege, R., Metzler, M., 2000. Electrophilic properties of patulin. N-Acetylcysteine and glutathione adducts. Chem. Res. Toxicol. 13 (5), 373−381.

Forstova, V., Belkova, B., Riddellova, K., Vaclavik, L., Prihoda, J., Hajslova, J., 2014. Acrylamide formation in traditional Czech leavened wheat-rye breads and wheat rolls. Food Control 38 (0), 221−226.

Franceschi, S., Bidoli, E., Baron, A.E., La Vecchia, C., 1990. Maize and risk of cancers of the oral cavity, pharynx, and esophagus in northeastern Italy. J. Natl Cancer Inst. 82 (17), 1407−1411.

Friedman, M., Levin, C.E., 2008. Review of methods for the reduction of dietary content and toxicity of acrylamide. J. Agric. Food Chem. 56 (15), 6113−6140.

Friedman, M.A., Dulak, L.H., Stedham, M.A., 1995. A lifetime oncogenicity study in rats with acrylamide. Fundam. Appl. Toxicol. 27 (1), 95−105.

Friedman, M., 2003. Chemistry, biochemistry, and safety of acrylamide. A review. J. Agric. Food Chem. 51 (16), 4504−4526.

Gama, E.M., da Silva Lima, A., Lemos, V.A., 2006. Preconcentration system for cadmium and lead determination in environmental samples using polyurethane foam/Me-BTANC. J. Hazard. Mater. 136 (3), 757−762.

Gardner, L.K., Lawrence, G.D., 1993. Benzene production from decarboxylation of benzoic acid in the presence of ascorbic acid and a transition-metal catalyst. J. Agric. Food Chem. 41 (5), 693−695.

Gelderblom, W.C.A., Kriek, N.P.J., Marasas, W.F.O., Thiel, P.G., 1991. Toxicity and carcinogenicity of the Fusanum monilzforine metabolite, fumonisin B1, in rats. Carcinogenesis 12 (7), 1247−1251.

Gelderblom, W.A., Marasas, W.O., Vleggaar, R., Thiel, P., Cawood, M.E., 1992. Fumonisins: isolation, chemical characterization and biological effects. Mycopathologia 117 (1−2), 11−16.

Gelderblom, W.C., Smuts, C.M., Abel, S., Snyman, S.D., Cawood, M.E., van der Westhuizen, L., Swanevelder, S., 1996. Effect of fumonisin B1 on protein and lipid synthesis in primary rat hepatocytes. Food Chem. Toxicol. 34 (4), 361−369.

Georgiadou, M., Dimou, A., Yanniotis, S., 2012. Aflatoxin contamination in pistachio nuts: a farm to storage study. Food Control 26 (2), 580−586.

Glaser, N., Stopper, H., 2012. Patulin: mechanism of genotoxicity. Food Chem. Toxicol. 50 (5), 1796−1801.

Gopinandhan, T.N., Kannan, G.S., Panneerselvam, P., Velmourougane, K., Raghuramulu, Y., Jayarama, J., 2008. Survey on ochratoxin A in Indian green coffee destined for export. Food Addit. Contam. Part B 1 (1), 51−57.

Guengerich, F.P., Kim, D.H., Iwasaki, M., 1991. Role of human cytochrome P-450 IIE1 in the oxidation of many low molecular weight cancer suspects. Chem. Res. Toxicol. 4 (2), 168−179.

Hamlet, C.G., Sadd, P.A., Liang, L., 2008. Correlations between the amounts of free asparagine and saccharides present in commercial cereal flours in the United Kingdom and the generation of acrylamide during cooking. J. Agric. Food Chem. 56 (15), 6145−6153.

Harrison, L.R., Colvin, B.M., Greene, J.T., Newman, L.E., Cole, J.R., 1990. Pulmonary edema and hydrothorax in swine produced by fumonisin B1, a toxic metabolite of Fusarium moniliforme. J. Vet. Diagn. Invest. 2 (3), 217−221.

Hashimoto, K., Aldridge, W.N., 1970. Biochemical studies on acrylamide, a neurotoxic agent. Biochem. Pharmacol. 19 (9), 2591−2604.

Hassen, W., Golli, E.E., Baudrimont, I., Mobio, A.T., Ladjimi, M.M., Creppy, E.E., Bacha, H., 2005. Cytotoxicity and Hsp 70 induction in Hep G2 cells in response to zearalenone and cytoprotection by sub-lethal heat shock. Toxicology 207 (2), 293−301.

Henderson, A.P., Barnes, M.L., Bleasdale, C., Cameron, R., Clegg, W., Heath, S.L., Lindstrom, A.B., Rappaport, S.M., Waidyanatha, S., Watson, W.P., Golding, B.T., 2005. Reactions of benzene oxide with thiols including glutathione. Chem. Res. Toxicol. 18 (2), 265−270.

Hendricks, K., 1999. Fumonisins and neural tube defects in South Texas. Epidemiology 10 (2), 198−200.

Hoffler, U., El-Masri, H.A., Ghanayem, B.I., 2003. Cytochrome P450 2E1 (CYP2E1) is the principal enzyme responsible for urethane metabolism: comparative studies using CYP2E1-null and wild-type mice. J. Pharmacol. Exp. Ther. 305 (2), 557−564.

Huwe, J.K., Larsen, G.L., 2005. Polychlorinated dioxins, furans, and biphenyls, and polybrominated diphenyl ethers in a U.S. meat market basket and estimates of dietary intake. Environ. Sci. Technol. 39 (15), 5606−5611.

IARC, 1974. International Agency for Research on Cancer, International Agency for Research on Cancer Monographs on the Evaluation of the Carcinogenic Risk of Chemicals to Man: Some Anti-thyroid and Related Substances, Nitrofurans and Industrial Chemicals, 7. World Health Organization, Lyon, France.

IARC, 1987. Benzene (Suppl. 7). In: IARC Monographs on the Evaluation of Carcinogenic Risks to Humans, pp. 1−42.

IARC, 1993a. International Agency for Research on Cancer. Evaluation of Carcinogenic Risks of Chemical to Humans. Some Naturally-occurring Substances: Food Items and Constituents. Heterocyclic Aromatic Amines and Mycotoxins. IARC Monographs, Lyon, France, 359−362.

IARC, 1993b. International Agency for Research on Cancer. In proceedings of the meeting of the IARC working group on beryllium, cadmium, mercury and exposures in the glass manufacturing industry. Scand. J. Work, Environ. Health 19, 360−374.

IARC, 1994. Acrylamide, IARC Monographs on the Evaluation of the Carcinogenic Risk of Chemicals to Humans. Lyon, France, 389−433.

IARC, 1999. International Agency for Research on Cancer. Overall Evaluation of Carcinogenicity to Humans. In: Monographs on the Evaluation of Carcinogenic Risk to Humans, vol. 1 (Lyon, France); 1−36.

IARC, 2002. International Agency for Research on Cancer. Monograph on the Evaluation of Carcinogenic Risks to Humans. In: Some Traditional Herbal Medicines, Some Mycotoxins, Naphthalene and Styrene. Summary of Data Reported and Evaluation, vol. 82. World Health Organization, Lyon, France, 171−175.

IARC, 2007. International Agency for Research on Cancer, International Agency for Research: Alcoholic Beverage Consumption and Ethyl Carbamate (Urethane), vol. 96. World Health Organization. Lyon, France. http://monographs.iarc.fr/ENG/Meetings/vol96-summary.pdf, 1−5.

Jarup, L., 2003. Hazards of heavy metal contamination. Br. Med. Bull. 68 (1), 167−182.

JECFA, 2002. Joint FAO/WHO Expert Committee on Food Additives. Health Implications of Acrylamide in Food. World Health Organization.

Jiang, L., Cao, J., An, Y., Geng, C., Qu, S., Jiang, L., Zhong, L., 2007. Genotoxity of acrylamide in human hepatoma G2 (HepG2) cells. Toxicol. In Vitro 21 (8), 1486−1492.

Johnson, K.A., Gorzinski, S.J., Bodner, K.M., Campbell, R.A., Wolf, C.H., Friedman, M.A., Mast, R.W., 1986. Chronic toxicity and oncogenicity study on acrylamide incorporated in the drinking water of Fischer 344 rats. Toxicol. Appl. Pharmacol. 85 (2), 154−168.

Juan, C., Raiola, A., MaÃes, J., Ritieni, A., 2014. Presence of mycotoxin in commercial infant formulas and baby foods from Italian market. Food Control 39 (0), 227−236.

Kabak, B., 2009. Ochratoxin A in cereal-derived products in Turkey: occurrence and exposure assessment. Food Chem. Toxicol. 47 (2), 348−352.

Khan, S., Cao, Q., Zheng, Y.M., Huang, Y.Z., Zhu, Y.G., 2008. Health risks of heavy metals in contaminated soils and food crops irrigated with wastewater in Beijing, China. Environ. Pollut. 152 (3), 686−692.

Kim, C.-W., Yun, J.-W., Bae, I.-H., Lee, J.-S., Kang, H.-J., Joo, K.-M., Jeong, H.-J., Chung, J.-H., Park, Y.-H., Lim, K.-M., 2010. Determination of spatial distribution of melamine cyanuric acid crystals in rat kidney tissue by histology and imaging matrix-assisted laser desorption/ionization quadrupole time-of-flight mass spectrometry. Chem. Res. Toxicol. 23 (1), 220−227. http://dx.doi.org/10.1021/tx900354z.

Kjallstrand, J., Petersson, G., 2001. Phenolic antioxidants in alder smoke during industrial meat curing. Food Chem. 74 (1), 85−89.

Klich, M.A., Pitt, J.I., 1988. Differentiation of *Aspergillus flavus* from *A. parasiticus* and other closely related species. Trans. Br. Mycol. Soc. 91 (1), 99−108.

Klich, M.A., Mullaney, E.J., Daly, C.B., Cary, J.W., 2000. Molecular and physiological aspects of aflatoxin and sterigmatocystin biosynthesis by *Aspergillus tamarii* and *A. ochraceoroseus*. Appl. Microbiol. Biotechnol. 53 (5), 605−609.

Kobayashi, E., Okubo, Y., Suwazono, Y., Kido, T., Nishijo, M., Nakagawa, H., Nogawa, K., 2002. Association between total cadmium intake calculated from the cadmium concentration in household rice and mortality among inhabitants of the cadmium-polluted Jinzu River basin of Japan. Toxicol. Lett. 129 (1−2), 85−91.

Krska, R., Welzig, E., Boudra, H., 2007. Analysis of *Fusarium toxins* in feed. Anim. Feed Sci. Technol. 137 (3−4), 241−264.

Lachenmeier, D.W., Schehl, B., Kuballa, T., Frank, W., Senn, T., 2005. Retrospective trends and current status of ethyl carbamate in German stone-fruit spirits. Food Addit. Contam. 22 (5), 397−405.

Lachenmeier, D.W., Reusch, H., Sproll, C., Schoeberl, K., Kuballa, T., 2008. Occurrence of benzene as a heat-induced contaminant of carrot juice for babies in a general survey of beverages. Food Addit. Contam. Part A 25 (10), 1216−1224.

Lan, Q., Zhang, L., Li, G., Vermeulen, R., Weinberg, R.S., Dosemeci, M., Rappaport, S.M., Shen, M., Alter, B.P., Wu, Y., Kopp, W., Waidyanatha, S., Rabkin, C., Guo, W., Chanock, S., Hayes, R.B., Linet, M., Kim, S., Yin, S., Rothman, N., Smith, M.T., 2004. Hematotoxicity in workers exposed to low levels of benzene. Science 306 (5702), 1774−1776.

Lawley, R., Curtis, L., Davis, J., 2012. The Food Safety Hazard Guidebook. RSC Publishing, p. 399.

Le, T., Yan, P., Xu, J., Hao, Y., 2013. A novel colloidal gold-based lateral flow immunoassay for rapid simultaneous detection of cyromazine and melamine in foods of animal origin. Food Chem. 138 (2−3), 1610−1615.

Le Kim, Y.-K., Koh, E., Chung, H.-J., Kwon, H., 2000. Determination of ethyl carbamate in some fermented Korean foods and beverages. Food Addit. Contam. 17 (6), 469−475.

Li, Y.C., Ledoux, D.R., Bermudez, A.J., Fritsche, K.L., Rottinghaus, G.E., 1999. Effects of fumonisin B1 on selected immune responses in broiler chicks. Poult. Sci. 78 (9), 1275−1282.

Li, F.-Q., Yoshizawa, T., Kawamura, O., Luo, X.-Y., Li, Y.-W., 2001. Aflatoxins and fumonisins in corn from the high-incidence area for human hepatocellular carcinoma in Guangxi, China. J. Agric. Food Chem. 49 (8), 4122−4126.

Li, Y., Zhou, Y.-C., Yang, M.-H., Ou-Yang, Z., 2012. Natural occurrence of citrinin in widely consumed traditional Chinese food red yeast rice, medicinal plants and their related products. Food Chem. 132 (2), 1040−1045.

Liu, G., Han, Z., Nie, D., Yang, J., Zhao, Z., Zhang, J., Li, H., Liao, Y., Song, S., De Saeger, S., Wu, A., 2012. Rapid and sensitive quantitation of zearalenone in food and feed by lateral flow immunoassay. Food Control 27 (1), 200–205.

LoPachin, R.M., 2004. The changing view of acrylamide neurotoxicity. Neurotoxicology 25 (4), 617–630.

Lu, J., Xiao, J., Yang, D.-J., Wang, Z.-T., Jiang, D.-G., Fang, C.-R., Yang, J., 2009. Study on migration of melamine from food packaging materials on markets. Biomed. Environ. Sci. 22 (2), 104–108.

Lund, K.H., Petersen, J.H., 2006. Migration of formaldehyde and melamine monomers from kitchen- and tableware made of melamine plastic. Food Addit. Contam. 23 (9), 948–955.

Lundqvist, C., Zuurbier, M., Leijs, M., Johansson, C., Ceccatelli, S., Saunders, M., Schoeters, G., Tusscher, G.T., Koppe, J.G., 2006. The effects of PCBs and dioxins on child health. Acta Paediatr. 95, 55–64.

Maaroufi, K., Chekir, L., Ekue Creppy, E., Ellouz, F., Bacha, H., 1996. Zearalenone induces modifications of haematological and biochemical parameters in rats. Toxicon 34 (5), 535–540.

Manova, A., Humenikova, S., Strelec, M., Beinrohr, E., 2007. Determination of chromium(VI) and total chromium in water by in-electrode coulometric titration in a porous glassy carbon electrode. Microchim. Acta 159 (1–2), 41–47.

Marasas, W.F., Kellerman, T.S., Gelderblom, W.C., Coetzer, J.A., Thiel, P.G., van der Lugt, J.J., 1988. Leukoencephalomalacia in a horse induced by fumonisin B1 isolated from Fusarium moniliforme. Onderstepoort J. Vet. Res. 55 (4), 197–203.

Marasas, W.F.O., Miller, J.D., Riley, R.T., Visconti, A., 2001. Fumonisins—occurrence, toxicology, metabolism and risk assessment. In: Summerell, B.A., Leslie, J.F., Backhouse, D., Bryden, W.L., Burgess, L.W. (Eds.), Fusarium. Paul E. Nelson Memorial Symposium. APS Press, St. Paul, Minn, pp. 332–359.

Masohan, A., Parsad, G., Khanna, M.K., Chopra, S.K., Rawat, B.S., Garg, M.O., 2000. Estimation of trace amounts of benzene in solvent-extracted vegetable oils and oil seed cakes. Analyst 125 (9), 1687–1689.

McMurry, S., Lochmiller, R., Vestey, M., Qualls Jr., C., Elangbam, C., 1991. Acute effects of benzene and cyclophosphamide exposure on cellular and humoral immunity of cotton rats, Sigmodon hispidus. Bull. Environ. Contam. Toxicol. 46 (6), 937–945.

Meharg, A.A., Deacon, C., Campbell, R.C.J., Carey, A.-M., Williams, P.N., Feldmann, J., Raab, A., 2008. Inorganic arsenic levels in rice milk exceed EU and US drinking water standards. J. Environ. Monit. 10 (4), 428–431.

Moake, M.M., Padilla-Zakour, O.I., Worobo, R.W., 2005. Comprehensive review of patulin control methods in foods. Compr. Rev. Food Sci. Food Saf. 4 (1), 8–21.

Montes, R., Segarra, R., Castillo, M.-A., 2012. Trichothecenes in breakfast cereals from the Spanish retail market. J. Food Compos. Anal. 27 (1), 38–44.

Mottram, D.S., Wedzicha, B.L., Dodson, A.T., 2002. Food chemistry: acrylamide is formed in the Maillard reaction. Nature 419 (6906), 448–449.

Musser, S.M., Plattner, R.D., 1997. Fumonisin Composition in Cultures of Fusarium moniliforme, Fusarium proliferatum, and Fusarium nygami. J. Agric. Food Chem. 45 (4), 1169–1173.

Naruszewicz, M., Zapolska-Downar, D., Kosmider, A., Nowicka, G., Kozlowska-Wojciechowska, M., Vikstrom, A.S., Tornqvist, M., 2009. Chronic intake of potato chips in humans increases the production of reactive oxygen radicals by leukocytes and increases plasma C-reactive protein: a pilot study. Am. J. Clin. Nutr. 89 (3), 773–777.

Newberne, P.M., Butler, W.H., 1969. Acute and chronic effects of aflatoxin on the liver of domestic and laboratory animals: a review. Cancer Res. 29 (1), 236–250.

Olanca, B., Cakirogullari, G.C., Ucar, Y., Kirisik, D., Kilic, D., 2014. Polychlorinated dioxins, furans (PCDD/Fs), dioxin-like polychlorinated biphenyls (dl-PCBs) and indicator PCBs (ind-PCBs) in egg and egg products in Turkey. Chemosphere 94 (0), 13–19.

Oymak, T., Tokalioglu, S., Yilmaz, V., Kartal, S., Aydin, D., 2009. Determination of lead and cadmium in food samples by the coprecipitation method. Food Chem. 113 (4), 1314–1317.

Ozden, S., Akdeniz, A.S., Alpertunga, B., 2012. Occurrence of ochratoxin A in cereal-derived food products commonly consumed in Turkey. Food Control 25 (1), 69–74.

Park, K.-K., Liem, A., Stewart, B.C., Miller, J.A., 1993. Vinyl carbamate epoxide, a major strong electrophilic, mutagenic and carcinogenic metabolite of vinyl carbamate and ethyl carbamate (urethane). Carcinogenesis 14 (3), 441–450.

Park, J., Kamendulis, L.M., Friedman, M.A., Klaunig, J.E., 2002. Acrylamide-induced cellular transformation. Toxicol. Sci. 65 (2), 177–183.

Peers, F.G., Linsell, C.A., 1973. Dietary aflatoxins and liver cancer–a population based study in Kenya. Br. J. Cancer 27 (6), 473–484.

Peijnenburg, W., Baerselman, R., de Groot, A., Jager, T., Leenders, D., Posthuma, L., Van Veen, R., 2000. Quantification of metal bioavailability for lettuce (*Lactuca sativa* L.) in field soils. Arch. Environ. Contam. Toxicol. 39 (4), 420–430.

Pestka, J.J., Smolinski, A.T., 2005. Deoxynivalenol: toxicology and potential effects on humans. J. Toxicol. Environ. Health Part B 8 (1), 39–69.

Pestka, J.J., Zhou, H.-R., Moon, Y., Chung, Y.J., 2004. Cellular and molecular mechanisms for immune modulation by deoxynivalenol and other trichothecenes: unraveling a paradox. Toxicol. Lett. 153 (1), 61–73.

Pestka, J.J., 2007. Deoxynivalenol: toxicity, mechanisms and animal health risks. Anim. Feed Sci. Technol. 137 (3–4), 283–298.

Peterson, S.W., Ito, Y., Horn, B.W., Goto, T., 2001. *Aspergillus bombycis*, a new aflatoxigenic species and genetic variation in its sibling species, A. nomius. Mycologia 93 (4), 689–703.

Petzinger, E., Weidenbach, A., 2002. Mycotoxins in the food chain: the role of ochratoxins. Livest. Prod. Sci. 76 (3), 245–250.

Pietri, A., Rastelli, S., Mulazzi, A., Bertuzzi, T., 2012. Aflatoxins and ochratoxin A in dried chestnuts and chestnut flour produced in Italy. Food Control 25 (2), 601–606.

Pique, E., Vargas-Murga, L., Gomez-Catalan, J., Lapuente, Jd., Llobet, J.M., 2013. Occurrence of patulin in organic and conventional apple-based food marketed in Catalonia and exposure assessment. Food Chem. Toxicol. 60 (0), 199–204.

Placinta, C.M., D'Mello, J.P.F., Macdonald, A.M.C., 1999. A review of worldwide contamination of cereal grains and animal feed with Fusarium mycotoxins. Anim. Feed Sci. Technol. 78 (1–2), 21–37.

Pozzi, C.R., Correa, B., Xavier, J.G., Direito, G.M., Orsi, R.B., Matarazzo, S.V., 2001. Effects of prolonged oral administration of fumonisin B1 and aflatoxin B1 in rats. Mycopathologia 151 (1), 21–27.

Rheeder, J.P., Marasas, W.F.O., Thiel, P.G., Sydenham, E.W., Shepard, G.S., Van Schalkwyk, D.J., 1992. Fusarium moniliforme and fumonisins in corn in relation to human oesophageal cancer in Transkei. Phytopathology 82, 353–357.

Rheeder, J.P., Marasas, W.F.O., Vismer, H.F., 2002. Production of fumonisin Analogs by Fusarium species. Appl. Environ. Microbiol. 68 (5), 2101–2105.

Safe, S., Hutzinger, O., 1984. Polychlorinated biphenyls (PCBs) and polybrominated biphenyls (PBBs): biochemistry, toxicology, and mechanism of action. Critical Reviews in Toxicology 13 (4), 319–395.

Schehl, B., Senn, T., Lachenmeier, D., Rodicio, R., Heinisch Jr., 2007. Contribution of the fermenting yeast strain to ethyl carbamate generation in stone fruit spirits. Appl. Microbiol. Biotechnol. 74 (4), 843–850.

Schlatter, J., 2004. Toxicity data relevant for hazard characterization. Toxicol. Lett. 153 (1), 83–89.

Schothorst, R.C., Jekel, A.A., 2003. Determination of trichothecenes in beer by capillary gas chromatography with flame ionisation detection. Food Chem. 82 (3), 475–479.

Senthil Kumar, K., Kannan, K., Paramasivan, O.N., Shanmuga Sundaram, V.P., Nakanishi, J., Masunaga, S., 2001. Polychlorinated dibenzo-p-dioxins, dibenzofurans, and polychlorinated biphenyls in human tissues, meat, fish, and wildlife samples from India. Environ. Sci. Technol. 35 (17), 3448–3455.

Seo, J.-A., Lee, Y.-W., 1999. Natural occurrence of the C Series of fumonisins in moldy corn. Appl. Environ. Microbiol. 65 (3), 1331−1334.

Shank, R.C., Bhamarapravati, N., Gordon, J.E., Wogan, G.N., 1972. Dietary aflatoxins and human liver cancer. IV. Incidence of primary liver cancer in two municipal populations of Thailand. Food Cosmet. Toxicol. 10 (2), 171−179.

Shephard, G.S., Marasas, W.F.O., Leggott, N.L., Yazdanpanah, H., Rahimian, H., Safavi, N., 2000. Natural occurrence of fumonisins in corn from Iran. J. Agric. Food Chem. 48 (5), 1860−1864.

Shifrin, V.I., Anderson, P., 1999. Trichothecene mycotoxins trigger a ribotoxic stress response that activates c-Jun N-terminal kinase and p38 mitogen-activated protein kinase and induces apoptosis. J. Biol. Chem. 274 (20), 13985−13992.

Shipp, A., Lawrence, G., Gentry, R., McDonald, T., Bartow, H., Bounds, J., Macdonald, N., Clewell, H., Allen, B., Van Landingham, C., 2006. Acrylamide: review of toxicity data and dose-response analyses for cancer and noncancer effects. Crit. Rev. Toxicol. 36 (6−7), 481−608.

Shundo, L., de Almeida, A.P., Alaburda, J., Lamardo, L.C.A., Navas, S.A., Ruvieri, V., Sabino, M., 2009. Aflatoxins and ochratoxin A in Brazilian paprika. Food Control 20 (12), 1099−1102.

Signes-Pastor, A.J., Deacon, C., Jenkins, R.O., Haris, P.I., Carbonell-Barrachina, A.A., Meharg, A.A., 2009. Arsenic speciation in Japanese rice drinks and condiments. J. Environ. Monit. 11 (11), 1930−1934.

Snyder, C.A., 1987. In: Benzene, Snyder, R. (Eds.), Ethel Browning's Toxicity and Metabolism of Industrial Solvents. Elsevier, New York, pp. 3−37.

Snyder, R., 2000. Overview of the toxicology of benzene. J. Toxicol. Environ. Health Part A 61 (5−6), 339−346.

Snyder, R., 2002. Benzene and leukemia. Crit. Rev. Toxicol. 32 (3), 155−210.

Soriano, J.M., Dragacci, S., 2004. Occurrence of fumonisins in foods. Food Res. Int. 37 (10), 985−1000.

Squire, R.A., 1981. Ranking animal carcinogens: a proposed regulatory approach. Science 214 (4523), 877−880.

Sridhara Chary, N., Kamala, C.T., Samuel Suman Raj, D., 2008. Assessing risk of heavy metals from consuming food grown on sewage irrigated soils and food chain transfer. Ecotoxicol. Environ. Saf. 69 (3), 513−524.

Stadler, R.H., Blank, I., Varga, N., Robert, F., Hau, J., Guy, P.A., Robert, M.-C., Riediker, S., 2002. Food chemistry: acrylamide from Maillard reaction products. Nature 419 (6906), 449−450.

Sydenham, E.W., Thiel, P.G., Marasas, W.F.O., Shephard, G.S., Van Schalkwyk, D.J., Koch, K.R., 1990. Natural occurrence of some Fusarium mycotoxins in corn from low and high esophageal cancer prevalence areas of the Transkei, Southern Africa. J. Agric. Food Chem. 38 (10), 1900−1903.

Tareke, E., Rydberg, P., Karlsson, P., Eriksson, S., Tornqvist, M., 2002. Analysis of acrylamide, a carcinogen formed in heated foodstuffs. J. Agric. Food Chem. 50 (17), 4998−5006.

Thiel, P., Marasas, W.O., Sydenham, E., Shephard, G., Gelderblom, W.A., 1992. The implications of naturally occurring levels of fumonisins in corn for human and animal health. Mycopathologia 117 (1−2), 3−9.

Tilson, H.A., Jacobson, J.L., Rogan, W.J., 1990. Polychlorinated biphenyls and the developing nervous system: cross-species comparisons. Neurotoxicol. Teratol. 12 (3), 239−248.

Tsujihata, M., 2008. Mechanism of calcium oxalate renal stone formation and renal tubular cell injury. Int. J. Urol. 15 (2), 115−120.

Turner, N.W., Subrahmanyam, S., Piletsky, S.A., 2009. Analytical methods for determination of mycotoxins: a review. Anal. Chim. Acta 632 (2), 168−180.

Tyan, Y.-C., Yang, M.-H., Jong, S.-B., Wang, C.-K., Shiea, J., 2009. Melamine contamination. Anal. Bioanal. Chem. 395 (3), 729−735.

Tyl, R.W., Friedman, M.A., 2003. Effects of acrylamide on rodent reproductive performance. Reprod. Toxicol. 17 (1), 1−13.

Tyl, R.W., Marr, M.C., Myers, C.B., Ross, W.P., Friedman, M.A., 2000. Relationship between acrylamide reproductive and neurotoxicity in male rats. Reprod. Toxicol. 14 (2), 147−157.

Van den Berg, M., Birnbaum, L.S., Denison, M., De Vito, M., Farland, W., Feeley, M., Fiedler, H., Hakansson, H., Hanberg, A., Haws, L., Rose, M., Safe, S., Schrenk, D., Tohyama, C., Tritscher, A., Tuomisto, J., Tysklind, M., Walker, N., Peterson, R.E., 2006. The 2005 world health organization reevaluation of human and mammalian toxic equivalency factors for dioxins and dioxin-like compounds. Toxicol. Sci. 93 (2), 223−241.

Van Egmond, H.P., 1989. Current situation on regulations for mycotoxins. Overview of tolerances and status of standard methods of sampling and analysis. Food Addit. Contam. 6 (2), 139−188.

van Leeuwen, F.X.R., Feeley, M., Schrenk, D., Larsen, J.C., Farland, W., Younes, M., 2000. Dioxins: WHO's tolerable daily intake (TDI) revisited. Chemosphere 40 (9−11), 1095−1101.

Wania, F., MacKay, D., 1996. Peer reviewed: tracking the distribution of persistent organic pollutants. Environ. Sci. Technol. 30 (9), 390A−396A.

Watzek, N., Bohm, N., Feld, J., Scherbl, D., Berger, F., Merz, K.H., Lampen, A., Reemtsma, T., Tannenbaum, S.R., Skipper, P.L., Baum, M., Richling, E., Eisenbrand, G., 2012. N7-Glycidamide-Guanine DNA adduct formation by orally ingested acrylamide in rats: a Dose-"Response study encompassing human diet-related exposure levels". Chem. Res. Toxicol. 25 (2), 381−390.

WHO, 2011. World Health Organization, Toxicological and Health Aspects of Melamine and Cyanuric Acid. http://www.whqlibdoc.who.int/publications/2009/9789241597951_eng.pdf.

Wild, C.P., Pionneau, F.A., Montesano, R., Mutiro, C.F., Chetsanga, C.J., 1987. Aflatoxin detected in human breast milk by immunoassay. Int. J. Cancer 40 (3), 328−333.

Wittkowski, R., Baltes, W., Jennings, W.G., 1990. Analysis of liquid smoke and smoked meat volatiles by headspace gas chromatography. Food Chem. 37 (2), 135−144.

Wong, V.W.T., Rainbow, P.S., 1986. Apparent and real variability in the presence and metal contents of metal-lothioneins in the crab *Carcinus maenas* including the effects of isolation procedure and metal induction. Comp. Biochem. Physiol. Part A: Physiol. 83 (1), 157−177.

Wu, Q.-J., Lin, H., Fan, W., Dong, J.-J., Chen, H.-L., 2006. Investigation into benzene, trihalomethanes and formaldehyde in chinese lager beers. J. Inst. Brew. 112 (4), 291−294.

Wu, J., Zhao, R., Chen, B., Yang, M., 2011. Determination of zearalenone in barley by high-performance liquid chromatography coupled with evaporative light scattering detection and natural occurrence of zearalenone in functional food. Food Chem. 126 (3), 1508−1511.

Yousef, M.I., El-Demerdash, F.M., 2006. Acrylamide-induced oxidative stress and biochemical perturbations in rats. Toxicology 219 (1−3), 133−141.

Zaied, C., Abid, S., Hlel, W., Bacha, H., 2013. Occurrence of patulin in apple-based-foods largely consumed in Tunisia. Food Control 31 (2), 263−267.

Zhu, M.J., Mendonca, A., Min, B., Lee, E.J., Nam, K.C., Park, K., Du, M., Ismail, H.A., Ahn, D.U., 2004. Effects of electron beam irradiation and antimicrobials on the volatiles, color, and texture of ready-to-eat Turkey breast roll. J. Food Sci. 69 (5), C382−C387.

Zinedine, A., Manes, J., 2009. Occurrence and legislation of mycotoxins in food and feed from Morocco. Food Control 20 (4), 334−344.

Zyzak, D.V., Sanders, R.A., Stojanovic, M., Tallmadge, D.H., Eberhart, B.L., Ewald, D.K., Gruber, D.C., Morsch, T.R., Strothers, M.A., Rizzi, G.P., Villagran, M.D., 2003. Acrylamide formation mechanism in heated foods. J. Agric. Food Chem. 51 (16), 4782−4787.

FOOD STANDARDS AND PERMISSIBLE LIMITS

3

Food standards are legislative instruments, often called "food laws," which are enacted to protect consumers against unsafe products, adulteration, and fraud; to ensure quality compliance; as well as to protect honest food producers, processors, distributors, and traders. They also facilitate the movement of goods within and between countries by providing a common lexicon for food quality and safety. Standards are designed for many purposes. The main concern is quality and safety, but any standard/regulation cannot be set to the zenith level of quality; therefore, above-average quality is generally taken into consideration in enforcement on a wider scale. Quality level and safety are of prime focus during the formulation of laws or standards to ensure the availability of nutritious food and fair trade practices across countries. Governmental agencies help prominently in the standard formulation process and describe fully what is required to meet a certain standard. Standards, grades, and their definitions and specifications act as reference points in aspects of quality, safeguarding the interest of farmers, processors, distributors, traders, and ultimately consumers. Standards help to ensure that food is wholesome and contains whatever the label claims, and they minimize the chances of deceit in terms of quality and economic value and associated risk. Permissible limits are often defined in the majority of food standards to identify maximum or minimum levels acceptable and thus provide an aid for avoiding foods having risk or any health hazard. Establishment of these limits or levels depends on the intended objective of examination by law enforcement agencies and bodies involved in enforcing the laws or standards. Scientific developments in the field of agriculture had led to increased use of modern-day chemicals: insecticides, pesticides, herbicides, veterinary drugs, and many exogenous chemicals with undesirable compounds, for example, plant toxicants, and mycotoxins, which can penetrate the food supply chain. Their residual effects on foods have magnified the associated risk, and they need to be scientifically monitored, with precise limits for labeling foods as safe for human consumption.

3.1 FOOD SAFETY REGULATORY BODIES

There are various food safety regulatory bodies and international organizations throughout the world, which are responsible for dealing with aspects of food safety and food trade among different countries, and for maintaining biodiversity. Some important bodies are listed below:

1. Codex Alimentarius Commission (CAC)
2. World Health Organization (WHO)
3. Food and Agricultural Organization (FAO)
4. World Trade Organization (WTO)

Rapid Detection of Food Adulterants and Contaminants. http://dx.doi.org/10.1016/B978-0-12-420084-5.00003-2
Copyright © 2016 Elsevier Inc. All rights reserved.

5. World Organization for Animal Health (WOAH)
6. European Union (EU) standards
7. US Food and Drug Administration (FDA)
8. International Plant Protection Organization (IPPO)
9. Convention on Biodiversity (CBD)
10. International Commission on Microbiological Specifications for Foods (ICMSF)

3.1.1 AMERICAN FOOD STANDARD BODIES

The FDA is a federal agency of the US Department of Health and Human Services, one of the US federal executive departments. The FDA is responsible for protecting and promoting public health through the regulation and supervision of food safety, tobacco products, dietary supplements, prescription and over-the-counter pharmaceutical drugs (medications), vaccines, biopharmaceuticals, blood transfusions, medical devices, electromagnetic radiation emitting devices, cosmetics, animal foods and feeds, and veterinary products. The FDA Food Safety Modernization Act of 2010 (FSMA) is a recent food law that came into existence in 2011. It aims to ensure the US food supply is safe by shifting the focus of federal regulators from responding to contamination to preventing it. The FSMA has given the FDA new authority to regulate the way foods are grown, harvested, and processed. The law grants the FDA a number of new powers, including mandatory recall authority, which the agency has sought for many years. The FSMA requires the FDA to undertake more than a dozen rule-makings and issue at least 10 guidance documents, as well as a host of reports, plans, strategies, standards, and notices, among other tasks. The law was prompted after many reported incidents of food-borne illnesses during the 2000s. Tainted food has cost the food industry billions of dollars in recalls, lost sales, and legal expenses. Earlier the US Department of Agriculture (USDA) and the FDA were the main actors in food regulation in the United States. These two federal agencies encompass all phases of the food regulatory system: They evaluate, investigate, regulate, inspect, and sanction. The USDA's work focused on the safety of meat, poultry, and some egg products, whereas the FDA regulated all other foodstuffs, such as whole eggs, seafood, fruits, vegetables, grain products, and milk.

3.1.2 EUROPEAN FOOD SAFETY AUTHORITY (EFSA)

The European Food Safety Authority (EFSA) is the keystone of European Union (EU) risk assessment regarding food and feed safety. In close collaboration with national authorities and in open consultation with its stakeholders, EFSA provides independent scientific advice and clear communication on existing and emerging risks. The EFSA is an independent European agency funded by the EU budget, which operates separately from the European Commission, European parliament, and EU member states.

EU Food Law is a market leading publication on food policy and legislation. It covers the latest legislative developments in the food industry, including health claims, nutrition claims, food safety, food labeling legislation, genetically modified food, food and drink advertising, traceability, food law enforcements, contaminants, and food hygiene proposals, among others. *EU Food Law* follows legislative developments from food authorities including the EFSA and the European Commission.

3.1.3 **CODEX ALIMENTARIUS COMMISSION (CAC)**

The CAC (*codex alimentarius* is Latin for "food code") is an intergovernmental body established in 1963 to set guidelines and standards to ensure "fair trade practices" and consumer protection in relation to the global food trade. The following are the basic objectives of the CAC:

- To protect the health of consumers and ensure fair practices in the food trade
- To promote the coordination of all food standards work undertaken by international governmental and nongovernmental organizations
- To determine priorities, initiate and guide the preparation of draft standards through and with the aid of appropriate organizations
- To finalize standards elaborated above and publish them in Codex Alimentarius either as regional or worldwide standards together with international standards already finalized by other bodies, wherever this is practicable
- To amend published standards, as appropriate, in the light of new developments

The Codex Alimentarius is a collection of internationally adopted food standards and related texts presented in a uniform manner. These food standards and related texts aim at protecting consumers' health and ensuring fair practices in the food trade. The Codex Alimentarius includes standards for all the principal foods, whether processed, semiprocessed, or raw, for distribution to consumers. Materials for further processing into foods should be included to the extent necessary to achieve the purposes of the Codex Alimentarius as defined. The Codex Alimentarius includes provisions with respect to food hygiene, food additives, residues of pesticides and veterinary drugs, contaminants, labeling and presentation, methods of analysis and sampling, and import and export inspection and certification.

3.1.4 **INTERNATIONAL STANDARDIZATION ORGANIZATION (ISO)**

The ISO is an international quality management accreditation agency. ISO 22000 is a standard developed by them dealing with food safety management. ISO 22000:2005 specifies requirements for a food safety management system whereby an organization in the food chain needs to demonstrate its ability to control food safety hazards in order to ensure that food is safe at the time of human consumption. It enables any establishment:

1. to plan, implement, operate, maintain, and update a food safety management system aimed at providing products that, according to their intended use, are safe for the consumer;
2. to demonstrate compliance with applicable statutory and regulatory food safety requirements;
3. to evaluate and assess customer requirements and demonstrate conformity with those mutually agreed customer requirements that relate to food safety, in order to enhance customer satisfaction;
4. to effectively communicate food safety issues to their suppliers, customers, and relevant interested parties in the food chain;
5. to ensure that the organization conforms to its stated food safety policy;
6. to demonstrate such conformity to relevant interested parties; and
7. to seek certification or registration of its food safety management system by an external organization, or make a self-assessment or self-declaration of conformity to ISO 22000:2005.

3.1.5 INDIAN FOOD STANDARD LEGISLATION AND REGULATORY BODIES

In 2006, India passed the Food Safety and Standard Act (FSSA), a modern integrated food law to serve as a single reference point in relation to the regulation of food products, including all types of food commodities. The passing of this Indian act is a significant first step; much more has to happen to eliminate the confusing overlap with old laws and regulations. The Food Safety and Standards Authority of India (FSSAI) still needs to make considerably substantive changes in infrastructure and appropriate stewardship for it to match with international standards of the United States and Europe.

1. By the mid-1990s, the food processing sector laws were framed in a veritable grid of regulation, including a multitude of state laws, as well as the following national laws:
   a. The Infant Milk Substitutes Feeding Bottles & Infants Food Act, 1992 (Rules: 1993) (regulation of production, supply)
   b. Milk and Milk Products Order, 1992
   c. Environmental Protection Act, 1986
   d. Pollution Control Act, 1986
   e. Meat Food Products Order (MFPO), 1973
   f. Insecticide Act, 1968
   g. Solvent Extracted Oil Control (SEO) Order, 1967
   h. Export (Quality Control and Inspection) Act, 1963
   i. Essential Commodities Act, 1955
   j. Fruits Product Order (FPO), 1955
   k. Prevention of Food Adulteration (PFA) Act, 1954 (Rules: 1955)
   l. Vegetable Oil Product Control (VOP) Order, 1947
   m. Agriculture Produce Act, 1937
2. Some food laws that can be declared voluntarily by the manufacturers of finished products include the following:
   a. AGMARK Standards (AGMARK)
   b. Codex Alimentarius Standards
   c. Bureau of Indian Standards Act, 1986
   d. Consumer Protection Act, 1986

3.1.5.1 Prevention of Food Adulteration Act, 1954

The Prevention of Food Adulteration (PFA) Act was promulgated by Parliament in 1954 to make provision for the prevention of adulteration of food, along with the Prevention of Food Adulteration Rules (1955), which was incorporated in 1955 as an extension to the PFA Act. Broadly, the PFA Act covers food standards, general procedures for sampling, analysis of food, powers of authorized officers, nature of penalties, and other parameters related to food. It deals with parameters relating to food additives, preservative, and coloring issues; packing and labeling of foods; prohibition and regulations of sales; and so on. Like the FPO, amendments to PFA Rules are incorporated with a recommendation made by the Central Committee of Food Standards, which has been setup by the central government under the Ministry of Health and Family Welfare, comprising members from different regions of the country. The provisions of the PFA Act and Rules are implemented by the state governments and local bodies as provided in the rules. The PFA Act had been in place for over five decades, and there was a need for change for various reasons, including the changing requirements of our food industry.

The PFA Act will be repealed from the date to be notified by the central government as per the Food Safety and Standards Act, 2006. Until that date, when new standards are specified, the requirements and other provisions of the PFA Act (1954) and Rules (1955) shall continue to be in force as a transitory provision for food standards. The PFA Division has contemplated the following draft rules/regulations for the Food Safety and Standards Act of India (FSSAI), 2006, which are in the process of being further scrutinized in consultation with state governments.

3.1.5.2 Milk and Milk Product Amendment Regulations, 2009

Consequent to the delicensing of the dairy sector in 1991 under the Industrial Development & Regulation Act, the Department of Animal Husbandry, Dairy and Fisheries promulgated the Milk and Milk Product Order (MMPO; 1992) under section 3 of the Essential Commodities Act, 1955. The objective of the order is to maintain and increase the supply of liquid milk of a desired quality in the interest of the general public, and also for regulating the production, processing, and distribution of milk and milk products. As per the provisions of this order, any person/dairy plant handling more than 10,000 L of milk per day or 500 MT of milk solids per annum needs to be registered with the registering authority appointed by the central government. There is no restriction on the setting up of new dairy units and expanding milk processing capacity; noting the requirement of registration is for enforcing the prescribed Sanitary and Hygienic Conditions, Quality and Food Safety Measures as specified in Schedule V of the MMPO 1992. To comply with the provisions of Paragraph 5(5)(B) of MMPO 1992, two inspection agencies—the National Productivity Council and the Export Inspection Council (EIC) of India—were notified for annual inspection of registered dairy units, on a rotating basis. Since their inception the central and state registering authorities have registered 803 dairy units with a combined milk processing capacity of 881.50 lakhs liters/day in the cooperative, private, and government sectors upto 31.03.2008. Further, the central registering authority has granted 12 new registrations with a milk processing capacity of 25.0 lakhs liters/day (9 dairy units for milk processing and the remaining 3 units for marketing/trading), enhanced the milk processing capacity of 14 dairy units, and canceled the registration of 10 dairy units during 2008−2009. MMPO 1992 was renamed MMPR (2009), and now it has been subsumed as milk and milk products regulations under section 99 of the FSSA of 2006.

3.1.5.3 Fruit Product Order (FPO), 1955

The FPO (1955), promulgated under section 3 of the Essential Commodities Act, 1955, with an objective to manufacture fruit and vegetable products while maintaining sanitary and hygienic conditions on the premises and quality standards laid down in the order. It is mandatory for all manufacturers of fruit and vegetable products, including some nonfruit products such as nonfruit vinegar, syrup, and sweetened aerated water, to obtain a license under this order. This order was implemented earlier by the Ministry of Food Processing Industries (now by the FSSAI) through the Directorate of Fruit and Vegetable Preservation, headquartered in New Delhi. The Directorate has five regional offices with headquarters located in New Delhi, Mumbai, Kolkata, Chennai, and Guwahati, as well as a suboffice at Lucknow in northern region. The field officers of the regional offices undertake periodic inspections of the manufacturing units to ensure maintenance of hygienic conditions in the factories and draw random samples of products from them, as well as from markets, which are analyzed in the laboratories to test their conformity according to the specifications laid out under the FPO. The Central Fruit Product Advisory Committee, comprising officials of concerned government departments, technical experts, and representatives of the Central Food Technology Research Institute, Bureau of Indian Standards, the fruits and vegetable processing industry, and consumer organizations, recommends amendments to the FPO.

3.1.5.4 Edible Oils Packaging Order, 1998

To ensure the availability of safe and quality edible oils in packed form at predetermined prices to consumers, the central government promulgated on September 17, 1998, an Edible Oils Packaging (Regulation) Order under the Essential Commodities Act, 1955, to make compulsory the packaging of edible oils sold in retail, unless specifically exempted by the concerned state governments.

3.1.5.5 Vegetable Oil Products Order, 1998

Two earlier orders—Vegetable Oil Products (Control) Order, 1947, and Vegetable Oil Products (Standards of Quality) Order, 1975—have been replaced by a single order called the "Vegetable Oil Products (Regulation) Order, 1998" for proper regulation of the manufacture, distribution, and sale of vegetable oil products. The vegetable oil products industry is regulated by this order through the Directorate of Vanaspati, Vegetable Oils & Fats, Department of Food, Public Distribution, Ministry of Consumer Affairs, and Food & Public Distribution.

3.1.5.6 Meat Food Products Order (MFPO), 1973

Consumption of meat and meat products, and consumers' preference for these products, is gradually increasing. In the meat and meat processing sectors, poultry is the fastest growing animal protein in India. Indian consumers prefer to buy fresh meat from the wet market, rather than processed or frozen meats. A mere 6% of poultry meat produced (about 100,000 MT) is sold in processed form. Of this, only about 1% undergoes processing into value-added products (ready-to-eat/ready-to-cook products). Processing of large animals is largely for the purpose of exports. Meat and meat products are highly perishable and can transmit diseases from animals to humans. Processing of meat products is licensed under the MFPO 1973 now under Ministry of Food Processing Industries.

3.1.5.7 Solvent-Extracted Oil, De-Oiled Meal and Edible Flour (Control) Order, 1967

This order is basically a quality control order to ensure that solvent-extracted oils in particular do not reach consumers for consumption before they are refined and conform to the quality standards specified in the order for the intended purpose. Standards for the solvent (hexane) that is to be used for the extraction of oil from oil-bearing materials have also been specified to eliminate possible contamination of oil from the solvent used. This order has various major roles:

1. Governing the manufacture, quality, and movement of solvent extracted oils, de-oiled meal, and edible flour
2. Protecting consumers through quality assurance of solvent extracted oils, de-oiled meal, and edible flour
3. Eliminating the possibility of diversion of the oils for unintended uses
4. Prohibiting purchase, offers to purchase, use, or stock for use, any solvent not conforming to the quality standards for extraction of vegetable oils, specifying particulars to be declared on the label affixed to the container.

3.1.5.8 Food Safety and Standards Authority of India (FSSAI)

The FSSAI was established under Food Safety and Standards Act, 2006, which consolidates various acts and orders that hitherto handled food-related issues in various ministries and departments. The FSSAI was created to develop science-based standards for articles of food and to regulate their manufacture, storage, distribution, sale, and import to ensure the availability of safe and wholesome food for human consumption. The Ministry of Health & Family Welfare, Government of India, is the administrative ministry for the implementation of the FSSAI. Various central acts including the PFA Act 1954, FPO 1955, MFPO 1973, Vegetable Oil Products (Control) Order 1947, Edible Oils Packaging (Regulation) Order 1988, Solvent Extracted Oil, De-Oiled Meal and Edible Flour (Control) Order 1967, and the MMPO 1992, have been repealed after commencement of the FSSAI Act, 2006. The FSSAI, headquartered in New Delhi, and the state food safety authorities enforce provisions of this act. Food Safety & Standards Regulations (FSSR), 2011, is divided into six parts:

1. Licensing and Registration of Food Businesses
2. Packaging and Labeling
3. Food Product Standards and Food Additives
4. Prohibition and Restriction on Sales
5. Contaminants, Toxins, and Residues
6. Laboratory and Sampling Analysis

The following regulations are covered under FSSR, 2011.
Regulation 2.1 Dairy products and Analogues
Regulation 2.2 Fats, oils and fat emulsion
Regulation 2.3 Fruits and vegetables products
Regulation 2.4 Cereal and cereal products
Regulation 2.5 Meat and meat products
Regulation 2.6 Fish and fish products
Regulation 2.7 Sweets and confectionary
Regulation 2.8 Sweetening agents including honey
Regulation 2.9 Salt, spices, condiments and related products
Regulation 2.10 Beverages (other than dairy and fruit- and vegetable-based)
Regulation 2.11 Other food products and ingredients
Regulation 2.12 Proprietary food
Regulation 2.13 Irradiation of food
Regulation 3.0 Substances added to food
Regulation 3.1 Food additives
Regulation 3.2 Standards of additives

3.1.5.9 Export Inspection Council

The EIC was set up by the Government of India under section 3 of the Export (Quality Control and Inspection) Act, 1963 (the 22nd of 1963), in order to ensure sound development of India's export trade through quality control and inspection and for matters connected thereto. Basmati rice, honey, fish and fish products, poultry meat products, egg products, and milk products are major food commodities for which the EIC provides certification; it also acts as an advisory body to the central government, which is empowered under the act to identify commodities to be subjected to quality control and/or inspection before export, to establish standards of quality for such identified commodities, and to specify the type of quality control and/or inspection to be applied to such commodities.

3.1.5.10 Agricultural and Processed Food Products Export Development Authority (APEDA)

The APEDA was established by the Government of India under the Agricultural and Processed Food Products Export Development Authority Act passed by the parliament in December 1985. The act (the second of 1986) came into effect on February 13, 1986, through a notification issued in the *Gazette of India*: Extraordinary: Part-II (Sec. 3(ii): 13.2.1986). The APEDA replaced the Processed Food Export Promotion Council and is mandated with the responsibility of promoting exports and developing the following scheduled products: meat and meat products, poultry and poultry products, dairy products, confectionery, biscuits and bakery products, honey, jaggery and sugar products, cocoa and its products, chocolates of all kinds, alcoholic and nonalcoholic beverages, cereal and cereal products, groundnuts, peanuts and walnuts, pickles, *papads* and chutneys, guar gum, floriculture and floriculture products, and herbal and medicinal plants. In addition to this, APEDA has been entrusted with the responsibility to monitor the importation of sugar.

3.2 STANDARDS AND PERMISSIBLE LIMITS

Food standards and permissible limits are fixed by food standard development bodies and are enacted by food regulatory bodies. These two bodies may be the same or different in a particular country, but it varies from country to country. Limits on food quality and safety parameters are fixed based on reviews of various aspects of their enactment, feasibility, effects on health, tolerance limits specified by different health organizations, and the possibility of controlling such levels during manufacture and quality monitoring. These factors vary by country and also change with time, as well as with scientific advancements to monitor and control food quality and safety parameters. Food standards and permissible limits of all countries are difficult to present in a single chapter; however, those for some countries are presented as a summary or compilation in the following subsections.

3.2.1 PHYSICAL ADULTERANTS

Physical adulterants or contaminants are those that are added in food purposely or by chance during processing operations. Some of these adulterants and their ill effects are described in Chapter 1. Limits and standards are not easily available, but it is desirable that any food materials be free from any physical adulterants and contaminants. Indicative limits of some of physical adulterants are enumerated in Table 3.1.

3.2.2 CHEMICAL RESIDUES

Chemical residues come into the food chain through various mechanisms, starting with the soil during plant growth through to processing, preservation, storage, and transport. Chemical residues that are present in amounts that are more than the permissible limits and are consumed for a long period may prove very dangerous to health. Limits of various chemical residues in various foods are compiled in Tables 3.2–3.5.

Table 3.1 Limits and Standards of Some Physical Adulterants/Contaminants in Different Food Types (The Limits Specified Here are Indicative Only. One Should Consult the Relevant Standard and Guidelines of Particular Regulatory Body for Practical Purposes)

3.1(a). AGMARK Standards of Maximum Permissible Limits in the Percentage of Different Refractions in Fair Average Quality of Wheat

Foreign Matter	Other Food Grains	Damaged Grains	Slightly Damaged Grains	Shriveled and Broken Grains
0.75 Within this, poisonous weed seeds should not exceed 0.4%, of which dhatura and akra (*Vicia* spp.) should not be >0.025% and 0.2%, respectively.	2.0	2.0 Within this, ergot-affected grains should not exceed 0.05%.	6.0	7.0

3.1(b). AGMARK Standards of Grade-Wise Allowable Foreign Physical Contaminants in Wheat (Percentage by Weight)

Grade Designation	Foreign Matter[a]	Other Food	Other Wheat	Damaged Grains	Slightly Damaged Grains	Immature, Shriveled, and Broken Grains	Weevilled Grains[b]
I	1.0	1.6	5.0	1.0	2.0	2.0	1.0
II	1.0	3.0	15.0	2.0	4.0	4.0	3.0
III	1.0	6.0	20.0	4.0	6.0	10.0	6.0
IV	1.0	8.0	20.0	5.0	10.0	10.0	10.0

3.1(c). Standards of Wheat Prescribed in PFA Rules 1955

Foreign Matter, %	Other Food Grains, %	Damaged Grains, %	Weevilled Grains, %
≤1, of which ≤0.25 mineral matter and ≤0.10 impurities of animal origin	≤6; total foreign matters and other edible grains ≤12%	≤6, of which ≤3 are karnal bunt—affected and ≤0.05 is ergot-affected grains	≤10 by count

Continued

Table 3.1 Limits and Standards of Some Physical Adulterants/Contaminants in Different Food Types (The Limits Specified Here are Indicative Only. One Should Consult the Relevant Standard and Guidelines of Particular Regulatory Body for Practical Purposes)—cont'd

3.1(d). Codex Alimentarius Commission Standard for Wheat and Durum Wheat

Parameters	Wheat	Durum Wheat
Moisture	≤14.5% m/m	≤14.5% m/m
Organic extraneous matter	≤1.5% m/m	≤1.5% m/m
Inorganic extraneous matter	≤0.5% m/m	≤ ≥0.5% m/m
Test weight, kg/hL	≥68	≥70
Shrunken and broken kernels, maximum	≤5.0% m/m	≤6.0% m/m
Edible grains other than wheat and durum wheat	≤2.0% m/m	≤3.0% m/m
Damaged kernels	≤6.0% m/m	≤4.0% m/m
Insect-bored kernels	≤1.5% m/m	≤2.5% m/m
Filth (impurities of animal origin, dead insects, etc.)	≤0.1% m/m	≤0.1% m/m
Ergot (*sclerotium* of the fungus *Claviceps purpurea*)	0 ≤ 0.05% m/m	≤0.05% m/m

3.1(e). Maximum Limits of Physical Adulterants in Grade A and Common Paddy

S. No.	Refractions	Maximum Limits, %
1	Foreign matter	
	a) Inorganic	1.0
	b) Organic	1.0
2	Damaged, discolored, sprouted, and weevilled grains	3.0
3	Immature, shrunken, and shriveled grains	3.0
4	Admixture of lower class	8.0
5	Moisture	17.0

Note: Within the overall limit of 1.0% for organic foreign matter, poisonous seeds can compose ≤0.5%, of which dhatura and akra seeds (*Vicia* spp.) are ≤0.025% and 0.2%, respectively. Paddy shall be in sound merchantable condition, dry, clean, wholesome, of good food value, uniform in color and size of grains, and free from molds, weevils, obnoxious smell, *Argemone maxicana, Lathyrus sativus* (*kesari*), and admixture of deleterious substances.

3.1(f). AGMARK Standards of Grades and Quality Definition of Paddy

Grades	Special Characteristics (Maximum Limit of Tolerance)			General Characteristics
	Foreign Matter, %	Admixture, %	Damaged, Immature, Weevilled, %	
I	1.0	5.0	1.0	It should be the dried, mature grains with a husk of *Oryza sativa* L., with uniform size, shape, and color; hard, clean, wholesome, and free from molds, weevils, obnoxious smell, discoloration, admixture of deleterious substances, and all other impurities. Moisture ≤14%
II	2.0	10.0	2.0	
III	4.0	15.0	5.0	
IV	7.0	30.0	10.0	

3.1(g). Limits of Physical Adulterants/Contaminants for Procurement of Maize by the Food Corporation of India

S. No.	Refractions	Maximum Limits, %
1	Foreign matters[c]	1
2	Other food grains	2
3	Damaged grains	1.5
4	Slightly damaged, discolored, and touched grains	4.5
5	Shriveled and immature grains	3
6	Weevilled grains	1
7	Moisture	14

3.1(h). AGMARK Standard for Maize

Grade Designation	Maximum Limits of Tolerance, %							
	Moisture	Foreign Matter		Other Edible Grains	Admixture of Different Varieties	Damaged Grains	Immature, Shriveled Grains	Weevilled Grains (by Count)
		Organic	Inorganic					
I	12	0.10	0	0.50	5	1	2	2
II	12	0.25	0.1	1	10	2	4	4
III	14	0.50	0.25	2	15	3	6	6
IV	14	0.75	0.25	3	15	4	6	8

In addition, maize should be mature grain of *Zea mays* L.; sweet, hard, clean, wholesome, uniform in size, shape, color, and in sound merchantable condition. It should be free from rodent hair and excreta, added coloring matter, molds, weevils, obnoxious substances, discoloration, poisonous seeds, and all other impurities except to the extent indicated in the schedule. Uric acid and aflatoxin should not exceed 100 mg and 30 μg/kg, respectively. It should also comply with the restrictions in regard to pesticide/insecticide residue (Rule 65), poisonous metals (Rule 57), naturally occurring toxic substances (Rule 57-B), and other provisions prescribed under the Prevention of Food Adulteration Rules 1955 and as amended from time to time.

Continued

Table 3.1 Limits and Standards of Some Physical Adulterants/Contaminants in Different Food Types (The Limits Specified Here are Indicative Only. One Should Consult the Relevant Standard and Guidelines of Particular Regulatory Body for Practical Purposes)—cont'd

3.1(i). Codex Alimentarius Commission Standard for Maize

Parameters	Maize (corn), % m/m
Moisture content	≤15.5
Organic extraneous matter	≤1.5
Inorganic extraneous matter	≤0.5
Filth (impurities of animal origin, including dead insects)	≤0.1
Kernels of other colors	≤5%
• Yellow maize. Maize grains that are yellow and/or light red in color are considered to be yellow maize. Maize grains that are yellow and dark red in color, provided the dark red color covers <50% of the surface of the grain, are also considered to be yellow maize.	
• White maize. Maize grains that are white and/or light pink in color are considered to be white maize. White maize also means maize grains that are white and pink in color, provided the pink color covers <50% of the surface of the grain.	≤2%
• Red maize. Maize grains that are pink and white or dark red and yellow in color are considered to be red maize, provided the pink or dark red color covers ≥50% of the surface of the grain.	≤5%
• Mixed maize	
Kernels of other shape	≤5.0% by weight of maize of other shapes
• Flint maize	≤5%
• Dent maize	5.0—95% by weight of flint maize
• Flint and dent maize	
Defects	
• Blemished grains (grains that are damaged by insects or vermin, stained, diseased, discolored, germinated, frost damaged, or otherwise materially damaged)	Maximum 7%, of which diseased grains must not exceed 0.5%
• Broken kernels	Maximum 6.0%
• Other grains	Maximum 2.0%

3.1(j). AGMARK Grade Specifications and Definition of Quality for *Kharif jowar*

Grade Designation	Moisture	Foreign Matter		Other Grains	Damaged Grains	Immature/Shriveled Grains	Weevilled Grains
		Organic	Inorganic	Maximum Limits of Tolerance, % by Weight			
I	12.00	0.10	0	1.00	1.00	2.0	0.5
II	12.00	0.25	0.10	1.50	2.00	4.0	1.0
III	14.00	0.50	0.25	2.00	3.00	6.0	2.0
IV	14.00	0.75	0.25	4.00	5.00	8.0	6.0

Jowar should be the dried mature grains of *Sorghum vulgare pers* raised in the *kharif* season; be sweet, hard, clean, wholesome, uniform in size, shape, color, and in sound merchantable condition; be free from added coloring matter, molds, weevils, obnoxious substances, discoloration, poisonous seeds and all other impurities except to the extent indicated in the schedule; uric acid and aflatoxin should not exceed 100 mg and 30 μg/kg, respectively; be free from rodent hair and excreta; and comply with the restrictions in regard to pesticide/insecticide residue (Rule 65), poisonous metals (Rule 57), naturally occurring toxic substances (Rule 57-B), and other provisions prescribed under the Prevention of Food Adulteration Rules, 1955, and as amended from time to time.

In foreign matter, the impurities of animal origin should not be more than 0.10%, whereas the ergot-affected grains should not exceed 0.05% by weight in damaged grains.

Definitions of some terminologies: *Foreign matter* means any extraneous matter than other food grains comprising (1) inorganic matter (includes metallic pieces, dust, sand, gravel, stones, dirt, pebbles, lumps or earth, clay, mud and animal filth etc.); (2) organic matter consisting of husk, straws, weeds and other inedible grains, etc.

Other edible grain means any edible grains (including oil seeds) other than the one that is under consideration. *Damaged grain* means grain that is sprouted or internally damaged as a result of heat, microbe, moisture, or weather, namely, ergot-affected grains and karnal bunt grains. *Immature and shriveled grain* means grain that is not properly developed. *Weevilled grain* means grain that is partially or wholly bored by insects injurious to grains but does not include germ-eaten grains and egg-spotted grains. *Poisonous, toxic, and/or harmful seeds* mean any seed that, if present in quantities above the permissible limit, may have a damaging or dangerous effect on health, organoleptic properties, or technological performance, such as dhatura (*D. fastuosa* Linn. and *D. stramonium* Linn.), corn cockle (*Agrostemma githago* L. *Machai Lallium remulenum* Linn.), and akra (*Vicia* species).

3.1(k). Codex Alimentarius Commission Standard for Sorghum Grains

Moisture Content: 14.5% m/m maximum, but lower limits vary destination to destination.

Defects: The product should not contain more than 8% total defects, including extraneous matter, inorganic extraneous matter, and filth, as contained in the standards, and blemished grains, diseased grains, broken kernels, husks, and other grains.

Toxic or noxious seeds: The products covered by the provisions of this standard should be free from *Crotalaria* spp., corn cockle (*Agrostemma githago* L.), castor beans (*Ricinus communis* L.), jimson weed (*Datura* spp.), and other seeds that are commonly recognized as harmful to health.

3.1(l). PFA Standard of Mustard and Rapeseed

Description: Mustard (*rai, sarson*) whole means the dried seeds of *Brassica alba. L. Boiss* (white rai), *Brassica compestris L. cv dichotoma* (black sarson), *Brassica compestris L. cv.* (yellow sarson), syn. *Brassica compestris L., cv. glauca* (yellow sarson), *Brassica compestris L. cv. toria* (toria), *Barassica juncea,* (L.) *Cosset czern* (rai, lotni), and *Brassica nigra* (L.) Koch (Benarasi rai).

Extraneous matter: ≤7.0% (insect-damaged matter ≤5%), and it should be free from seeds of *Argemone maxicana* Linn. and added coloring.

Continued

Table 3.1 Limits and Standards of Some Physical Adulterants/Contaminants in Different Food Types (The Limits Specified Here are Indicative Only. One Should Consult the Relevant Standard and Guidelines of Particular Regulatory Body for Practical Purposes)—cont'd

3.1(m). AGMARK Standards of Mustard and Rapeseed Comprising *Brassica compestris* cv. Sarson/Toria dichotoma and *Brassica juncea/Brassica nigra* (Rai) grown in India

Grade Designation	Maximum Allowed, % by Weight				
	Foreign Matter	Dead, Badly Discolored, and Damaged	Unripe, Shriveled, and Slightly Damaged	Small Atrophied Seeds	Admixture[d] of Other Varieties of Mustard
Special	1.0	1.0	1.5	5.0	5.0
Standard	2.0	1.5	3.0	10.0	10.0
General	3.0	2.0	4.0	20.0	15.0

3.1(n). AGMARK Standards of Pulses (Whole Grain)

Pulses	Grades	Moisture, %	Maximum Limits of Tolerance , % by Weight		Other Edible Grains	Damaged Grains	Weevilled Grains (by Count)
			Foreign Matter				
			Organic	Inorganic			
Urd whole (blackgram)	Special	10.0	0.10	0	0.1	0.5	2.0
	Standard	12.0	0.50	0.10	0.5	2.0	4.0
	General	14.0	0.75	0.25	3.0	5.0	6.0
Moong (whole)	Special	10.0	0.10	0	0.1	0.5	2.0
	Standard	12.0	0.50	0.10	0.5	2.0	4.0
	General	14.0	0.75	0.25	3.0	5.0	6.0
Masoor (lentil) whole	Special	10.0	0.10	0	0.1	0.5	2.0
	Standard	12.0	0.50	0.10	0.5	2.0	4.0
	General	14.0	0.75	0.25	3.0	5.0	6.0
Arhar/tur (red gram) whole	Special	10.0	0.10	0	0.5	0.5	3.0
	Standard	12.0	0.50	0.10	0.5	2.0	5.0
	General	14.0	0.75	0.25	2.0	5.0	10.0
Kabuli chana	Special	10.0	0.10	0	0.5	0.5	3.0
	Standard	12.0	0.50	0.10	2.0	2.0	6.0
	General	16.0	0.75	0.25	4.0	5.0	10.0

Pulses	Grades	Moisture, %	Organic, %	Inorganic, %		
Chana whole (bengal gram)	Special	10.0	0.10	0	0.5	3.0
	Standard	12.0	0.50	0.10	2.0	6.0
	General	16.0	0.75	0.25	5.0	10.0
Peas (whole)	Special	10.0	0.10	0	0.5	3.0
	Standard	12.0	0.50	0.10	2.0	6.0
	General	16.0	0.75	0.25	5.0	10.0
French bean/rajma	Special	10.0	0.10	0	2.0	2.0
	Standard	12.0	0.50	0.10	4.0	6.0
	General	14.0	0.75	0.25	5.0	10.0
Lobia	Special	10.0	0.10	0	2.0	2.0
	Standard	12.0	0.50	0.10	4.0	6.0
	General	14.0	0.75	0.25	5.0	10.0

In foreign matter, the impurities of animal origin should not be more than 0.10% by weight. In general, whole-grain pulses should be dried, mature seeds of pulses; be sweet, clean, wholesome, uniform in size, shape, color, and in sound merchantable conditions; be free from living and dead insects, fungus infestation, added coloring matter, molds, obnoxious smell, discoloration; be free from rodent hair and excreta; be free from toxic or noxious seeds, namely, *Crotolaria* spp., corn cockle (*Agrostemma githago* L.), castor beans (*Ricinus communis* L.), Jimson weed (*Dhatura* spp.), *Argemone mexicana, khesari*, and other seeds that are commonly recognized as harmful to health; and uric acid and aflatoxin should not exceed 100 mg and 30 μg/kg, respectively.

3.1(o). AGMARK Standards of Pulses (Split/Husked)

Pulses	Grades	Moisture, %	Foreign Matter Organic, %	Inorganic, %	Other Edible Grains, %	Damaged Grains, %	Broken and Fragmented Grains, %	Weevilled Grains, %
Urd split (Husked)	Special	10.0	0.10	0	0.1	0.5	0.5	1.0
	Standard	12.0	0.50	0.10	0.5	2.0	2.0	2.0
	General	14.0	0.75	0.25	3.0	5.0	5.0	3.0
Urd split (Unhusked)	Special	10.0	0.10	0	0.1	0.5	2.0	1.0
	Standard	12.0	0.50	0.10	0.5	2.0	4.0	2.0
	General	14.0	0.75	0.25	3.0	5.0	6.0	3.0
Moong split (Husked)	Special	10.0	0.10	0	0.1	0.5	1.0	1.0
	Standard	12.0	0.50	0.10	0.5	2.0	3.0	2.0
	General	14.0	0.75	0.25	3.0	5.0	6.0	3.0

Continued

Table 3.1 Limits and Standards of Some Physical Adulterants/Contaminants in Different Food Types (The Limits Specified Here are Indicative Only. One Should Consult the Relevant Standard and Guidelines of Particular Regulatory Body for Practical Purposes)—cont'd

3.1(o). AGMARK Standards of Pulses (Split/Husked)

Pulses	Grades	Moisture, %	Foreign Matter		Other Edible Grains, %	Damaged Grains, %	Broken and Fragmented Grains, %	Weevilled Grains, %
			Organic	Inorganic				
Moong split (Unhusked)	Special	10.0	0.10	0	0.1	0.5	0.5	1.0
	Standard	12.0	0.50	0.10	0.5	2.0	2.0	2.0
	General	14.0	0.75	0.25	3.0	5.0	5.0	3.0
Masoor (lentil) split (husked)	Special	10.0	0.10	0	0	0.5	0.1	1.0
	Standard	12.0	0.50	0.10	0.5	2.0	0.5	2.0
	General	14.0	0.75	0.25	2.0	5.0	1.0	3.0
Arhar/tur (red gram) split (husked)	Special	10.0	0.10	0	0	0.5	2.0	1.0
	Standard	12.0	0.50	0.10	0.2	2.0	5.0	2.0
	General	14.0	0.75	0.25	0.5	5.0	8.0	3.0
Chana split (husked)/dal chana	Special	10.0	0.10	0	0.5	0.5	0.5	1.0
	Standard	12.0	0.50	0.10	1.0	2.0	2.0	2.0
	General	16.0	0.75	0.25	2.0	5.0	5.0	3.0

In foreign matter, the impurities of animal origin should not be more than 0.10%. In general, dal should consist of unhusked and split of same category; be sweet, clean, wholesome, uniform in size, shape, color, and in sound merchantable condition; be free from living and dead insects, fungus infestation, added coloring matter, molds, obnoxious smell, discoloration; be free from rodent hair and excreta; be free from toxic or noxious seeds, namely, *Crotolaria* spp., corn cockle (*Agrostemma githago* L.), castor bean (*Ricinus communis* L.), Jimson weed (*Dhatura* spp.), *Argemone mexicana*, khesari, and other seeds that are commonly recognized as harmful to health; uric acid and aflatoxin should not exceed 100 mg and 30 µg/kg, respectively.

[a]Within this, mineral matters and impurities of animal origin should not be more than 0.25% and 0.1%, respectively.

[b]If a lot of wheat, paddy, or maize has ≤1%, >1−4%, >4−7%, or >7−15% of weevilled grain, it is designated as A, B, C, and D grades, respectively, by the Central Warehousing Corporation of India.

[c]Foreign matter ≤1%, in which mineral matter ≤0.25, impurities of animal origin ≤0.1%, moisture ≤16.0%, other edible grains ≤3%, damaged grains ≤5%, and weevilled grains ≤10% by count.

[d]This does not apply to Brassica juncea or Brassica nigra if mixed with Brassica campestris sarsons/tonne dichotoma. In general the seeds should be of shape, size, color, and pungency characteristic of the varietyform; be mature, hard, wholesome, and well dried; have moisture not exceeding 6%; not have any trace of Argemone seeds; be free from molds or insect damage and deleterious substances; not bear the grains of any other species; and be in a sound merchantable condition.

Table 3.2 Tolerance Limits of Some Insecticides, Pesticides, and Chemical Residues in Different Foods

S. No.	Insecticides	Food	Tolerance Limit (mg/kg or ppm)[a]
1	Aldrin, dieldrin (the limits apply to aldrin and dieldrin singly or in any combination and are expressed as dieldrin)	Food grains	0.01
		Milled food grains	0
		Milk and milk products	0.15 (On a fat basis)
		Fruits and vegetables	0.1
		Meat	0.2
		Eggs	0.1 (On a shell-free basis)
2	Carbaryl	Fish	0.2
		Food grains	1.5
		Milled food grains	0
		Okra and leafy vegetables	10.0
		Potatoes	0.2
		Other vegetables	5.0
		Cottonseed (whole)	1.0
		Maize cob (kernels)	1.0
		Rice	2.50
		Maize	0.50
		Chilies	5.00
3	Chlordane (residue to be measured as *cis*- plus *trans*- chlordane)	Food grains	0.02
		Milled food grains	0
		Milk and milk products	0.05 (On a fat basis)
		Vegetables	0.2
		Fruits	0.1
		Sugar beet	0.3
4	DDT (the limits apply to DDT, DDD, and DDE singly or in any combination)	Milk and milk products	1.25 (On a fat basis)
		Fruits and vegetables, including potato	3.5
		Meat, poultry, and fish	7.0 (On a whole-product basis)
		Eggs	0.5 (On a shell-free basis)
5	DDT (singly)	Carbonated water	0.001
6	DDD (singly)	Carbonated water	0.001
7	DDE (singly)	Carbonated water	0.001
8	Diazinon	Food grains	0.05
		Milled food grains	0
		Vegetables	0.5
9	Dichlorvos (content of di-chloroacetaldehyde must be reported where possible)	Food grains	1.0
		Milled food grains	0.25
		Vegetables	0.15
		Fruits	0.1

Continued

Table 3.2 Tolerance Limits of Some Insecticides, Pesticides, and Chemical Residues in Different Foods—cont'd

S. No.	Insecticides	Food	Tolerance Limit (mg/kg or ppm)[a]
10	Dicofol	Fruits and vegetables	5.0
		Tea (dry manufactured)	5.0
		Chilies	1.0
11	Dimethoate (residue to be determined as dimethoate and expressed as imethoate)	Fruits and vegetables	2.0
		Chilies	0.5
12	Endosulfan (residues are measured and reported as total of endosulfan A and B and endosulfan-sulfate)	Fruits and vegetables	2.0
		Cottonseed	0.5
		Cottonseed oil (crude)	0.2
		Bengal gram	0.20
		Pigeon pea	0.10
		Fish	0.20
		Chilies	1.0
		Cardamom	1.0
13	Endosulfan A	Carbonated water	0.001
14	Endosulfan B	Carbonated water	0.001
15	Endosulfan-sulfate	Carbonated water	0.001
16	Fenitrothion	Food grains	0.02
		Milled food grains	0.005
		Milk and milk products	0.05 (On a fat basis)
		Fruits	0.5
		Vegetables	0.3
		Meat	0.03
17	Heptachlor (combined residues of heptachlor and its epoxide to be determined and expressed as heptachlor)	Food grains	0.01
		Milled food grains	0.002
		Milk and milk products	0.15 (On a fat basis)
		Vegetables	0.05
18	Hydrogen cyanide	Food grains	37.5
		Milled food grains	3.0
19	Hydrogen phosphide	Food grains	0
		Milled food grains	0
20	Inorganic bromide (determined and expressed as total bromide from all sources)	Food grains	25.0
		Milled food grains	25.0
		Fruits	30.0
		Dried fruits	30.0
		Spices	400.00
21	Hexachlorocycle hexane and its isomers		
	(a) Alpha (b) isomer	Rice grain, unpolished	0.10
		Rice grain, polished	0.05
		Milk (whole)	0.02
		Fruits and vegetables	1.00

Table 3.2 Tolerance Limits of Some Insecticides, Pesticides, and Chemical Residues in Different Foods—cont'd

S. No.	Insecticides	Food	Tolerance Limit (mg/kg or ppm)[a]
		Fish	0.25
		Carbonated water	0.001
	(b) Beta (b) isomer	Rice grain, unpolished	0.10
		Rice grain, polished	0.05
		Milk (whole)	0.02
		Fruits and vegetables	1.00
		Fish	0.25
		Carbonated water	0.001
	(c) Gamma (b) isomer (known as lindane)	Food grains (except rice)	0.10
		Milled food grains	0
		Rice grain, unpolished	0.10
		Rice grain, polished	0.05
		Milk	0.01
		Milk products	0.20
		Milk products (having <2% fat)	0.20 (On whole basis)
		Fruits and vegetables	1.00
		Fish	0.25
		Eggs	0.10 (On shell-free basis)
		Meat and poultry	2.00 (On whole basis)
		Carbonated water	0.001
	(d) Delta (b) isomer	Rice grain, unpolished	0.10
		Rice grain, polished	0.05
		Milk (whole)	0.02
		Fruits and vegetables	1.00
		Fish	0.25
		Carbonated water	0.001
22	Malathion (to be determined and expressed as combined residues of malathion and malaoxon)	Food grains	4.0
		Milled food grains	1.0
		Fruits	4.0
		Vegetables	3.0
		Dried fruits	8.0
		Carbonated water	0.001
23	Parathion (combined residues of parathion and paraoxon to be determined and expressed as parathion)	Fruits and vegetables	0.5

Continued

Table 3.2 Tolerance Limits of Some Insecticides, Pesticides, and Chemical Residues in Different Foods—cont'd

S. No.	Insecticides	Food	Tolerance Limit (mg/kg or ppm)[a]
24	Parathion methyl (combined residues of parathion methyl and its oxygen analog to be determined and expressed as parathion methyl)	Fruits	0.2
		Vegetables	1.0
25	Phosphamidon residues (expressed as the sum of phosphamidon and its desethyl derivative)	Food grains	0.05
		Milled food grains	0
		Fruits and vegetables	0.2
26	Pyrethrins (sum of pyrethrins I and II and other structurally related insecticide ingredients of pyrethrum)	Food grains	0
		Milled food grains	0
		Fruits and vegetables	1.0
27	Chlorienvinphos (residues to be measured as alpha and beta isomers of chlorienvinphos)	Food grains	0.025
		Milled food grains	0.006
		Milk and milk products	0. 2 (Fat basis)
		Meat and poultry	0.2 (Carcass fat)
		Vegetables	0.05
		Groundnuts	0.05 (Shell-free basis)
		Cotton seed	0.05
28	Chlorobenzilate	Fruits	1.0
		Dry fruits, almonds and walnuts	0.2 (Shell-free basis)
29	Chlorpyrifos	Food grains	0.05
		Milled food grains	0.01
		Fruits	0.5
		Potatoes and onions	0.01
		Cauliflower and cabbage	0.01
		Other vegetables	0.2
		Meat and poultry	0.1 (Carcass fat)
		Milk and milk products	0.01 (Fat basis)
		Cotton seed	0.05
		Cottonseed oil (crude)	0.025
		Carbonated water	0.001
30	2,4-Dichlorophenoxyacetic acid	Food grains	0.01
		Milled food grains	0.003
		Potatoes	0.2
		Milk and milk products	0.05
		Meat and poultry	0.05
		Eggs	0.05 (Shell-free basis)
		Fruits	2.0

Table 3.2 Tolerance Limits of Some Insecticides, Pesticides, and Chemical Residues in Different Foods—cont'd

S. No.	Insecticides	Food	Tolerance Limit (mg/kg or ppm)[a]
31	Ethion (residues to be determined as ethion and its oxygen analog and expressed as ethion)	Tea (dry manufactured)	5.0
		Cucumber and squash	0.5
		Other vegetables	1.0
		Cotton seed	0.5
		Milk and milk products	0.5 (Fat basis)
		Meat and poultry	0.2 (Carcass fat basis)
		Eggs	0.2 (Shell-free basis)
		Food grains	0.025
		Milled food grains	0.006
		Peaches	1.0
		Other fruits	2.0
		Dry fruits	0.1 (Shell-free basis)
32	Formothion (determined as dinethoate and its oxygen analog and expressed as dimethoate, except in the case of citrus fruits where it is to be determined as formothion)	Citrus fruits	0.2
		Other fruits	1.0
		Vegetable	2.0
		Peppers and tomatoes	1.0
33	Monocrotophos	Food grains	0.025
		Milled food grains	0.006
		Citrus fruits	0.2
		Other fruits	1.0
		Carrot, turnip, potato, and sugar beet	0.05
		Onion and peas	0.1
		Other vegetables	0.2
		Cottonseed	0.1
		Cottonseed oil (raw)	0.05
		Meat and poultry	0.02
		Milk and milk products	0.02
		Eggs	0.02 (Shell-free basis)
		Coffee (raw beans)	0.1
		Chillies	0.2
		Cardamom	0.5
34	Paraquat dichloride (determined as paraquat cations)	Food grains	0.1
		Milled food grains	0.025
		Potato	0.2
		Other vegetables	0.05
		Cotton seed	0.2

Continued

Table 3.2 Tolerance Limits of Some Insecticides, Pesticides, and Chemical Residues in Different Foods—cont'd

S. No.	Insecticides	Food	Tolerance Limit (mg/kg or ppm)[a]
35	Phosalone	Cottonseed oil (edible refined)	0.05
		Milk (whole)	0.01
		Fruits	0.05
		Pears	2.0
		Citrus fruits	1.0
		Other fruits	5.0
		Potatoes	0.1
		Other vegetables	1.0
		Rapeseed/mustard oil (crude)	0.05
36	Trichlorfon	Food grains	0.05
		Milled food grains	0.0125
		Sugar beet	0.05
		Fruits and vegetables	0.1
		Oil seeds	0.1
		Edible oil (refined)	0.05
		Meat and poultry	0.1
		Milk (whole)	0.05
37	Thiometon (residues determined as thiometon, its sulfoxide, and sulfone, expressed as thiometon)	Food grains	0.025
		Milled food grains	0.006
		Fruits	0.5
		Potatoes, carrots, and sugar beets	0.05
		Other vegetables	0.5
38	Acephate	Safflower seed	2.0
		Cotton seed	2.0
39	Methamido-phos (a metabolite of acephate)	Safflower seed	0.1
		Cotton seed	0.1
40	Aldicarb (sum of aldicarb, its sulfoxide, and sulfone, expressed as aldicarb)	Potatoes	0.5
		Chewing tobacco	0.1
41	Atrazine	Maize	0
		Sugarcane	0.25
42	Carbendazim	Food grains	0.50
		Milled food grains	0.12
		Vegetables	0.50
		Mango	2.00
		Banana (whole)	1.00
		Other fruits	5.00
		Cotton seed	0.10

Table 3.2 Tolerance Limits of Some Insecticides, Pesticides, and Chemical Residues in Different Foods—cont'd

S. No.	Insecticides	Food	Tolerance Limit (mg/kg or ppm)[a]
		Groundnut	0.10
		Sugar beet	0.10
		Dry fruits	0.10
		Eggs	0.10 (Shell-free basis)
		Meat and poultry	0.10 (Carcass fat basis)
		Milk and milk products	0.10 (Fat basis)
43	Benomyl	Food grains	0.50
		Vegetables	0.50
		Mango	2.00
		Banana (whole)	1.00
		Other fruits	5.00
		Cotton seed	0.10
		Groundnut	0.10
		Sugar beet	0.10
		Dry fruits	0.10
		Eggs	0.10 (Shell-free basis)
		Meat and poultry	0.10 (Carcass fat basis)
		Milk and milk products	0.10 (Fat basis)
44	Captan	Fruit and vegetables	15.00
45	Carbofuran (sum of carbofuran and 3-hydroxy carbofuran expressed as carbofuran)	Food grains	0.10
		Milled food grains	0.03
		Fruits and vegetables	0.10
		Oil seeds	0.10
		Sugarcane	0.10
		Meat and poultry	0.10 (Carcass fat basis)
		Milk and milk products	0.05 (Fat basis)
46	Copper oxychloride (determined as copper)	Fruit	20.00
		Potatoes	1.00
		Other vegetables	20.00
47	Cypermethrin (sum of isomers) (fat-soluble residue)	Wheat grains	0.05
		Milled wheat grains	0.01
		Brinjal	0.20
		Cabbage	2.00
		Bhindi	0.20
		Oil seeds (except groundnut)	0.20
		Meat and poultry	0.20 (Carcass fat basis
		Milk and milk products	0.01 (Fat basis)
48	Decamethrin/deltamethrin	Cotton seed	0.10
		Food grains	0.50
		Milled food grains	0.20
		Rice	0.05

Continued

Table 3.2 Tolerance Limits of Some Insecticides, Pesticides, and Chemical Residues in Different Foods—cont'd

S. No.	Insecticides	Food	Tolerance Limit (mg/kg or ppm)[a]
49	Edifenphos	Rice	0.02
		Rice bran	1.00
		Meat and poultry	0.02 (Carcass fat basis)
		Milk and milk products	0.01(Fat basis)
50	Fenthion (sum of fenthion, its oxygen analog, and their sulfoxides and sulfones, expressed as fenthion)	Food grains	0.10
		Milled food grains	0.03
		Onions	0.10
		Potatoes	0.05
		Beans	0.10
		Peas	0.50
		Tomatoes	0.50
		Other vegetables	1.00
		Musk melon	2.00
		Meat and poultry	2.00 (Carcasses fat basis)
		Milk and milk products	0.05 (Fat basis)
51	Fenvalerate (fat-soluble residue)	Cauliflower	2.00
		Brinjal	2.00
		Okra	2.00
		Cotton seed	0.20
		Cotton seed oil	0.10
		Meat and poultry	1.00 (Carcass fat basis)
		Milk and milk product	0.01 (Fat basis)
52	Dithiocarbamates (the residue tolerance limit is determined and expressed as mg/CS2/kg and refer separately to the residues arising from any or each group of dithiocarbamates) (a) Dimethyl dithiocarbamate residue resulting from the use of ferbam or ziram, and (b) ethylene bis-dithiocarbamates resulting from the use of mancozeb, maneb, or zineb (including zineb derived from nabam plus zinc sulfate) (c) Mancozeb	Food grains	0.20
		Milled food grains	0.05
		Potatoes	0.10
		Tomatoes	3.00
		Cherries	1.00
		Other fruits	3.00
		Chilies	1.0

Table 3.2 Tolerance Limits of Some Insecticides, Pesticides, and Chemical Residues in Different Foods—cont'd

S. No.	Insecticides	Food	Tolerance Limit (mg/kg or ppm)[a]
53	Phenthoate	Food grains	0.05
		Milled food grains	0.01
		Oilseeds	0.03
		Edible oils	0.01
		Eggs	0.05 (Shell-free basis)
		Meat and poultry	0.05 (Carcass fat basis)
		Milk and milk products	0.01 (Fat basis)
54	Phorate (sum of phorate, its oxygen analog, and their sulfoxides and sulfones, expressed as phorate)	Food grains	0.05
		Milled food grains	0.01
		Tomatoes	0.10
		Other vegetables	0.05
		Fruits	0.05
		Oil seeds	0.05
		Edible oils	0.03
		Sugarcane	0.05
		Eggs	0.05 (Shell-free basis)
		Meat and poultry	0.05 (Carcass fat basis)
		Milk and milk products	0.05 (Fat basis)
55	Simazine	Maize	0
		Sugarcane	0.25
56	Pirimiphos-methyl	Rice	0.50
		Food grains (except rice)	5.00
		Milled food grains (except rice)	1.00
		Eggs	0.05 (Shell-free basis)
		Meat and poultry	0.05 (Carcass fat basis)
		Milk and milk products	0.05 (Fat basis)
57	Alachlor	Cottonseed	0.05
		Groundnut	0.05
		Maize	0.10
		Soybeans	0.10
58	Alfa nephthyl acetic acid	Pineapple	0.50
59	Bitertanol	Wheat	0.05
		Groundnut	0.10
60	Captafol	Tomato	5.00
61	Cartaphydrochloride	Rice	0.50
62	Chlormequatchloride	Grape	1.00
		Cottonseed	1.00
63	Chlorothalonil	Groundnut	0.10
		Potato	0.10

Continued

Table 3.2 Tolerance Limits of Some Insecticides, Pesticides, and Chemical Residues in Different Foods—cont'd

S. No.	Insecticides	Food	Tolerance Limit (mg/kg or ppm)[a]
64	Diflubenzuron	Cottonseed	0.20
65	Dodine	Apple	5.00
66	Diuron	Cottonseed	1.00
		Banana	0.10
		Maize	0.50
		Citrus	1.00
		Grapes	1.00
67	Ethephon	Pineapple	2.00
		Coffee	0.10
		Tomato	2.00
		Mango	2.00
68	Fluchloralin	Cottonseed	0.05
		Soybeans	0.05
69	Malic hydrazide	Onion	15.00
		Potato	50.00
70	Metalyxyl	Bajra	0.05
		Maize	0.05
		Sorghum	0.05
71	Methomyl	Cottonseed	0.10
72	Methyl chloro-phenoxy-acetic acid	Rice	0.05
		Wheat	0.05
73	Oxadiazon	Rice	0.03
74	Oxydemeton methyl	Food grains	0.02
75	Permethrin	Cucumber	0.50
		Cottonseed	0.50
		Soybeans	0.05
		Sunflower seed	1.00
76	Quinolphos	Rice	0.01
		Pigeon pea	0.01
		Cardamom	0.01
		Tea	0.01
		Fish	0.01
		Chilies	0.2
77	Thiophenatemethyl	Apple	5.00
		Papaya	7.00
78	Triazophos	Chilies	0.2
		Rice	0.05
		Cottonseed oil	0.1
		Soybean oil	0.05
79	Profenofos	Cottonseed oil	0.05
80	Fenpropathrin	Cottonseed oil	0.05

Table 3.2 Tolerance Limits of Some Insecticides, Pesticides, and Chemical Residues in Different Foods—cont'd

S. No.	Insecticides	Food	Tolerance Limit (mg/kg or ppm)[a]
81	Fenarimol	Apple	5.0
82	Hexaconazole	Apple	0.1
83	Iprodione	Rapeseed	0.5
		Mustard seed	0.5
		Rice	10.0
		Tomato	5.0
		Grapes	10.0
84	Tridemorph	Wheat	0.1
		Grapes	0.5
		Mango	0.05
85	Penconazole	Grapes	0.2
86	Propiconazole	Wheat	0.05
87	Myclobutanil	Groundnut seed	0.1
		Grapes	1.0
88	Sulfosulfuron	Wheat	0.02
89	Trifluralin	Wheat	0.05
90	Ethoxysulfuron	Rice	0.01
91	Metolachlor	Soybean oil	0.05
92	Glyphosphate	Tea	1.0
93	Linuron	Pea	0.05
94	Oxyfluorfen	Rice	0.05
		Groundnut oil	0.05
95	Carbosulfan	Rice	0.2
96	Tricyclazole	Rice	0.02
97	Imidacloprid	Cotton seed oil	0.05
		Rice	0.05
98	Butachlor	Rice	0.05
99	Chlorimuron-ethyl	Wheat	0.05
100	Diclofop-methyl	Wheat	0.1
101	Metribuzin	Soybean oil	0.1
102	Lambdacyhalothrin	Cottonseed oil	0.05
103	Fenazaquin	Tea	3.0
104	Pendimethalin	Wheat	0.05
		Rice	0.05
		Soybean oil	0.05
		Cottonseed oil	0.05
105	Pretilachlor	Rice	0.05
106	Fluvalinate	Cottonseed oil	0.05
107	Metasulfuron-methyl	Wheat	0.1
108	Methabenzthiazuron	Wheat	0.5
109	Imazethapyr	Soybean oil	0.1
		Groundnut oil	0.1

Continued

Table 3.2 Tolerance Limits of Some Insecticides, Pesticides, and Chemical Residues in Different Foods—cont'd

S. No.	Insecticides	Food	Tolerance Limit (mg/kg or ppm)[a]
110	Cyhalofop-butyl	Rice	0.5
111	Triallate	Wheat	0.05
112	Spinosad	Cottonseed oil	0.02
		Cabbage	0.02
		Cauliflower	0.02
113	Thiamethoxam	Rice	0.02
114	Fenobucarb	Rice	0.01
115	Thiodicarb	Cottonseed oil	0.02
116	Anilophos	Rice	0.1
117	Fenoxy-prop-*p*-ethyl	Wheat	0.02
		Soybean seed	0.02
118	Glufosinate-ammonium	Tea	0.01
119	Clodinafop-propanyl	Wheat	0.1
120	Dithianon	Apple	0.1
121	Kitazin	Rice	0.2
122	Isoprothiolane	Rice	0.1
123	Acetamiprid	Cottonseed oil	0.1
124	Cymoxanil	Grapes	0.1
125	Triadimefon	Wheat	0.5
		Pea	0.1
		Grapes	2.0
126	Fosetyl-A1	Grapes	10
		Cardamom	0.2
127	Isoproturon	Wheat	0.1
128	Propargite	Tea	10.0
129	Difenoconazole	Apple	0.01
130	b-Cyfluthrin	Cottonseed	0.02
131	Ethofenprox	Rice	0.01
132	Bifenthrin	Cottonseed	0.05
133	Benfuracarb	Red gram	0.05
		Rice	0.05
134	Quizalofop-ethyl	Soybean seed	0.05
135	Flufenacet	Rice	0.05
136	Buprofezin	Rice	0.05
137	Dimethomorph	Grapes	0.05
		Potatoes	0.05
138	Chlorfenopyr	Cabbage	0.05
139	Indoxacarb	Cotton seed	0.1
		Cottonseed oil	0.1
		Cabbage	0.1

[a]*These limits are indicative only. For practical purposes, one must consult particular standards/norms of the regulatory authority at that point in time.*

Table 3.3 Permissible Limits of Some Heavy Metals in Different Types of Foods

Metal Contaminants	Article of Foods	Permissible Limits, ppm by Weight[a]
1. Lead	**1.** Beverages	
	Concentrated soft drinks (but not including concentrates used in the manufacture of soft drinks)	0.5
	Fruits and vegetable juice (including tomato juice, but not including lime juice and lemon juice)	1.0
	Concentrates used in the manufacture of soft drinks, lime juice, and lemon juice	2.0
	Baking powder	10
	Edible oils and fats	0.5
	Infant milk substitute and infant foods	0.2
	Turmeric, whole and powder	10.0
	2. Other foods	
	Anhydrous dextrose and dextrose monohydrate, edible oils and fats, refined white sugar (sulfated ash content not exceeding 0.03%	0.5
	Ice cream, iced lollies, and similar frozen confections	1.0
	Canned fish, canned meats, edible gelatin, meat extracts and hydrolyzed protein, dried or dehydrated vegetables (other than onions)	5.0
	All types of sugar, sugar syrup, invert sugar, and direct consumption of colored sugars with sulfated ash content exceeding 1.0%	5.0
	Raw sugars except those sold for direct consumption or used for manufacturing purposes other than the manufacture of refined sugar	5.0
	Edible molasses, caramel liquid, and solid glucose and starch conversion products with a sulfated ash content exceeding 1.0%	5.0
	Cocoa powder	5.0 (Dry fat-free substance)
	Yeast and yeast products	5.0 (Dry matter basis)
	Tea, dehydrated onions, dried herbs and spices, flavorings, alginic acid, alignates, agar, carrageen, and similar products derived from seaweed	10.0 (Dry matter basis)
	Liquid pectin, chemicals not otherwise specified, used as ingredients or in the preparation or processing of food	10.0
	Food coloring other than caramel	10.0 (Dry coloring matter basis)
	Solid pectin	50.0
	Hard-boiled sugar confectionery	2.0
	Iron-fortified common salt	2.0
	Corned beef, luncheon meat, cooked ham, chopped meat, canned chicken, canned mutton, and goat meat and other related meat products	2.5
	Brewed vinegar and synthetic vinegar	0

Continued

Table 3.3 Permissible Limits of Some Heavy Metals in Different Types of Foods—cont'd

Metal Contaminants	Article of Foods	Permissible Limits, ppm by Weight[a]
	3. Foods not specified	2.5
2. Copper	**1.** Beverages	7.0
	Soft drinks excluding concentrates and carbonated water	
	Carbonated water	1.5
	Toddy	5.0
	Concentrates for soft drinks	20.0
	2. Other foods	
	Chicory (dried or roasted), coffee beans, flavorings/pectin liquid	30.0
	Coloring matter	30.0 (Dry coloring matter basis)
	Edible gelatin	30.0
	Tomato ketchup	50.0 (Dried total solids)
	Yeast and yeast products	60.0 (Dry matter)
	Cocoa powder	70.0 (Fat-free substance)
	Tomato puree, paste, powder, juice and cocktails	100.0 (Dried basis)
	Tea	150.0
	Pectin (solid)	300.0
	Hard-boiled sugar confectionery	5.0
	Iron-fortified common salt	2.0
	Turmeric, whole and powder	5.0
	Juice of orange, grape, apple, tomato, pineapple, and lemon	5.0
	Pulp and pulp products of any fruit	5.0
	Infant milk substitute and infant foods	15.0 (But not <2.8)
	Brewed vinegar and synthetic vinegar	0
	Caramel	20
	3. Foods not specified	30.0
3. Arsenic	**1.** Milk	0.1
	2. Beverages	
	Soft drinks intended for consumption after dilution, except carbonated water	0.5
	Carbonated water	0.25
	Infant milk substitute and infant foods	0.05
	Turmeric, whole and powder	0.1
	Juice of orange, grape, apple, tomato, pineapple, and lemon	0.2
	Pulp and pulp products of any fruit	0.2

Table 3.3 Permissible Limits of Some Heavy Metals in Different Types of Foods—cont'd

Metal Contaminants	Article of Foods	Permissible Limits, ppm by Weight[a]
	Preservatives, antioxidants, emulsifying and stabilizing agents, and synthetic food colors	3.0 (Dry matter basis)
	Ice cream, iced lollies, and similar frozen confections	0.5
	Dehydrated onions, edible gelatine, liquid pectin	2.0
	Chicory (dried or roasted)	4.0
	Dried herbs, finings and clearing agents, solid pectin all grades, spices	5.0
	Food coloring other than synthetic coloring	5.0 (Dry matter basis)
	Hard-boiled sugar confectionery	1.0
	Iron-fortified common salt	1.0
	Brewed vinegar and synthetic vinegar	0.1
	3. Foods not specified	1.1
4. Tin	**1.** Processed and canned products	250.0
	Hard-boiled sugar confectionery	5.0
	Jam, jellies, and marmalade	250
	Juice of orange, apple, tomato, pineapple, and lemon	250
	Pulp and pulp products of any fruit	250
	Infant milk substitute and infant foods	5.0
	Turmeric, whole and powder	0
	Corned beef, luncheon meat, cooked ham, chopped meat, canned chicken, canned mutton and goat meat	250
	2. Foods not specified	250
5. Zinc	**1.** Ready-to-drink beverages	5.0
	Juice of orange, grape, tomato, pineapple, and lemon	5.0
	Pulp and pulp products of any fruit	5.0
	Infant milk substitute and infant foods	50.0 (But not <25.0)
	Edible gelatin	100.0
	Turmeric, whole and powder	25.0
	Fruit and vegetable products	50.0
	Hard-boiled sugar confectionery	5.0
	Foods not specified	50.0
6. Cadmium	Infant milk substitute and infant foods	0.1
	Turmeric, whole and powder	0.1
	Other foods	1.5
7. Mercury	Fish	0.5
	Other foods	1.0
8. Methyl mercury	All foods (calculated as the element)	0.25
9. Chromium	Refined sugar	20 ppb

Continued

Table 3.3 Permissible Limits of Some Heavy Metals in Different Types of Foods—cont'd		
Metal Contaminants	**Article of Foods**	**Permissible Limits, ppm by Weight[a]**
10. Nickel	All hydrogenated, partially hydrogenated, interesterified vegetable oils and fats such as vanaspati, table margarine, bakery and industrial margarine, bakery shortening, fat spread, and partially hydrogenated soybean oil	1.5

[a]*Limits specified here are indicative only. One should consult the relevant standards and guidelines of a particular regulatory body for practical purposes at that point in time.*

Table 3.4 Tolerance Limits of Some Antibodies in Any Food Materials		
S. No.	**Antibiotics**	**Tolerance Limit (mg/ kg, ppm)[a]**
1	Tetracycline	0.1
2	Oxytetracycline	0.1
3	Trimethoprim	0.05
4	Oxolinic acid	0.3

[a]*Limits specified here are indicative only. One should consult the relevant standards and guidelines of a particular regulatory body for practical purposes at that point in time.*

3.2.3 MICROBIOLOGICAL CONTAMINANTS

Microbial contaminants mostly are introduced through sources such as the environment, processing mediums, machinery or manpower, and the raw materials themselves, but in the majority of cases microbes grow during storage as a result of either contamination from surrounding foods/materials or fermentation/rotting of the food itself. Microbial or bacterial loads beyond permissible limits or bacteria are very dangerous to health, even if some of them are present in one count. Possible microbiological adulterants/contaminants and their ill effects on health are discussed in Chapter 1; permissible limits for some of these in various important foods are presented in Tables 3.6—3.9.

3.2.4 TOXINS

Toxins are toxic metabolites produced by various molds or fungi when they grow on food products before or after harvest, or during transportation or storage. Some molds, such as *Aspergillus* and *Penicillium* species, can invade food after harvest and produce mycotoxins, whereas others, such as *Fusarium* species, infect food and produce mycotoxins before harvest. Mycotoxins remain in food long after the mold producing them has died; they can therefore be present in foods that are not visibly moldy. Further, many (but not all) mycotoxins are stable and able to survive the usual conditions of

Table 3.5 Antibiotic Residues and Other Pharmacologically Active Substances in Fish and Fishery Products

S. No.	Antibiotics and Other Pharmacologically Active Substances	
1	All nitrofurans, including	
2	Furaltadone	
3	Furazolidone	
4	Furylfuramide	
5	Nifuratel	
6	Nifuroxime	
7	Nifurprazine	
8	Nitrofurnatoin	
9	Nitrofurazone	
10	Chloramphenicol	
11	Neomycin	
12	Nalidixic acid	
13	Sulfamethoxazole	
14	*Aristolochia* spp. and preparations thereof	Prohibited in any unit processing seafoods including shrimp, prawns, or any other variety of fish and fishery products
15	Chloroform	
16	Chloropromazine cholchicine	
17	Dapsone	
18	Dimetridazole	
19	Metronidazole	
20	Ronidazole	
21	Ipronidazole	
22	Other nitromidazoles	
23	Clenbuterol	
24	Diethylstibestrol	
25	Sulfanoamide drugs (except approved sulfadimethoxine, sulfabromomethazine, and sulfaethoxypyridazine)	
26	Fluoroquinolones	
27	Glycopeptides	

cooking or processing. Mycotoxins are undesirable because of their adverse effects on both human and animal health. Food grains, especially rye, bajra, sorghum, and wheat, have a tendency to become infected with the ergot fungus *Claviceps purpurea*. Consumption of ergot-infected grains can lead to ergotism. Mycotoxins produced by certain molds, *Aspergillus flavus*, and *Aspergillus parasiticus* are known as aflatoxins. These fungi develop in many foods, particularly maize, sorghum, and groundnuts, under improper storage conditions and produce aflatoxins, of which B1 and G1 are the most potent hepatotoxins, in addition to being carcinogenic. Moisture content of foods above 16% and temperatures ranging from 11 to 37 °C favor toxin formation.

Table 3.6 Microbiological Limits Specified by the PFA for Different Food Products

Prevention of Food Adulteration (PFA) Rules, 1956, specify microbiological requirements for pathogens such as *Esherichia coli*, *Staphylococcus auerus*, *Salmonella* and *Shigella*, *Vibrio cholerae*, *Vibrio parahaemolyticus*, *Clostridium perfringens*, *Clostridium botulinum*, and *Listeria monocytogenes*. (Limits specified here are indicative only. One should consult the relevant standards and guidelines of particular regulatory body for practical purposes at that point in time.)

Infant milk food

Infant formula

Milk cereal-based complementary food

Processed cereal-based complementary food and similar foods

- Bacterial count per gram: (not more than) 10,000
- Coliform count absent in 0.1 g
- Yeast and mold count absent in 0.1 g
- *Salmonella* and *Shigella* absent in 25 g
- *E. coli* absent in 0.1 g
- *S. aureus* absent in 0.1 g

Mineral water

- Yeast and mold counts: absent
- *Salmonella* and *Shigella*: absent
- *E. coli* or thermotolerant coliforms: absent in 250 mL
- Total coliform bacteria: absent in 250 mL
- Fecal streptococci and *S. aureus*: absent in 250 mL
- *P. aeruginosa*: absent in 250 mL
- Sulfite-reducing anaerobes: absent in 50 mL
- *V. cholerae*: absent in 50 mL
- *V. parahaemolyticus*: absent in 250 mL

Packaged drinking water (other than mineral water)

- Yeast and mold counts: absent in 250 mL
- *Salmonella* and *Shigella*: absent in 250 mL
- *E. coli* or thermotolerant coliforms: absent in 250 mL
- Total coliform bacteria: absent in 250 mL

- Fecal streptococci and *S. aureus*: absent in 250 mL
- *P. aeruginosa*: absent in 250 mL
- Sulfite-reducing anaerobes: absent in 50 mL
- *V. cholerae*: absent in 250 mL
- *V. parahaemolyticus*: absent in 250 mL
- Aerobic microbial count: The total viable colony count shall not exceed 100 per mL at 20 °C–22 °C in 72 h on agar–agar or on agar–gelatin mixture, and 20 per mL at 37 °C in 24 h on agar–agar

Meat and meat products: corned beef, luncheon meat, cooked ham, chopped meat, canned chicken, canned mutton and goat meat

- Total plate count:1000/g maximum
- *E. coli*: absent in 25 g
- *Samonella*: absent in 25 g
- *S. aureus*: absent in 25 g
- *C. perfringens* and *C. botulinum*: absent in 25 g
- *L. monocytogenes*: absent in 25 g
- Yeast and mold count: 1000/g maximum

Frozen mutton, goat, beef, and buffalo meat

- Total plate count: 10000/g maximum
- *E. coli*: 100/g
- *Samonella*: absent in 25 g
- *S. aureus*: 100/g maximum
- *C. perfringens* and *C. botulinum*: 30/g maximum
- *L. monocytogenes*: absent in 25 g
- Yeast and mold count: 1000/g maximum

Microbiological criteria for solvent-extracted soy flour, solvent-extracted groundnut flour, solvent-extracted sesame flour, solvent-extracted coconut flour, and solvent-extracted cotton seed flour

- Total bacterial count: not more than 50,000/g
- Coliform bacteria: not more than 10/g
- *Salmonella* bacteria: none in 25 g

Table 3.7 Indicative Microbiological Limits for Meat and Meat Products[a]

S. No.	Product	Total Plate Count	Escherichia coli	Staphylococcus aureus	Salmonella and Shigella	Vibrio cholerae	Vibrio parahaemolyticus	Clostridium prefringens
1	Frozen shrimp or prawns							
	Raw	<5 lakh/g	<20/g	<100/g	Absent in 25 g	Absent in 25 g	Absent in 25 g	
	Cooked	<1 lakh/g	Absent in 25 g	Absent in 25 g	Absent in 25 g	Absent in 25 g	Absent in 25 g	
2	Frozen lobsters							
	Raw	<5 lakh/g	<20/g	<100/g	Absent in 25 g	Absent in 25 g	Absent in 25 g	
	Cooked	1 lakh/g	Absent in 25 g	Absent in 25 g	Absent in 25 g	Absent in 25 g	Absent in 25 g	
3	Frozen squid	<5 lakh/g	<20/g	<100/g	Absent in 25 g	Absent in 25 g	Absent in 25 g	—
4	Frozen fish	<5 lakh/g	<20/g	<100/g	Absent in 25 g	Absent in 25 g	Absent in 25 g	—
5	Frozen fish fillets or minced fish flesh or mixtures thereof	<5 lakh/g	<20/g	<100/g	Absent in 25 g	Absent in 25 g	Absent in 25 g	—

6	Dried shark fins	<5 lakh/g	<20/g	<100/g	Absent in 25 g	Absent in 25 g	Absent in 25 g	—
7	Salted fish/dried salted fish	<5 lakh/g	Absent in 25 g	Absent in 25 g	Absent in 25 g	Absent in 25 g	Absent in 25 g	—
8	Canned fish	Nil	Absent in 25 g	Absent in 25 g	Absent in 25 g	Absent in 25 g	Absent in 25 g	Absent in 25 g
9	Canned shrimp	Nil	Absent in 25 g	Absent in 25 g	Absent in 25 g	Absent in 25 g	Absent in 25 g	—
10	Canned sardines or sardine-type products	Nil	Absent in 25 g	Absent in 25 g	Absent in 25 g	Absent in 25 g	Absent in 25 g	—
11	Canned salmon	Nil	Absent in 25 g	Absent in 25 g	Absent in 25 g	Absent in 25 g	Absent in 25 g	—
12	Canned crab meat	Nil	Absent in 25 g	Absent in 25 g	Absent in 25 g	Absent in 25 g	Absent in 25 g	—
13	Canned tuna and bonito	Nil	Absent in 25 g	Absent in 25 g	Absent in 25 g	Absent in 25 g	Absent in 25 g	—

[a] *Readers should consult actual limits prescribed by a particular standards and regulatory body at that point in time.*

Table 3.8 Microbiological Limits for Fruits and Vegetables Products[a]

S. No.	Products	Parameters	Limits
1	Thermally processed fruits and vegetable products	1. Total plate count 2. Incubation at 37 °C for 10 days and 55 °C for 7 days	1. Not more than 50/mL 2. No change in pH
2	1. Dehydrated fruits and vegetable products 2. Soup powders 3. Desiccated coconut powder 4. Table olives 5. Raisins 6. Pistachio nuts 7. Dates 8. Dry fruits and nuts	Total plate count	Not more than 40,000/g
3	Carbonated beverages, ready-to-serve beverages including fruit beverages	1. Total plate count 2. Yeast and mold count 3. Coliform count	<50 CFU/mL <2.0 CFU/mL Absent in 100 mL
4	Tomato products 1. Tomato juices and soups 2. Tomato puree and paste 3. Tomato ketchup and tomato sauce	1. Mold count 2. Yeast and spores 1. Mold count 1. Mold count 2. Yeast and spores 3. Total plate count	Positive in >40% of the field examined <125/l (60 cmm) Positive in >60% of the field examined Positive in >40% of the field examined <125/l (60 cmm) <10000/ml

	Product	Parameter	Limit
5	Jam/marmalade/fruit jelly/fruit chutney and sauce	1. Mold count 2. Yeast and spores	Positive in >40% of the field examined
6	Other fruits and vegetables products	Yeast and mold count	<125/1 (60 cmm)
7	Frozen fruits and vegetable products	Total plate count	<40000/g
8	Preserves	Mold count	Absent in 25 g/mL
9	Pickles	Mold count	Absent in 25 g/mL
10	Fruit cereal flakes	Mold count	Absent in 25 g/mL
11	Candied and crystallized or glazed fruit and peel	Mold count	Absent in 25 g/mL
12	1. All fruits and vegetable products and ready-to-serve beverages including fruit beverages and synthetic products 2. Table olives 3. Raisins 4. Pistachio nuts 5. Dates 6. Dry fruits and nuts 7. vinegars	1. Flat sour organisms 2. *Staphylococcus aureus* 3. *Salmonella* 4. *Shigella* 5. *Clostridium botulinum* 6. *Esherichia coli* 7. *Vibrio cholera*	<1000 CFU/g for products having a pH <5.2 Absent in 25 g/mL Absent in 25 g/mL Absent in 25 g/mL Absent in 25 g/mL Absent in 1 g/mL Absent in 25 g/mL

[a]*Readers should consult the actual limits prescribed by a particular standards and regulatory body at that point in time.*

Table 3.9 Indicative Microbiological Limits for Milk and Milk Products[a]

S. No.	Requirement	Ice cream/ Frozen Dessert/Milk Lolly/Ice Candy/Dried Ice Cream Mix	Cheese/ Processed	Evaporated Milk	Sweetened Condensed Milk	Butter	Butter Oil/ Butter Fat and Ghee
1	Total plate count	<250,000/g	<50,000/g	<500/g	<500/g	<500/g	<500/g
2	Coliform count	<10/g	Absent in 0.1 g	Absent in 0.1 g	Absent in 0.1 g	<5/g	Absent in 0.1 g
3	*E. coli*	Absent in 1 g	Absent in 1 g	Absent in 1 g	Absent in 1 g	Absent in 1 g	Absent in 1 g
4	*Salmonella*	Absent in 25 g	Absent in 25 g	Absent in 25 g	Absent in 25 g	Absent in 25 g	Absent in 25 g
5	*Shigella*	Absent in 25 g	Absent in 25 g	Absent in 25 g	Absent in 25 g	Absent in 25 g	Absent in 25 g
6	*Staphylococcus aureus*	Absent in 1 g	Absent in 1 g	<100/g	<100/g	Absent in 1 g	Absent in 1 g
7	Yeast and mold count	Absent in 1 g	Absent in 1 g	Absent in 1 g	<10/g	<20/g	Absent in 1 g
8	Anaerobic spore count	Absent in 1 g	Absent in 1 g	Not more than 5/g	Absent in 1 g	Absent in 1 g	Absent in 1 g
9	*Listeria monocytogenes*	Absent in 1 g	Absent in 25 g	Absent in 1 g	Absent in 1 g	Absent in 1 g	Absent in 1 g

[a]Readers should consult the actual limits prescribed by a particular standards and regulatory body at that point in time.

Regulations have been established in many countries to protect consumers from the harmful effects of mycotoxins. Various factors play a role in the decision-making process of setting limits for mycotoxins. These include scientific factors such as the availability of toxicological data and survey data, knowledge about the distribution of mycotoxins in commodities, and analytical methodology. Economic and political factors such as commercial interests and sufficiency of food supply have an impact as well. At present, more than 100 countries (covering approximately 85% of the world's inhabitants) have specific regulations or detailed guidelines for mycotoxins in food. For guidance, permissible concentrations of toxins in food are shown in Tables 3.10–3.12.

Table 3.9 Indicative Microbiological Limits for Milk and Milk Products[a]

Yogurt/Dahi	Milk Powder/ Cream Powder	Edible Casein Products	UHT Milk/ UHT Flovored Milk	Pasteurized Milk	Sterilized Milk/ Sterilized Flavored Milk	Khoya/ Chhana/ Paneer	Chakka/ Srikhand
<1,000,000/g	<50,000/g	<50,000/g	Nil	<30,000/g	Nil	<50,000/g	<50,000/g
<10/g	<0.1/g	Absent in 0.1 g	Absent in 0.1 g	Absent in 0.1 g	Absent in 0.1 g	<90/g	<10/g
Nil in 1.0 g	Absent in 0.1 g	Absent in 1 g	Absent in1 g	Absent in 1 g	Absent in 1 g	Absent in 1 g	Nil in 1 g
Nil in 25 g	Absent in 25 g	Absent in 25 g	Absent in 25 g	Absent in 25 g	Absent in 25 g	Absent in 25 g	Nil in 25 g
Nil in 25 g	Absent in 25 g	Absent in 25 g	Absent in 25 g	Absent in 25 g	Absent in 25 g	Absent in 25 g	Nil in 25 g
<100/g	Absent in 0.1 g	Absent in 1 g	Absent in g	Absent in 1 g	Absent in 1 g	<100/g	<100/g
<100/g	Absent in 1 g	Absent in 1 g	Absent in 1 g	Absent in 1 g	Absent in 1 g	<250/g	Chakka: <10/g
Nil in 1 g	Absent in 1 g	Absent in 1 g	<5/g	Absent in 1 g	Not more than 5/g	Absent in 1 g	Nil in 1 g
Nil in 1 g	Absent in 1 g	Absent in 1 g	Absent in 1 g	Absent in 1 g	Absent in 1 g	Absent in 1 g	Nil in 1 g

Table 3.10 Indicative Limit of Naturally Occurring Toxic Substances Caused by Crop Contaminants

S. No.	Contaminants	Article of Food	Limits[a] (μg/kg)
1	Aflatoxin	All articles of food	30
2	Aflatoxin M1	Milk	0.5
3	Patulin	Apple juice and apple juice ingredients in other beverages	50
4	Ochratoxin A	Wheat, barley, and rye	20

[a]For practical use, readers are advised to consult particular standards and regulatory bodies for actual values given by them at that point in time.

Table 3.11 Indicative Toxic Substances Possibly Naturally Occurring in Food

S. No.	Substance	Maximum Limit, ppm[a]
1	Agaric acid	100
2	Hydrocyanic acid	5
3	Hypericine	1
4	Saffrole	10

[a]*For practical use, readers are advised to consult particular standards and regulatory bodies for actual values given by them at that point in time.*

Table 3.12 Indicative Guidelines and Regulations for Toxins in Food and Feed

Country/Region	Mycotoxin	Food/Feed	Action Level[a]
United States of America	Patulin	Apple juice, apple juice concentrate, and apple juice products	50 ppb
	Fumonisin B1 + B2 + B3	Degermed dry milled corn products	2 ppm
		Whole or partially degermed dry milled corn products	4 ppm
		Dry milled corn bran	4 ppm
		Cleaned corn intended for mass production	4 ppm
		Cleaned corn intended for popcorn	3 ppm
		Corn and corn by-products intended for:	
		Equids and rabbits	5 ppm
		Swine and catfish	20 ppm
		Breeding ruminants, poultry, and mink (includes lactating dairy cattle and hens laying eggs for human consumption)	30 ppm
		Ruminants ($\geq$3 months old being raised for slaughter and mink for pelt production)	60 ppm
		Poultry being raised for slaughter	100 ppm
		All other species or classes of livestock and pet animals	10 ppm
	Aflatoxin B1 + B2 + G1 + G2	Foods	20 ppb
		Brazil nuts, pistachio nuts, peanuts, and peanut products	20 ppb
	Aflatoxin M1	Milk	0.5 ppb
	Aflatoxin B1 + B2 + G1 + G2	Corn and peanut products intended for:	
		Finishing beef cattle	300 ppb
		Finishing swine $\geq$100 lb	200 ppb

Table 3.12 Indicative Guidelines and Regulations for Toxins in Food and Feed—cont'd

Country/Region	Mycotoxin	Food/Feed	Action Level[a]
		Breeding beef cattle, breeding swine, or mature poultry	100 ppb
		Corn, peanut products, and other animal feeds and feed ingredients (excluding cottonseed meal intended for immature animals)	100 ppb
		Corn, peanut products, cottonseed meal, and other animal feed ingredients intended for dairy animals, for animal species, or when the intended use is unknown	20 ppb
		Cottonseed meal for beef cattle, swine, poultry	300 ppb
		DON finished wheat products	1 ppm
European Union	Ochratoxin A	Raw cereal grains	5 ppb
		All products derived from cereals	3 ppb
		Dried vine fruit (currants, raisins, sultanas)	10 ppb
	Patulin	Apple juice and other foods derived from apples	50 ppb
		Solid apple products (e.g., purees, compotes)	25 ppb
		Apple juice and solid apple products intended for infant foods	10 ppb
	Aflatoxin B1	Spices (*Capsicum* and *Piper* spp., nutmeg, ginger, and turmeric)	5 ppb
	Aflatoxin B1 + B2 + G1 + G2	Spices (*Capsicum* and *Piper* spp., nutmeg, ginger, and turmeric)	10 ppb
	Aflatoxin B1	Groundnuts, nuts, dried fruit, and processed products thereof for direct consumption or as an ingredient	2 ppb
		Groundnuts to be subjected to sorting or other physical treatment before human consumption or use as ingredient	8 ppb
		Nuts and dried fruit to be subjected to sorting or other physical treatment before human consumption or use as ingredient	5 ppb
		Cereals and processed products thereof intended for direct consumption or as an ingredient	2 ppb
	Aflatoxin B1 + B2 + G1 + G2	Groundnuts, nuts, dried fruit, and processed products thereof intended for direct consumption or as an ingredient	4 ppb
		Groundnuts to be subjected to sorting or other physical treatment before human consumption or use as ingredient	15 ppb

Continued

Table 3.12 Indicative Guidelines and Regulations for Toxins in Food and Feed—cont'd

Country/Region	Mycotoxin	Food/Feed	Action Level[a]
Codex Alimentarius Commission		Nuts and dried fruit to be subjected to sorting or other physical treatment before human consumption or use as ingredient	10 ppb
		Cereals and processed products thereof intended for direct consumption or as an ingredient	4 ppb
	Aflatoxin M1	Milk (raw, for manufacture of milk-based products, and heat-treated)	0.05 ppb
	Patulin	Apple juice and apple juice ingredients in other beverages	50 ppb
	Aflatoxin B1 + B2 + G1 + G2	Peanuts intended for further processing	15 ppb
	Aflatoxin M1	Milk	0.5 ppb

[a]For practical use, readers are advised to consult particular standards and regulatory bodies for actual values given by them at that point in time.

BASIC DETECTION TECHNIQUES

The competitiveness of food production is now more dependent on safety, quality, and production procedures than on quantity and price. Food safety is related to protection of human health from microbiological, chemical, and physical hazards. These hazards are difficult to detect by consumers and thus mostly are implemented by food-regulating bodies of respective countries. There are numerous detection techniques, but their acceptance varies from country to country depending on their own standards, the methodologies available, the level of accuracy, and so on. This chapter deals with some basic conventional microbiological, molecular, microscopic techniques, among others, to give readers an idea about detection before moving on to the chapters describing rapid/modern detection techniques.

4.1 MICROBIOLOGICAL METHODS
4.1.1 ENUMERATION OF MICROBES IN FOOD

Enumeration of microbial populations forms the basis for determining the quality, spoilage, and safety of food. As the name implies, enumeration of microbes in general means a complete ordered listing of the individual viable microbes in any given sample. Enumeration of microorganisms is especially important in the areas of public health, for example, in dairy microbiology, food microbiology, and water microbiology, to determine their safety based on the numbers of microorganisms in those products and their effect on safe human consumption. It is also useful in understanding the ecological role of microbes in a food metrics environment or other physiological or biochemical studies.

Bacteria are so small and numerous that it is very difficult to count them. Different methods are used to determine the number of microorganisms present in a given sample. This can be accomplished by a direct or indirect method of counting the microorganisms or a direct or indirect method of measuring microbial biomass. One method to enumerate bacteria is to dilute the sample to a certain point, so that their number is reduced to a very small amount and can easily be counted. Microbial counting has always proven its potential utility in basic sciences and is preferably applied to determine the number of microbes available for physiological or biochemical studies. If the number of microbes present in a culture is known, the amount of protein or DNA can also be calculated. Moreover, microbial enumeration is also routinely applied in areas related to public health.

4.1.1.1 Direct Count of Cells/Measurement

Microbial cells are counted directly under a microscope.

Rapid Detection of Food Adulterants and Contaminants. http://dx.doi.org/10.1016/B978-0-12-420084-5.00004-4
Copyright © 2016 Elsevier Inc. All rights reserved.

4.1.1.1.1 Direct Microscopic Count

Petroff-Hausser or Neubauer counting chambers can be used as a direct method to determine the number of microbial cells in a culture or liquid medium. A chamber is designed in such a way that a known volume is enclosed by the coverslip, slide, and ruled lines. In this procedure, the number of cells in a given volume of culture liquid is counted directly in 10–20 microscope fields. The average number of cells per field is calculated, and the number of bacterial cells per milliliter of original sample then is computed.

The main advantage of this method is that enumeration is fast, although it is not possible to distinguish living from dead cells. The method is therefore not very useful for determining the number of viable cells in a culture.

4.1.1.1.2 Direct Count with Fluorescent Dyes

Several dyes have been used as a rapid method for enumerating bacteria. Direct counting of micro-organisms can be improved using fluorescent dyes. Some fluorescent dyes commonly used for counting bacterial cells are acridine orange (AO), 4,6-diamidino-2-phenylindole (DAPI) for total bacteria, and 5-cyano-2,3-ditolyl tetrazolium chloride (CTC) for actively respiring bacteria. Among these, AO is the most widely used. At appropriately low AO concentrations, actively growing bacteria fluoresce red-orange as a result of the prevalence of RNA and inactive bacteria fluoresce green owing to the predominance of DNA (Hobbie et al., 1977). However, several factors such as growth media, the RNA-to-DNA ratio in a cell, AO staining concentration and procedure, the method of cell fixation (if any), as well as cell taxonomy may affect the AO color reaction. Direct counting using AO is therefore not a good method to assess the activity of natural and undefined bacterial communities. In addition, bacterium-sized particles, which may also be stained or autofluorescent, interfere with bacterial counts (Ghiorse and Balkwill, 1983). DAPI is reported to be superior to AO in enumerating the total number of bacteria when background fluorescence is high (Porter and Feig, 1980). The redox dye CTC is also gaining popularity for enumerating actively growing bacteria (Rodriguez et al., 1992). Actively respiring bacteria readily reduce it to insoluble, highly fluorescent CTC-formazan, which emits in the red spectrum, with excitation by ultraviolet light, whereas nonrespiring bacteria and abiotic material emit in the blue or blue-green spectrum.

These stains are useful in estimating the total number of microorganisms in a diverse niche, where metabolically diverse populations coexist, making it difficult to enumerate the microbial populations using viable counting procedures. Staining is often used for enumerating populations of a sample in which counts are low and viable plate counts are known to severely underestimate the numbers. It has been widely used in the dairy industry for milk and milk products; it has also been applied to beverages, foods, clinical specimens, and in environmental research such as marine microbiology. The technique involves trapping bacterial cells on the surface of filters, staining them with a fluorescent dye, and visualizing them using a microscope. It is also known as the direct epifluorescent filter technique. Other dyes—auramine and rhodamine—emit a bright yellow or orange color under a fluorescent microscope when they bind to the cell wall of mycobacteria. These stains are gradually replacing acid-fast stains.

4.1.1.2 Indirect Count of Cells/Measurement

Using the indirect method, the microbial population in a given sample is grown and appropriately diluted or concentrated on a suitable medium and counted. Then it is computed to estimate the microbial population in the entire original sample.

4.1.1.2.1 Standard Plate Count (Viable Counts)

A viable cell is defined as a cell that is able to divide and form colonies. A viable cell count is usually carried out either by spreading a test sample onto the surface of agar plates (spread plate method), or mixing it with molten agar, followed by pouring it onto plates (pour plate method) and allowing it to solidify. The plates are then incubated under optimized conditions so that colonies form. After incubation, the colonies are counted. Enumeration/counting in food microbiology refers to determining the number of colony-forming units (CFUs), or viable microbial cells present in a unit volume or weight of a sample. The number of bacteria present in the sample is represented by individual CFUs.

However, the number of colonies developing on the plates should neither be too small nor too large. Statistically, the most valid method is to count colonies only on plates with 30 and 300 colonies. Therefore, the test sample needs to be suitably diluted so as to obtain the appropriate number of colonies.

The numbers of CFUs are divided by the product of the dilution factor and the volume of the diluted suspension on the plate to calculate the number of bacteria per milliliter present in the original sample. The advantage of this method is that it is relatively easy to perform. Its drawbacks, however, are that it is time-consuming and, because of the variable requirements (pH, temperature, nutrient) of different microbial communities, it is difficult to develop a universal set of growth conditions to cater the needs of all kinds of microbial populations. It is therefore practically impossible to enumerate all microorganisms using viable plating. Selective procedures for the enumeration of specific microbial populations can be designed, however.

4.1.1.2.2 Turbidometric Measurement

Bacterial populations can be determined measuring the turbidity or optical density of culture broth. The instruments used for measuring turbidity or optical density work on the principle of light transmitted or absorbed. A light beam is passed through a sample and the intensity of the transmitted light reaching a detector is measured. The light beam consists of a stream of photons. When photons interact with an analyte (the analyte is the molecule being studied), there is a possibility that the analyte will absorb the photon. This absorption reduces the number of photons in the beam of light, thereby reducing the intensity of the light beam. This equipment contains a light source and a light detector separated by a sample compartment. The cells suspended in the culture interrupt the passage of light, allowing less light to reach the photoelectric cell. The amount of light energy transmitted through the suspension is measured using the spectrophotometer as percentage of transmission (%T; 0–100%). The density of a cell suspension is expressed as absorbance or optical density. Before turbidometric measurements are made, the spectrophotometer must be adjusted to 100% transmittance (0% absorbance). This is done using a sample of uninoculated medium or as a blank. The percentage transmittance of various dilutions of the bacterial culture is then measured and the values converted to optical density, based on the formula in Eqn (4.1):

$$\text{Absorbance or Optical Density} = -\log\frac{\%T}{100} \tag{4.1}$$

The more turbid a solution, the less light will be transmitted, and vice versa. The turbidity (which is directly proportional to the number of cells) is used as an indicator of bacterial concentration in the sample. This method of enumeration is fast and is usually preferred when a large number of microbial cultures are to be counted. Measuring the turbidity of cells is much faster than the standard plate count, although it must be correlated initially with cell number. This is achieved by determining the turbidity of different concentrations of a given species of microorganism in a particular medium and then using the standard plate count to determine the number of viable organisms per milliliter of a sample. A standard curve can then be drawn, in which a specific turbidity or optical density reading is matched to a specific number of viable organisms. The number of viable organisms may subsequently be read directly from the standard curve, without necessitating time-consuming standard counts.

The advantage of this method is that enumeration is fast and is usually preferred when a large number of cultures are to be counted. On the other hand, disadvantages include the shielding effect of cells, because of which the measurement is no longer accurate.

4.1.1.2.3 The Most Probable Number

The most probable number (MPN) technique is a statistical method of estimating the concentration of bacteria. It estimates the population density of viable microorganisms in a test sample. Random distribution of a microbial population throughout the test sample is the key requirement for the application of the MPN method. The test samples are first serially diluted to a level where there are no more viable microorganisms. Thereafter, replicates (3, 5, or 10) of multiple serial dilutions are inoculated on suitable media. The set of replicates (number) showing the presence of viable microorganisms gives an estimate of the number of viable microorganisms present in the sample. Published MPN tables, which have been developed based on the assumption that numbers of viable microorganisms present in replicate samples are distributed around a mean number in accordance with Poisson distribution, are referred to make an estimate and know its confidence limits. The more replicate tubes used, the better will be the precision of the estimate. The MPN technique is widely used in food and water microbiology to estimate numbers of coliforms as an indication of fecal contamination.

The coliform test was named by the Public Health Service in 1914 to *Enterobacteriaceae* family of microbes. It is a bacterial indicator that is often used to test the sanitary quality of foods and water. Coliforms are defined as rod-shaped, gram-negative, non-spore-forming microbes, and some of them are able to ferment lactose with the production of acid and gas when incubated at 35−37 °C. The technique includes three successive steps, namely, a presumptive test, a confirmed test, and a completed test. In the presumptive test, lactose broth tubes are inoculated with a series of decimal dilutions to give an estimate of the MPN. The MPN confirmation procedure is carried out by transferring positive presumptive tubes (gas production within 24−48 h) to 2% brilliant green bile lactose broth to test for gas production within 48 h at 35 °C for total coliforms and at 44.5 °C for fecal coliforms. Positive tubes are used to calculate the MPN. The presence of coliform bacteria in samples is established in completed tests using a protocol with eosin methylene blue agar plates.

4.2 MICROSCOPIC ANALYSIS OF FOOD SAMPLES

Optic or electronic microscopic analyses are well-known techniques in the area of food science. The suitability of this technique has been proven. The first study, based on the application of optical microscopic analysis as a characterization tool for the origin of honey, came into light in the 1950s when Maurizio studied the quantity of pollens in honeys. Maurizio and Louveaux (1965) used optical microscopic analysis to characterize European melliferous plants establishing their "pollinic identity card." In the following year, Louveaux (1966) used this technique for 40 Canadian honeys. Further, in 1978, a manual/book entitled *Methods of Mellissopalynology* was published (Louveaux et al., 1978) as an official method for determining the floral origin of honeys.

Nowadays, analysis of food samples combined with microscopic analysis and physicochemical methods is used to identify the floral origin of honeys. This technique was also used to study the effect of bactericide on some Brazilian (Cortopassi-Laurino and Gelli, 1991) and Greek monofloral honeys (Thrasyvoulou and Manikis, 1995). Kerkvliet et al. (1995) used the method developed by Louveaux and Maurizio to detect the adulteration of honeys with cane sugar and cane sugar products using optical microscopy. This study is based on the detection of parenchymal cells, sclerous rings, and other cane sugar constitutive cells.

4.3 BIOCHEMICAL METHODS

Biochemical methods target specific biochemical properties of adulterants/contaminants. One of the key factors in identification and classification of microbes are enzymes, which govern biochemical reactions that occur both outside and inside cells. Selective growth media for isolating a particular microbe based on biochemical properties is constituted. The selective agent may be a protein source, often a hydrolyzate of casein; a fermentable sugar, such as lactose or glucose; and occasionally an indicator such as neutral red or bromo cresol purple. Several specific media such as tryptic soy agar, sorbitol MacConkey agar, rainbow agar, Biosynth, Staad, and Fluorocult (Brauns et al., 1991) have been used for selective isolation of the common human pathogen *Escherichia coli* O157:H7, which can cause diarrhea, hemorrhagic colitis, and hemolytic uremic syndrome. Strain can even be differentiated using differential selective media such as Fraser enrichment broth and University of Vermont medium modified enrichment broth.

Scientists from the National Dairy Research Institute (Karnal, India) recently developed a microtechnique for the detection of *Listeria monocytogenes* in milk within 24 h. It is based on the principle of targeting specific marker enzyme(s) of target bacteria, leading to the release of free chromogen following an "enzyme–substrate reaction" that is visually detected by a color change after initial enrichment of the bacterium in the selective medium. A semiquantitative method for the detection of aflatoxin (AF) M1 in milk has also been developed using spore-immobilized ampoules and a chromogenic substrate. It is reported to detect the presence of AFM1 at more than the permissible limit set by CODEX. A spore-based biosensor was also developed to detect *Enterococci*; an *Enterococci*-specific marker enzyme acts on germinogenic substances of the spores coated on a gold chip to release germinant, which then leads to germination-mediated concomitant enzyme activity, the product of which could be detected using fluorescence spectroscopy. These methods are promising, are inexpensive, and give both qualitative and quantitative information about the tested components. The only drawback is that they involve laborious sample preparation, requiring trained manpower, and many of them use polluting chemical reagents.

A cultural medium comprises nonpolar or weakly polar substances. If a bacterial population converts these substrates, the medium will produce highly charged end products. Impediometry measures the conductance changes of the cultural medium caused by bacterial metabolism. The conductance curve as a function of time coincides with the detection time, which is inversely proportional to the initial bacterial population in the cultural medium. The conductance method can be applied to various analyses in microbiology as well as in biochemistry.

Biochemical methods are being widely used for specific analyses of chemical compounds in various products. White (1993) used ultraviolet/visible spectrophotometry to determine proline content in American honeys, whereas Davies and Harris (1982) applied this method to determine geographic origin. They distinguished between English and Welsh honeys based on the ninhydrin reaction to produce colored samples using this method. Enzymatic processes have been used to determine parameters such as glycerol content (Fernandez-Muino and Sancho, 1996), which can be considered to be an indicator of the fermentation state of honey. D-Gluconic acid (Mato et al., 1997), galactose, lactose (Val et al., 1998), and L-malic acid (Mato et al., 1998) are other honey constituents that can be evaluated by enzymatic methods. All of the above methods work as honey characterization tools and collectively provide an indication of honey quality.

High-performance liquid chromatography (HPLC) is one of the most frequently used techniques for detecting adulterants in food products based on their biochemical properties. It is routinely used at the industrial and laboratory levels to evaluate quality. Blanch et al. (1998) proposed various approaches to evaluate the authenticity of olive oil and hazelnut oil by HPLC and HPLC-gas chromatography. This work of characterization was based on the hazelnut oil marker filbertone ((E)-5-methylhept-2-en-4-one), which is a known adulterant in olive oil and occurs naturally in olives (Ruiz del Castillo et al., 1998). Mouly et al. (1998) developed a method to detect flavones such as glycosides and polymethoxylates in citrus juices like orange and lemon juices. Guyot et al. (1998) detected phenolic compounds in different French ciders, whereas Bengoechea et al. (1997) made the same analyses on manufactured purees and concentrates from peach and apple fruits. Mohler-Smith and Nakai (1990) explained the application of HPLC data in the classification of cheese varieties using multivariate analysis of HPLC profiles. In addition, Mohler-Smith and Nakai included both proposed principal component analysis and linear discriminant analysis of 55 peak areas from each cheese chromatogram, with 90% correct classification. The utilization of linear discriminant analysis after principal component analysis improves the classification of the data set as opposed to the direct utilization of the discriminant analysis.

A huge number of chemical compounds are extensively analyzed using the HPLC method for characterization as well as for authentication of components such as flavonoids in fruit juices (Kawaii et al., 1999; Robards et al., 1997), organic acids in apple juices (Guyot et al., 1998; Blanco et al., 1996), phenolic pigments in black tea liquors (McDowell et al., 1995), proline isomers and amino acids in wines (Calabrese et al., 1995; Moreno-Arribas et al., 1998), and anthocyanins in jams (Garcia-Viguera et al., 1997) and juices.

4.4 ANALYTICAL METHODS

Atomic absorption spectrometry (AAS) and atomic emission spectrometry (AES) are very old known analytical techniques that are used to determine the physical properties of any chemical element (all mineral elements listed in the Mendeleev classification) that composes matter. This absorption or emission is treated as the fingerprint of the specific element being analyzed. Identification of the mineral content of food products using this technique was initiated in the 1930s. In the beginning, the amount of minerals in honeys was studied and the relationship between the degree of pigmentation and the mineral content was established. Later, McLellan (1975) used AAS to evaluate calcium, magnesium, potassium, and sodium contents in honeys; then, Petrovic et al. (1993) determined the selenium content in honeys from eastern Croatia. AAS/AES are very useful techniques for determining mineral content in food products (Julshamn et al., 1998). Applications of this technique in the detection of different adulterants in wines (Baxter et al., 1997), cheeses (Fresno et al., 1995), sugars (Leblebici and Volkan, 1998), and honeys (Prats-Moya et al., 1997; Vinas et al., 1997) are well documented.

Simpkins et al. (2000) used AAS and inductively coupled plasma AES techniques to detect adulteration of Australian orange juices and showed that, depending on the origin of the orange juice, the number and concentration of elements varies. They verified mixing of two qualities of orange juice with the help of principal component analysis. Sun et al. (1997) applied both discrimination analytical methods (ACP, high-content analysis, Bayes discrimination method, and Fischer discrimination method) and artificial neural network (back-propagation architecture) methods to inductively coupled plasma optical emission spectrometry data to classify wine samples from six different regions.

The isotope ratio mass spectrometry (IRMS) technique has also been used to detect adulteration in food products. This technique was used for the first time to detect an addition of potential industrial syrup (from C-4 plants) in food products (White et al., 1998). The procedure included measuring the isotope ratio of an analyte converted into a simple gas and isotopic representative of the original sample before entering the ion source of an IRMS system. The system quickly measures isotope ratios (2H/1H, 15N/14N, 13C/12C, 18O/16O, and 34S/32S) in H_2, N_2, CO_2, CO, and SO_2 gases, respectively. These analytical developments were followed by a study of the detection of adulteration of honey using continuous-flow IRMS (Hernandez, 1998).

4.5 MOLECULAR METHODS

Molecular methods have been successfully used as a tool for enhancing the sensitivity and specificity of authenticity testing and microbial contaminant identification. Polymerase chain reaction (PCR), which is an in vitro technique based on the principle of DNA polymerization reaction, in which a particular DNA sequence is amplified several fold, has brought revolutionary changes to the detection, identification, and characterization of microbial contaminants (Settanni et al., 2005). PCR-based technologies such as PCR restriction fragment length polymorphism and randomly amplified polymorphic DNA have been used frequently for the differentiation and identification of microbial isolates (Shao et al., 2008; Guo et al., 2008). Another variant of PCR is real-time PCR, which enables simultaneous detection and quantification. It has been successfully used for the specific detection and quantification of goat's milk adulteration with cows' milk in the range of 0.5—100% using a standard curve, along with an R^2 value better than 0.99 (Dąbrowska et al., 2010). Multiplex PCR (MPCR) offers the added advantage of simultaneous detection of multiple microorganisms using simultaneous amplification of more than one locus in a single reaction (Chamberlain et al., 1988). In MPCR, many specific primer sets are collected into a single PCR assay. MPCR is one of the fastest independent approaches for strain-specific detection in complex matrices. Like the majority of the other molecular techniques, MPCR commonly targets the 16S ribosomal RNA gene, the gene most broadly used to infer phylogenetic relationships among bacteria (Rosselló-Mora and Amann, 2001). This technique is widely applied in various fields of microbiology for rapid differentiation of pathogenic and nonpathogenic microorganisms. Techniques based on PCR have been reported to successfully detect pathogenic microorganisms such as *Escherichia coli* (Osek and Gallien, 2002), *Clostridium perfringens* (Baums et al., 2004), *Salmonella enterica* (Lim et al., 2003), *Staphylococcus aureus* (Kwon et al., 2004), and multiple species of single genera, such as *Campylobacter* spp. (Harmon et al., 1997; Houng et al., 2001; Cloak and Fratamico, 2002; Wang et al., 2002; Klena et al., 2004) and *Listeria* spp. (Brosch et al., 1996; Graves et al., 1991), without compromising accuracy. Similarly, nonpathogenic microorganisms such as starter cultures used in food fermentations (Aquilanti et al., 2006; Valmori et al., 2006) and probiotic strains (Ventura et al., 2001; Mullié et al., 2003;

Yost and Nattress, 2000) could be differentiated using MPCR. The main disadvantage of PCR technology is its inability to distinguish between dead and live cells, and DNA from dead cells can still be amplified after a significant time interval (Lei et al., 2008). Inhibitory compounds, such as large amounts of fat and protein in food samples, further limit the application of PCR technology. Therefore, efficient procedures are required for successful elimination of such inhibition for routine use of this technology. Nonetheless, environmental conditions affecting somatic cell numbers in milk, such as animal mastitis or changes from raw to heat-treated milk, may adversely affect PCR results, even though a few recent techniques have suggested that PCR is unaffected by these conditions (Lopez-Calleja et al., 2005).

At the moment, PCR lacks certain practicality for routine commercial use, but with further improvement it could become a very good option in the future. DNA-analytical methods based on PCR have become more and more important in ensuring the quality and safety of food because of their simplicity, specificity, and sensitivity for monitoring microorganisms (Allmann et al., 1995; Candrian, 1995) and for detecting food constituents (Allmann et al., 1993; Meyer et al., 1994; Meyer, 1995a; Meyer and Candrian, 1996; Hubner et al., 1997; Jankiewicz et al., 1997). The first detection method specifically developed for identification of a commercialized, genetically engineered plant was demonstrated for detection of the FlavrSavrä tomato (Meyer, 1995b). Specific PCR methods to distinguish between natural contamination by the cauliflower mosaic virus or *Agrobacterium* and the presence of these sequences in genetically modified organisms and more specific methods for newly approved genetically modified organisms are required. Interpretation of results will become more complex because of the increased number of different methods applied. Therefore, frame conditions and decision matrices (e.g., duplicate analysis, repetition, positive and negative controls, confirmation, control experiments) must be standardized.

4.6 IMMUNOASSAYS

In immunoassays, a specific substance is quantified at a very low concentration using an immunological reaction. Antigens present in association with organisms that are difficult to culture, such as the hepatitis B virus and Chlamydia *trachomatis*, have been detected using this technique. Some known techniques are radioimmunoassay, enzyme immunoassay, immunoprecipitation, and fluorescence immunoassays. Numerous studies report immunological techniques for food authentication using enzyme-linked immunosorbent assay (ELISA). The technique includes the cultivation of antibodies that are able to bind the protein of interest, hence enabling the detection of that protein or antigen via both qualitative and quantitative means. ELISA has been commonly used to detect the occurrence of AFB1 in animal feed (Kolosova et al., 2006; Lee et al., 2004; Tudorache and Bala, 2008) and AFM1 in infant milk products, liquid milk, and ultra-high-temperature processed milk samples (Kang'ethe and Lang'a, 2009; Gundinic and Filazi, 2009; Pathirana et al., 2010; Gomma and Deeb, 2010; Mohammadian et al., 2010; Rastogi et al., 2004).

ELISA has also been compared with HPLC in the detection of the minimum level of occurrence of AFM1 in pasteurized milk as well as in dairy products; limits of detection using ELISA and HPLC were found to be 2 and 10 pg/mL, respectively (Kim et al., 2000). Commercial ELISA kits for detection of AFB1 are also available in the market; for example, Romer Labs (Union, MO, USA) and R-Biopharm AG (Darmstadt, Germany) have commercialized ELISA kits with a detection limit of 40 and 50 ng/g, respectively (Huybrechts, 2011).

Development of immunochemical methods has led to the development of many rapid and sensitive methods for monitoring and quantifying AFB1 in contaminated food and feed (Tudorach and Bala, 2008). Liu et al. (2013) recently reported a highly sensitive and rapid competitive direct ELISA and a gold nanoparticle immunochromatographic strip method for screening for AFB1 in food and feed samples. The strip has been designed and reported to detect very low levels of AFB1 (2.0 ng/mL) in food and feed samples. A number of reports are available describing ELISA techniques for the detection of the presence of vegetable proteins in milk powder (Sanchez et al., 2002) and the differentiation of milk from different species (Bania et al., 2001; Moatsou and Anifantakis, 2003; Hurley et al., 2004). Yin et al. (2010) described an indirect competitive ELISA (icELISA) method that depended on the preparation of monoclonal antibodies for the detection of melamine in raw milk, milk powder, and animal feeds. The detection limits of melamine in milk, milk powder, and feeds were 0.1, 0.2, and 0.5 mg/L, respectively. The properties and specificity of this technique for melamine in milk were improved (Lei et al., 2011) using three haptens of melamine with a separate spacer. The limit of detection was 8.9 ng/mL. This icELISA represented a much lower cross-reactivity than cyromazine, a fly-killing insecticide widely used in vegetables and stables, compared with the ELISA results previously reported. Recoveries obtained from milk samples in this study were in agreement with those obtained using the HPLC-mass spectrometry method, indicating that the detection performance of the icELISA could meet the requirement of the residue limit set by the Codex Alimentarius Commission. This technique seems promising for the authentication of food products; to date, however, limited advances have been made in extending its authentication capabilities. Preparation and production of a specific antibody for a specific protein is the drawback of ELISA. Wang et al. (2011) introduced a fluorescence polarization immunoassay that depended on a polyclonal antibody for determination of melamine in milk. The antibody in the fluorescence polarization immunoassay represented about 21.2% cross-reactivity to the fly, and the same was confirmed using HPLC. The presence of whey in processed and raw milk has also been detected using immunochromatographic assay of glycomacropeptide (GMP), also known as caseinomacropeptide (Oancea, 2009). GMP is a bioactive, 64-amino-acid residue glycopeptide that is released enzymatically in whey from κ-casein by the action of chymosin during cheese making (Eigel et al., 1984). The addition of whey in milk (because of its low cost) does not cause any health hazard, but supplementation of infant formula with GMP enhances the absorption of trace minerals, leading to excess dietary intake (Kelleher et al., 2003).

4.7 ELECTRICAL METHODS

The electrical properties of liquid food in controlled circumstances mainly depend on different parameters such as measuring current, voltage, frequency, impulse, type of electric current, experimental conditions (particularly temperature), and different chemical components in the raw material and their degree of dissociation (Zywica et al., 2005; Jha et al., 2011). Liquids such as milk have high water and mineral content and are characterized by good ionic conductivity. The conductivity of milk is determined by two major parameters, namely, impedance (the main component of which is resistance) and admittance (the main component of which is conductance). The relationship among milk constituents and their electric properties has been investigated and applied for quality evaluation. Zhuang et al. (1997) found statistically significant ($P < 0.05$) correlation between the protein content of a commercial whey powder determined using electric conductance and Kjeldahl's method. Mabrook and Petty (2003a) used the method of electrical admittance spectroscopy to study the water and fat content of milk. Over the frequency range of 5 Hz to 1 MHz, the electrical circuit was dominated by a single time constant. To eliminate the effect of electrode polarization, the conductance of the milk was measured at high frequencies, where it showed a saturation value. The characteristics of all milk samples at 100 kHz and 8 °C revealed a linear decrease in conductance with increasing water content. Admittance data for full-fat or semifat (skim) milk showed an increase in milk conductance with decreasing fat content. Linear correlation between the conductance of milk and its water content has been applied in the detection of adulterations through the addition of water. The characteristics of all skim milk samples at 100 kHz and 8 °C revealed a linear decrease in conductance with increasing water content over the entire range of water concentrations. At the same time, the conductance of full-fat milk showed a decrease only at added water concentrations higher than 10%. At lower added water concentrations, the full-fat milk exhibited an anomalous conductivity maximum at 2–3% added water (Mabrook and Petty, 2003a). Mabrook and Petty (2003b) developed a novel method to detect water added to full-fat milk using single-frequency electrical conductance measurements. The electrical conductivity of milk has also been studied as a means to detect freshness and the adulteration of milk (Mabrook and Petty, 2003b). Conductance of milk is mainly based on the presence of ions (Na^+, K^+, Cl^-). Since the salt content of pure milk remains constant, it poses a constant conductance in controlled conditions of temperature and measurement. Further addition of an adulterant leads to a change in salt concentration, which results in variations in the conductance of milk samples with frequencies ranging from 20 Hz to 1 MHz. On this basis, Sadat et al. (2006) detected adulteration of detergents and synthetic milk in natural milk using alternating-current conductance from 20 Hz to 1 MHz. Because natural milk samples have been found to exhibit higher conductance at 100 kHz than synthetic milk, the addition of synthetic milk resulted in a decrease in conductance with respect to the concentration added to natural milk. The conductance has been found to decrease at the rate of 0.13 mS, with a 10% increase in the percentage of synthetic milk at 8 °C. Conductance of milk decreases with increase in temperature. Dielectric properties, or permittivities, are intrinsic properties that determine the interaction of electromagnetic energy with milk when subjected to dielectric heating. The dielectric properties of milk have also been successfully used for their quality evaluation. The open-ended coaxial line probe method associated with network and impedance analyzers is a useful technique to determine dielectric properties (i.e., dielectric constant, dielectric loss factor, and loss tangent) of materials, especially for liquid foods, according to the reflection coefficient at the material–probe interface. Guo et al. (2010) proposed the use of loss factor as an indicator in predicting milk concentration and freshness. The dielectric properties of water from diluted cow's milk with milk concentrations from 70% to 100% stored for 36 h at 22 °C and 144 h at 5 °C, along with electrical conductivity and pH value, were measured at room temperature for frequencies ranging from 10 to 4500 MHz using open-ended coaxial line probe technology. The raw milk had the lowest dielectric constant when the frequency was higher than about 20 MHz, and it had the highest loss factor at each frequency. The highest linear coefficient of determination (0.995) between the milk concentration and the loss factor was observed at 915 MHz. The change in tendency of the loss factor was inverse of the pH during milk storage, with the best linear correlation ($R^2 = 0.983$) at 1100 MHz. Das et al. (2011) successfully differentiated synthetic milk and original milk using an impedance sensor when the synthetic milk was reconstructed adding a minimum 15% of adulteration. The measurement was based on the change in constant value of the phase angle of the impedance sensor resulting from the change of ionic property of the adulterated milk. It was observed that the sensor can identify 5% of liquid whey adulteration, 10% of tap water adulteration, and 0.6 mg urea/mL milk. Change in the electrical properties of the medium as a result of the breakdown of complex molecules into charged catabolic products has enabled the wide application of electrical methods in dairy microbiology. Commercially available instruments like the Bactometer (Gnan and Luedecke, 1982; Suhren and Heeschen, 1987) and the Malthus system (Fung, 1994) measure the concentration of microorganisms growing in a medium by measuring the electrical properties of the medium. These instruments have been used for detecting specific microorganisms in milk (Easter and Gibson, 1989; Kowalik and Ziajka, 2005), milk powder (Neaves et al., 1988) and yogurt; for monitoring the quality of raw milk (Senyk et al., 1988); for estimating microorganisms in raw and pasteurized milk and cheese (Vasavada, 1993); and for predicting the shelf-life of milk (Visser and DeGroote, 1984; White, 1993). The requirement of a large bacterial population ($1 \times 10^6 - 1 \times 10^7$ CFUs/mL) for efficient detection is, however, the major drawback of this technology (Vasavada, 1993; Adams and Moss, 2000).

4.8 RAPID AND NONDESTRUCTIVE METHODS

Spectroscopic methods are being developed and are being used routinely in a few food industries as part of their quality control protocols. These methods are rapid, cost-effective, reliable, and (most of the time) noninvasive. The expansion of these tools to include screening for possible microbial contamination is becoming more feasible as a result of research advances during the past two decades. Infrared spectroscopy is a method that performs a chemical analysis of a sample, probing the absorption of radiation by organic chemical bonds in a wavelength range dominated by overtones and combinations thereof. Since infrared spectra are highly overlapped and complex, chemometrics (multivariate analysis) is often used to perform classification. Other advanced nondestructive techniques being applied for food quality and safety evaluation and detection of internal defects (e.g., microbes) are ultrasonic, X-ray, computed tomography, and magnetic resonance imaging. These methodologies, their theory, and practical applications are described in detail in Chapters 6 and 7.

REFERENCES

Adams, M.R., Moss, M.O., 2000. Food Microbiology. The Royal Society of Chemistry, Cambridge.

Allmann, M., Candrian, U., Lüthy, J., 1993. Polymerase chain reaction (PCR): a possible alternative to immunological methods assuring safety and quality of food. Detection of wheat-contamination in non-wheat food products. Z. Lebensm. Unters. Forsch. 196, 248−251.

Allmann, M., Höfelein, Ch, Köppel, E., Lüthy, J., Meyer, R., Niederhauser, C., Wegmüller, B., Candrian, U., 1995. Polymerase chain reaction (PCR) for the detection of pathogenic microorganisms in bacteriological monitoring of dairy products. Res. Microbiol. 146, 85−97.

Aquilanti, L., Dell'Aquila, L., Zannini, E., Zochetti, A., Clementi, F., 2006. Resident lactic acid bacteria in raw milk Canestrato Pugilese cheese. Lett. Appl. Microbial. 43, 161−167.

Bania, J., Ugorski, M., Polanowski, A., Adamczyk, E., 2001. Application of polymerase chain reaction for detection of goats' milk adulteration by milk of cow. J. Dairy Res. 68, 333−336.

Baums, C.G., Schotte, U., Amtsberg, G., Goethe, R., 2004. Diagnostic multiplex PCR for toxin genotyping of *Clostridium perfringens* isolates. Veter. Microbiol. 20, 11−16.

Baxter, M.J., Crews, H.M., Dennis, M.J., Goodall, I., Anderson, D., 1997. The determination of the authenticity of wine from its trace element composition. Food Chem. 60 (3), 443−450.

Bengoechea, L.M., Sancho, A.I., Bartolomé, B., Estrella, I., Gomez-Cordovés, C., Hernandez, M.T., 1997. Phenolic composition of industrially manufactured purées and concentrates from peach and apple fruits. J. Agric. Food Chem. 45, 4071–4075.

Blanch, G.P., Mar Caja, M., Ruiz del Castillo, M.L., Herraiz, M., 1998. Comparison of different methods for the evaluation of the authenticity of olive oil and hazelnut oil. J. Agric. Food Chem. 46, 3153–3157.

Blanco, D., Quintanilla, M.E., Mangas, J.J., Gutierrez, M.D., 1996. Determination of organic acids in apple juice by capillary liquid chromatography. J. Liq. Chromatogr. Relat. Technol. 19 (16), 2615–2621.

Brauns, L.A., Hudson, M.C., Oliver, J.D., 1991. Use of the polymerase chain reaction in detection of culturable and nonculturable Vibrio vulnificus cells. Appl. Environ. Microbiol. 57, 2651–2655.

Brosch, R., Brett, M., Catimel, B., Luchansky, J.B., Ojeniyi, B., Rocourt, J., 1996. Genomic fingerprinting of 80 strains from the WHO multicentre international typing study of listeria monocytogenes via pulsed-field gel electrophoresis (PFGE). Int. J. Food Microbiol. 32, 343–355.

Calabrese, M., Stancher, B., Riccobon, P., 1995. High-performance liquid chromatography determination of proline isomers in Italian wines. J. Agric. Food Chem. 69, 361–366.

Candrian, U., 1995. Polymerase chain reaction in food microbiology. J. Microbiol. Methods 23, 89–103.

Chamberlain, J.S., Gibbs, R.A., Ranier, J.E., 1988. Deletion screening of the Duchenne muscular dystrophy locus via multiplex DNA amplification. Nucleic Acids Res. 16, 11141–11156.

Cloak, O.M., Fratamico, P.M., 2002. A multiplex polymerase chain reaction for the differentiation of Campylobacter jejuni and Campylobacter coli from a swine processing facility and characterization of isolates by pulsed-field gel electrophoresis and antibiotic resistance profiles. J. Food Prot. 65, 266–273.

Cortopassi-Laurino, M., Gelli, D.S., 1991. Analyse pollinique, proprieties physico-chimiques et action Anti-bactérienne des miels d'abeilles africanisées Apis mellifera et de Méliponisés du Brésil. Apidologie 22, 61–73.

Dąbrowska, A., Wałecka, E., Bania, J., ślazko, M., Szołtysik, M., Chrzanowska, J., 2010. Quality of UHT goat's milk in Poland evaluated by real-time PCR. Small Rumin. Res. 94 (1), 32–37.

Das, S., Sivaramakrishna, M., Biswas, K., Goswami, B., 2011. Performance study of a 'constant phase angle based' impedance sensor to detect milk adulteration. Sens. Actuators, A 167, 273–278.

Davies, A.M.C., Harris, R.G., 1982. Free amino acid analysis of honeys from England and Wales: application to the determination of the geographical origin of honeys. J. Apic. Res. 21 (3), 168–173.

Easter, M.C., Gibson, D.M., 1989. Detection of microorganisms by electrical measurements. Prog. Ind. Microbiol. 26, 57–100.

Eigel, M.N., Buttler, J.E., Ernstorm, C.A., Farrell, H.M., Harwalker, V.R., Jenness, R., Whitney, R.M., 1984. Nomenclature of proteins of cow's milk: fifth revision. J. Dairy Sci. 67, 1599–1631.

Fernandez-Muino, M.A., Sancho, M.T., 1996. Direct enzymatic analysis of glycerol in honey: a simplified method. J. Sci. Food Agric. 71, 141–144.

Fresno, J.M., Prieto, B., Urdiales, R., Sarmiento, R.M., 1995. Mineral content of some Spanish cheese varieties. Differentiation by source of milk and by variety from their content of main and trace elements. J. Sci. Food Agric. 69, 339–345.

Fung, D.Y.C., 1994. Rapid methods and automation in food microbiology: a review. Food Rev. Int. 10, 357–375.

Garcia-Viguera, C., Zafrilla, P., Tomas-Barberan, F.A., 1997. Determination of authenticity of fruit jams by HPLC analysis of anthocyanins. J. Sci. Food Agric. 73, 207–213.

Ghiorse, W.C., Balkwill, D.L., 1983. Enumeration and morphological characterization of bacteria indigenous to subsurface environments. Dev. Ind. Microbiol. 24, 213–224.

Gnan, S., Luedecke, L.O., 1982. Impedance measurements in raw milk as an alternative to the standard plate count. J. Food Protect. 45, 4–7.

Gomma, G.M., Deeb, A.M.M., 2010. Occurrence of aflatoxin M1 in milk collected from Kafr El-Sheikh, Egypt 1. Zag. Vet. 38, 144–150.

Graves, L.M., Swaminathan, B., Reeves, M.W., Wenger, J., 1991. Ribosomal DNA fingerprinting of *Listeria monocytogenes* using a digoxigenin-labelled DNA probe. Eur. J. Epidemiol. 7, 77–82.

Gundinic, U., Filazi, A., 2009. Detection of aflatoxin M1 concentrations in UHT milk consumed in Turkey markets by ELISA. Pak. J. Biol. Sci. 12 (8), 653–656.

Guo, W., Zhu, X., Liu, H., Yue, R., Wang, S., 2010. Effects of milk concentration and freshness on microwave dielectric properties. J. Food Eng. 99, 344–350.

Guo, Z.X., Weng, S.P., He, J.G., 2008. Development of an RT-PCR detection method for mud crab reovirus. J. Virol. Methods 151, 237–241.

Guyot, S., Marnet, N., Laraba, D., Sanoner, P., Drilleau, J.F., 1998. Reversed-phase HPLC following thiolysis for quantitative estimation and characterization of the four classes of phenolic compounds in different tissue zones of a french cider apple variety (*Malus domestica* Var. Kermerrien). J. Agric. Food Chem. 46, 1698–1705.

Harmon, K.M., Ransom, G.M., Wesley, I.V., 1997. Differentiation of *Campylobacter jejuni* and *Campylobacter coli* by polymerase chain reaction. Mol. Cell. Probes 11, 195–200.

Hernandez, H., 1998. Detection of adulteration of honey: application of continuous-flow IRMS. VAM Bull 18 (Spring), 12–14.

Hobbie, J.E., Daley, R.J., Jasper, S., 1977. Use of nuclepore filters for counting bacteria by fluorescence microscopy. Appl. Environ. Microbiol. 33, 1225–1228.

Houng, H.S., Sethabutr, O., Nirdnoy, W., Katz, D.E., Pang, L.W., 2001. Development of a ceuE-based multiplex polymerase chain reaction (PCR) assay for direct detection and differentiation of *Campylobacter jejuni* and *Campylobacter coli* in Thailand. Diagn. Microbiol. Infect. Dis. 40, 11–19.

Hubneret, P., Burgener, M., Liuthy, J., 1997. Application of molecular biology for the identifcation of fish. In: Amado, R., Battaglia, R. (Eds.), Proceedings of the Ninth European Conference on Food Chemistry, Authenticity and Adulteration of Food ± the Analytical Approach, 24–26 September, Interlaken, vol. 1, ISBN 3-9521414-0-2, pp. 49–54.

Hurley, I.P., Coleman, R.C., Ireland, H.E., Williams, J.H.H., 2004. Measurement of bovine IgG by indirect competitive ELISA as a means of detecting milk adulteration. J. Dairy Sci. 87, 543–549.

Huybrechts, B., 2011. Evaluation of Immunoassay Kits for Aflatoxin Determination in Corn and Rice. CODA-CERVA Veterinary And Agrochemical Research Centre, NRL, Belgium.

Jankiewicz, A., Hubner, P., Bi ogl, K.W., Dehne, L.I., Vieths, S., Baltes, W., Liuthy, J., 1997. Celery allergy: PCR as a tool for the detection of trace amounts of celery in processed foods. In: Amado, R., Battaglia, R. (Eds.), Proceedings of the Ninth European Conference on Food Chemistry, Authenticity and Adulteration of Food ± the Analytical Approach, 24–26 September 1997, Interlaken, vol. 1, ISBN 3-9521414-0-2, pp. 131–136.

Jha, S.N., Narsaiah Narsaiah, K., Basediya, A.L., Sharma, R., Jaiswal, P., Kumar, R., Bhardwaj, R., 2011. Measurement techniques and application of electrical properties for non destructive quality evaluation of foods. J. Food Sci. Technol. 48 (4), 387–411.

Julshamn, K., Maage, A., Wallin, H.C., 1998. Determination of magnesium and calcium in foods by atomic absorption spectrometry after microwave digestion: NMKL collaborative study. J. Assoc. Off. Anal. Chem. 81 (6), 1202–1208.

Kang'ethe, E.K., Lang'a, K.A., 2009. Aflatoxin B1 and M1 contamination of animal feeds and milk from urban centers in Kenya. Afr. Health Sci. 9 (4), 218–226.

Kawaii, S., Tomono, Y., Katase, E., Ogawa, K., Yano, M., 1999. HL-60 differentiating activity and flavonoid content of the readily extractable fraction prepared from citrus juices. J. Agric. Food Chem. 47, 128–135.

Kelleher, S.L., Chatterton, D., Neilsen, K., Lönnerdal, B., 2003. Glycomacropeptide and lactalbumin supplementation of infant formula affects growth and nutritional status in infant rhesus monkeys. Amer. J. Clin. Nutr. 77, 1261–1268.

Kerkvliet, J.D., Shrestha, M., Tuladhar, K., Manandhar, H., 1995. Microscopic detection of adulteration of honey with cane sugar and cane sugar products. Apidologie 26, 131−139.

Kim, E.K., Shon, D.H., Ryu, D., Park, J.W., Hwang, H.J., Kim, Y.B., 2000. Occurrence of aflatoxin M1 in Korean dairy products determined by ELISA and HPLC. Food Addit. Contam. 17, 59−64.

Klena, J.D., Parker, C.T., Knibb, K., Ibbitt, J.C., Devane, P.M., Horn, S.T., Miller, W.G., Konkel, M.E., 2004. Differentiation of *Campylobacter coli*, *Campylobacter jejuni*, *Campylobacter lari*, and *Campylobacter upsaliensis* by a multiplex PCR developed from the nucleotide sequence of the lipid A gene lpxA. J. Clin. Microbiol. 42, 5549−5557.

Kolosova, A.Y., Shim, W.B., Yang, Z.Y., Eremin, S.A., Chung, D.H., 2006. Anal. Bioanal. Chem. 384, 286−294.

Kowalik, J., Ziajka, S., 2005. Assessment of the growth of *Listeria monocytogenes* in milk on the basis of PMP70 program and individual research. Medycyna Wet. 61, 940−942.

Kwon, N.H., Kim, S.H., Park, K.T., Bae, W.K., Kim, J.Y., Lim, J.Y., Ahn, J.S., Lyoo, K.S., Kim, J.M., Jung, W.K., Noh, K.M., Bohach, G.A., Park, Y.H., 2004. Application of extended single-reaction multiplex polymerasechain reaction for toxin typing of *Staphylococcus aureusisolates* in South Korea. Int. J. Food Microbiol. 97, 137−145.

Leblebici, J., Volkan, M., 1998. Sample preparation for Arsenic, Copper, Iron and Lead. Determination in sugar. J. Agric. Food Chem. 46, 173−177.

Lee, N.A., Wang, S., Allan, R.D., Kennedy, I.R., 2004. A rapid aflatoxin B1 ELISA: development and validation with reduced matrix effects for peanuts, corn, pistachio, and soybeans. J. Agric. Food Chem. 52, 2746−2755.

Lei, H., Su, R., Haughey, S.A., Wang, Q., Yang, J., Xu, Z., 2011. Development of a specifically enhanced enzyme-linked immunosorbent assay for the detection of melamine in milk. Molecules 16, 5591−5603.

Lei, I.F., Roffey, P., Blanchard, C., Gu, K., 2008. Development of a multiplex PCR method for the detection of six common foodborne pathogens. J. Food Drug Anal. 16, 37−43.

Lim, Y.H., Hirose, K., Izumiya, H., Arakawa, E., Takahashi, H., 2003. Multiplex polymerase chain reaction assay for selective detection of *Salmonella enterica serovar typhimurium*. Jpn. J. Infect. Dis. 56, 151−155.

Liu, B.H., Hsu, Y.T., Lu, C., Yih, F., 2013. Detecting aflatoxin B1 in foods and feeds by using sensitive rapid enzyme-linked immunosorbent assay and gold nanoparticle immunochromatographic strip. Food Control 30 (1), 184−189.

Lopez-Calleja, I., Gonzalez, I., Fajardo, V., Martin, I., Hernandez, P.E., Garcia, T., Martin, R., 2005. Application of polymerase chain reaction to detect adulteration of sheeps milk with goats milk. J. Dairy Sci. 88, 3115−3120.

Louveaux, J., 1966. Pollenanalyse einiger kanadischer honig. Z. Bienenforsch. 8, 195−202.

Louveaux, J., Maurizio, A., Vorwohl, G., 1978. Methods of melissopalynology. Bee World 59 (4), 139−157.

Mabrook, M.F., Petty, M.C., 2003a. A novel technique for the detection of added water to full fat milk using single frequency admittance measurements. Sens. Actuators, B 96, 215−218.

Mabrook, M.F., Petty, M.C., 2003b. Effect of composition on the electrical of milk. J. Food Eng. 60, 321−325.

Mato, I., Huidobro, J.F., Sanchez, M.P., Muniategui, S., Fernandez-Muino, M.A., Sancho, M.T., 1997. Enzymatic determination of total d-gluconic acid in honey. J. Agric. Food Chem. 45, 3550−3553.

Mato, I., Huidobro, J.F., Sanchez, M.P., Muniategui, S., Fernandez-Muino, M.A., Sancho, M.T., 1998. Enzymatic determination of L-malic acid in honey. Food Chem. 62 (4), 503−508.

Maurizio, A., Louveaux, J., 1965. Pollens de plantes mellifères d'Europe. Union des groupements api- coles français, Paris.

McDowell, I., Taylor, S., Gay, C., 1995. The phenolic pigment composition of black tea liquors−Part I: predicting quality. J. Agric. Food Chem. 69, 467−474.

McLellan, A.R., 1975. Calcium, magnesium, potassium, sodium in honey and in nectar secretion. J. Apic. Res. 14 (2), 57−61.

Meyer, R., 1995a. Detection of genetically engineered plants by polymerase chain reaction PCR using the flavr Savrä tomato as an example. Z. Lebensm. Unters. Forsch. 201, 583−586.

Meyer, R., 1995b. Detection of genetically engineered food by the polymerase chain reaction (PCR). Mitt. Gebiete Lebensm. Hyg. 86, 648−656.

Meyer, R., Candrian, U., 1996. PCR-based analysis for the identifcation and characterization of food components. Lebensm. Wiss. Technol. 29, 1−9.

Meyer, R., Candrian, U., Li uthy, J., 1994. Detection of pork in heated meat products by the polymerase chain reaction. J. AOAC Int. 77, 617−622.

Moatsou, G., Anifantakis, E., 2003. Recent developments in antibody-based analytical methods for the differentiation of milk from different species. Int. J. Dairy Technol. 56, 133−138.

Mohammadian, B., Khezri, M., Ghasemipour, N., Mafakheri, Sh, Poorghafour, L.P., 2010. Aflatoxin M1 contamination of raw and pasteurized milk produced in Sanandaj. Iran. Arch. Razi Inst. 65 (2), 99−104.

Mohler-Smith, A., Nakai, S., 1990. Classification of cheese varieties by multivariate analysis of HPLC profiles. Can. Inst. Food Sci. Technol. J. 23 (1), 53−58.

Moreno-Arribas, V., Pueyo, E., Polo, M.C., Martin-Alvarez, P.J., 1998. Changes in the amino acid composition of the different nitrogenous fractions during the aging of wine with yeast. J. Agric. Food Chem. 46, 4042.

Mouly, P., Gaydou, E.M., Auffray, A., 1998. Simultaneous separation of flavone glycosides and polymethoxylated flavones in citrus juices using liquid chromatography. J. Chromatogr. A 800, 171−179.

Mullié, C., Odou, M.F., Singer, E., Romond, M.B., Izard, D., 2003. Multiplex PCR using 16S rRNA gene-targeted primers for the identification of bifidobacteria from human origin. FEMS Microbiol. Lett. 222, 129−136.

Neaves, P., Waddell, M.J., Prentice, G.A., 1988. A medium for the detection of Lancefield group D cocci in skimmed milk powder by measurement of conductivity changes. J. Appl. Bacteriol. 65, 437−448.

Oancea, S., 2009. Identification of glycomacropeptide as indicator of milk and dairy drinks adulteration with whey by immunochromatographic assay. Roman. Biotechnol. Lett. 14 (1), 4146−4151.

Osek, J., Gallien, P., 2002. Molecular analysis of *Escherichia coli* O157 strains isolated from cattle and pigs by the use of PCR and pulsed-field gel electrophoresis methods. Vet. Med. Czech 47 (6), 149−158.

Pathirana, U.P.D., Wimalasiri, K.M.S., Silva, K.F.S.T., Gunarathne, S.P., 2010. Investigation of farm gate cow milk for aflatoxin M1. Trop. Agric. Res. 21 (2), 119−125.

Petrovic, Z.T., Mandic, M., Grgic, J., Grgic, Z., Besic, J., 1993. Selenium levels in some species of honey in eastern Croatia. Dtsch. Lebensm.-Rundsch 89, 46−48.

Porter, K.G., Feig, Y.S., 1980. The use of DAPI for identifying and counting aquatic microflora. Limnol. Oceanogr. 25, 943−948.

Prats-Moya, S., Grané-Teruel, N., Berenguer-Navarro, V., Martin-Carratala, M.L., 1997. Inductively coupled plasma application for the classification of 19 almond cultivars using inorganic element composition. J. Agric. Food Chem. 45, 2093−2097.

Rastogi, S., Dwivedi, P.D., Khanna, S.K., Mukul, D., 2004. Detection of aflatoxin M1 contamination in milk and infant milk products from Indian markets by ELISA. Food Control 15, 287−290.

Robards, K., Li, X., Antolovich, M., Boyd, S., 1997. Characterization of citrus by chromatographic analysis of flavonoids. J. Sci. Food Agric. 75, 87−101.

Rodriguez, G., Phipps, D., Ishiguro, K., Ridgeway, H.F., 1992. Use of a fluorescent redox probe for direct visualization of actively respiring bacteria. Appl. Environ. Microbiol. 58, 1801−1808.

Rosselló-Mora, R., Amann, R., 2001. The species concept for prokaryotes. FEMS Microbiol. Rev. 25, 39−67.

Ruiz del Castillo, M.L., Mar Caja, M., Herraiz, M., Blanch, G.P., 1998. Rapid recognition of olive oil adulterated with hazelnut oil by direct analysis of the enantiomeric composition of filberstone. J. Agric. Food Chem. 46, 5128−5131.

Sadat, A., Mustajab, P., Khan, I.A., 2006. Determining the adulteration of natural milk with synthetic milk using ac conductance measurement. J. Food Eng. 77, 472−477.

Sanchez, L., Perez, M.D., Puyol, P., Calvo, M., Brett, G., 2002. Determination of vegetal proteins in milk powder by enzyme linked immunosorbent assay: interlaboratory study. J. AOAC Int. 85, 1390–1397.

Senyk, G.F., Goodall, C., Kozlowski, S.M., Bandler, D.K., 1988. Selection of tests for monitoring the bacteriological quality of refrigerated raw milk. J. Dairy Sci. 71, 613–619.

Settanni, L., Van Sinderen, D., Rossi, J., Corsetti, A., 2005. Rapid differentiation and in situ detection of 16 sourdough lactobacillus species by multiplex PCR. Appl. Environ. Microbiol. 71, 3049–3059.

Shao, Y., Feldman-Cohen, L.S., Osuna, R., 2008. Functional characterization of the *Escherichia coli* Fis-DNA binding sequence. J. Mol. Biol. 376, 771–785.

Simpkins, W.A., Louie, H., Wu, M., Harrison, M., Goldberg, D., 2000. Trace elements in Australian orange juice and other products. Food Chem. 71, 423–433.

Suhren, G., Heeschen, W., 1987. Impedance assays and bacteriological testing of milk and milk products. Milchwissenschaft 42, 619–627.

Sun, L.X., Danzer, K., Thiel, G., 1997. Classification of wine samples by means of artificial neural networks and discrimination analytical methods. Fresenius' J. Anal. Chem. 359, 143–149.

Thrasyvoulou, A., Manikis, J., 1995. Some physicochemical and microscopic characteristics of greek unifloral honeys. Apidologie 26, 441–452.

Tudorache, M., Bala, C., 2008. Sensitive aflatoxin B1 determination using a magnetic particles-based enzyme-linked immunosorbent assay. Sensors 8 (12), 7571–7580.

Val, A., Huidobro, J.F., Sanchez, M.P., Muniategui, S., Fernandez-Muino, M.A., Sancho, M.T., 1998. Enzymatic determination of galactose and lactose in honey. J. Agric. Food Chem. 46, 1381–1385.

Valmori, S., Settani, L., Suzzi, G., Gardini, F., Vernocchi, P., Corsetti, A., 2006. Application of novel polyphasic approach to study the lactobacilli composition of sourdoughs from the Abruzzo region (Central Italy). Lett. Appl. Microbiol. 43, 343–349.

Vasavada, P.C., 1993. Rapid methods and automation in dairy microbiology. J. Dairy Sci. 76, 3101–3113.

Ventura, M., Reniero, R., Zink, R., 2001. Specific identification and targeted characterization of *Bifidobacterium lactis* from different environmental isolates by a combined multiplex-PCR approach. Appl. Environ. Microbiol. 67, 2760–2765.

Vinas, P., Lopez-Garcia, I., Lanzon, M., Hernandez-Cordoba, M., 1997. Direct determination of lead, cadmium, zinc; copper in honey by electrothermal atomic absorption spectrometry using hydrogen peroxide as a matrix modifier. J. Agric. Food Chem. 45, 3952–3956.

Visser, I.J.R., De-Groote, J.M.F.H., 1984. The malthus microbiological analyser as an aid in the detection of post pasteurization contamination of pasteurized milk. Neth. Milk Dairy J. 38, 151–156.

Wang, G., Clark, C.G., Taylor, T.M., Pucknell, C.C., Price, B.L., Woodward, D.L., Rodgers, F.G., 2002. Colony multiplex PCR assay for identification and differentiation of *Campylobacter jejuni, C. coli, C. lari, C. upsaliensis*, and *C. fetus* subsp. fetus. J. Clin. Microbiol. 40, 4744–4747.

Wang, Q., Haughey, S.A., Sun, Y., Eremin, S.A., Li, Z., Liu, H., 2011. Development of a fluorescence polarization immunoassay for the detection of melamine in milk and milk powder. Anal. Bioanal. Chem. 399, 2275–2284.

White, C., 1993. Rapid methods for estimation and prediction of shelf-life of milk and dairy products. J. Dairy Sci. 76, 3126–3132.

White, J.W., Winters, K., Martin, P., Rossmann, A., 1998. Stable carbon isotope ratio analysis of honey: validation of internal standard procedure for worldwide application. J. Assoc. Off. Anal. Chem. 81 (3), 610–619.

Yin, W.W., Liu, J.T., Zhang, T.C., Li, W.H., Liu, W., Meng, M., 2010. Preparation of monoclonal antibody for melamine and development of an indirect competitive ELISA for melamine detection in raw milk, milk powder and animal feeds. J. Agri. Food Chem. 58 (14), 8152–8157.

Yost, C.K., Nattress, F.M., 2000. The use of multiplex PCR reactions to characterize populations of lactic acid bacteria associated with meat spoilage. Lett. Appl. Microbiol. 31, 129–133.

Zhuang, Y., Zhou, W., Nguyen, M.H., Hourigan, J.A., 1997. Determination of protein content of whey powder using electrical conductivity measurement. Int. Dairy J. 7 (10), 647–653.

Zywica, R., Pierzynowska-Korniak, G., Wójcik, J., 2005. Application of food products electrical model parameters for evaluation of apple puree dilution. J. Food Eng. 67, 413–418.

BIOSENSOR

5

Biosensors are emerging as an attractive solution for rapid detection of food-borne pathogens, toxins, pesticides and drug residues, and heavy metal ions in foods. Biosensors combine the selectivity of biology with the processing power of modern microelectronics and optoelectronics to offer powerful new analytical tools with major applications in medicine, environmental diagnostics, and foods and their processing industries. It is gaining importance and popularity over conventional analytical techniques because of specificity, low cost, fast response time, portability, ease of use, and continuous real-time signal; conventional methods, on the other hand, are costly, laboratory-bound, and require trained personnel. Biosensors offer rapid and accurate detection with minimal sample preparation. They are also amenable for online analysis. This chapter describes type of biosensors and their conceptual understanding with practical applications.

5.1 CONCEPTS

The concept of the biosensor was pioneered by Clark and Lyons in 1962; they proposed that enzymes could be immobilized on electrical detectors to form enzyme electrodes. The first enzyme electrode was devised for monitoring glucose concentrations in blood using glucose oxidase (GOD). An oxidoreductase GOD was held next to a platinum electrode in a membrane sandwich. The platinum electrode polarized at $+0.6$ V and responded to the peroxide produced by the enzyme reaction with substrate.

$$\text{Glucose} + O_2 \rightarrow \text{gluconic acid} + H_2O_2 \tag{5.1}$$

A biosensor is an analytical device that consists of a biological component coupled to a transducer that converts biochemical activity into a quantifiable electrical/optical signal. It consists of five parts, as shown in Figure 5.1.

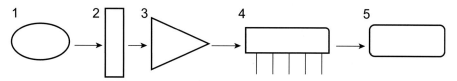

FIGURE 5.1

The schematic flow of a biosensor: (1) biological recognition element; (2) transducer; (3) signal conditioning circuit; (4) microprocessor; (5) display.

Rapid Detection of Food Adulterants and Contaminants. http://dx.doi.org/10.1016/B978-0-12-420084-5.00005-6
Copyright © 2016 Elsevier Inc. All rights reserved.

1. *"Biological recognition element"* (e.g., tissue, microorganisms, organelles, cell receptors, enzymes, antibodies, nucleic acids): converts the analyte into product with or without reactant.
2. *Transducer*: detects the occurrence of the reaction and converts it into an electrical, optical, or other signal. Transduction could be based on change in electrical, optical, mechanical, mass, acoustic, thermal, chemical, and magnetic properties.
3. *Signal conditioning circuit*: associated electronics or microprocessors that are primarily responsible for signal processing and amplification.
4. *Microprocessor*: signal is digitized and stored for further processing.
5. *Display*: graphical/numerical real-time display of the analyte concentration.

Biosensors combine the selectivity of biological reactions with the processing power of modern microelectronics and optoelectronics to offer powerful new analytical tools with major applications in medicine, diagnostics, and the environmental and food processing industries. An ideal biosensor should have many of the following features:

1. A biocomponent that is highly specific to analyte and stable to provide longer operational life
2. Minimal dependence of the reaction on parameters such as stirring, pH, and temperature
3. A response signal that is accurate, precise, and reproducible, with linearity over the useful analytical range
4. A high signal-to-noise ratio, with minimum noise
5. Cheap, small, portable, and simple to use

5.2 TYPE OF BIOSENSORS

There are numerous types of biosensors. They are mainly classified based on their biological recognition elements and transduction methods.

5.2.1 BIOLOGICAL RECOGNITION TYPE

Based on biological recognition elements, biosensors can be categorized into five major types (Monk and Walt, 2004):

1. *Enzyme-based biosensors*: These use immobilized enzymes. Enzymes are chosen in such a way that the reactions couple the target analyte to form the basis of detection.
2. *Immunoassay-based biosensors*: In this biosensor the specific antigen–antibody binding is monitored (e.g., indirectly via a fluorescent optical label or directly by observing a refractive index or reflectivity change, which requires no label).
3. *Nucleic acid–based biosensors*: These use the hybridization of single-stranded DNA (ssDNA) to form double-stranded DNA with complementary sequences. Of the two, one ssDNA is labeled with an optical indicator for detection.
4. *Organelle biosensors*: These use the effects of an analyte on an intact whole cell/organelle as the principle of detection.
5. *Biomimetic biosensors*: These use nonbiological materials to mimic biological selectivity.

5.2.2 **TRANSDUCER TYPE**

Based on the transduction methods, biosensors are categorized into four major types (Figure 5.2).

5.2.2.1 *Optical Biosensors*

Optical sensors are based on methods such as ultraviolet–visual absorption, fluorescence, phosphorescence, bioluminescence, chemiluminescence, reflectance, scattering and refractive index, caused by the interaction of the biocatalyst with the target analyte. These are the most commonly used sensors because of their high sensitivity and selectivity for bioanalytic purposes. Optical sensors can be grouped into the following types.

5.2.2.1.1 Surface Plasmon Resonance

Surface plasmon resonance biosensors detect changes in refractive index that occur when the target analyte interacts with a biorecognition element on the sensor. These sensors use electromagnetic wave to measure the change in the angle of the reflected light as a function of change of density of medium against time. This change produces a variation in the propagation constant of the surface plasmon wave, and the variation is measured to produce a reading (Mol and Fischer, 2010).

5.2.2.1.2 Fluorescent Biosensors

In this type of optical biosensor, a fluorescent compound in conjugation with antibodies helps in the detection of toxins or other contaminants. Fluorescence-based biosensors incorporate fluorochrome molecules (e.g., fluorescein isothiocyanate), which are used to produce light during the biorecognition event. Fluorescence needs an external source of light (short-wavelength light) to initiate electronic

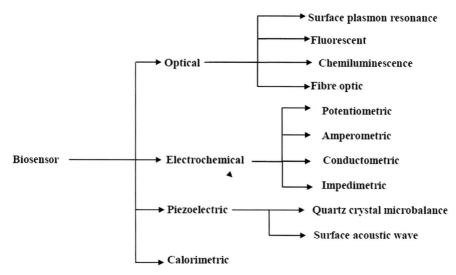

FIGURE 5.2

Types of biosensors based on the use of transducers.

transitions in an atom or molecule, which then generate luminescence (a longer-wavelength light) (Daly and McGrath, 2003).

5.2.2.1.3 Chemiluminescence

In chemiluminescent biosensors, the biochemical reaction between the analyte and the immobilized biomolecule generates luminescence. When a chemical reaction occurs, the atom or molecule relaxes from its excited state to its ground state, which results in light as a side product of the reaction. This emitted light can be detected using a photo multiplier tube (Zhang et al., 2005).

5.2.2.1.4 Fiber Optic

The basic principle of fiber optic biosensors lies behind fluorescently labeled analyte, which, when bound to the surface of the biosensor, is excited by the laser wave (635 nm). As a result, it generates fluorescent signals that are detected by the fluorescent detector in the real-time system (Bhunia, 2008).

5.2.2.2 *Electrochemical Biosensors*

Electrochemical biosensors are self-contained integrated devices that provide specific quantitative or semiquantitative analytical information through a biorecognition element that remains in direct contact with an electrochemical transduction element. These types of biosensors use an electrode as the transduction element for the detection of analyte. Although electrochemical methods has several advantages, such as low cost, easy miniaturization, and the ability to work with turbid samples, their sensitivity and selectivity are slightly limited compared with optical detection methods. Electrochemical biosensors can be classified into potentiometric, amperometric, conductometric, and impedimetric based on the observed parameters of potential, current, conductance, and impedance, respectively.

5.2.2.2.1 Potentiometric Biosensors

In potentiometric biosensors, the biological recognition element converts the recognition process into a potential signal to provide analytical information. It is equipped with two electrodes: a reference electrode for providing a constant half-cell potential and an indicator electrode, which aids in the development of a variable potential. The transduction depends on the potential difference generated during the recognition process between these two electrodes at nearly zero current. This potential difference or electromotive force is measured through a high-impedance voltmeter, which is proportional to the logarithm of the concentration of the substance being considered. The technique facilitates the detection of extremely minute concentration changes caused by the generation of a logarithmic concentration response (Grieshaber et al., 2008).

5.2.2.2.2 Amperometric Biosensors

Amperometric biosensors measure the current produced through the oxidation and reduction of an electroactive species in a biochemical reaction that is related to the concentration of analyte in solution. These biosensors utilize three types of electrode: a working electrode, a reference electrode controlling the potential of the working electrode, and a counter or auxiliary electrode, which helps measure current flow (Wang et al., 2008).

5.2.2.2.3 Conductometric Biosensors

These sensors are based on transduction methods that measure the change in electrical conductivity or current flow. The device uses two metal electrodes separated by a certain distance and an alternating current voltage applied across the electrodes. It causes a change in ionic concentration and leads to a change in current flow, which is measured using an ohmmeter (Perumal and Hashim, 2014).

5.2.2.2.4 Impedimetric Biosensors

With impedimetric biosensors, a controlled alternating voltage is applied over a range of frequencies, resulting in a current flowing through the biosensor, which is used to calculate the impedance. The device consists of three electrodes: the working electrode for current measurement, the counter electrode for supplying current to the cell, and the reference electrode for the measurement of voltage. In addition, it also contains a potentiostat for providing high-input impedance to maintain the voltage and a frequency response analyzer to supply an excitation waveform and to provide a high-precision, wide-band method of measuring the impedance (Barsoukov and Macdonald, 2005).

5.2.2.3 Piezoelectric Biosensors

Piezoelectricity is a linear interaction between the mechanical and electrical systems in a noncentric crystal or similar structure. These biosensors measure small changes in mass. They are relatively easy to use, are cost-effective, and offer label-free detection and real-time monitoring. There are some drawbacks, however, such as lack of sensitivity and specificity along with excessive interference. The two main types of mass-based sensors are (1) bulk wave or quartz crystal microbalance (QCM) and (2) surface acoustic wave (SAW).

5.2.2.3.1 Quartz Crystal Microbalance

QCM sensors are based on resonance frequency changes on a QCM following mass changes on the probe/transducer surface. When a piezoelectric sensor surface equipped with an antibody is placed in a solution containing analytes, the interaction of the agent and the antibody-coated surface causes an increase in the crystal mass and gives rise to a corresponding frequency shift. This shift can be measured electrically to determine the additional mass of the crystal (Velusamy et al., 2010). Quartz is the most commonly used piezoelectric material because of its low cost and stability against thermal, chemical, and mechanical stress.

5.2.2.3.2 Surface Acoustic Wave

SAW devices produce and detect acoustic waves on the surface of a piezoelectric crystal by means of interdigital transducers. Thus, the acoustic energy is sturdily restricted at the surface of the device in the range of the acoustic wavelength, regardless of the substrate thickness (Grate and Frye, 1996) For this reason, the wave is potentially very sensitive toward any change on the surface, such as mass loading, viscosity, and conductivity.

5.2.2.4 Calorimetric Biosensors

Calorimetric biosensors are based on changes in temperature that occur when a reaction takes place between the biorecognition element and a suitable analyte. This thermal change can be correlated to the amount of products formed or reactants consumed. The calorimetric device uses either a thermistor (usually metal oxide) or thermopile (usually ceramic semiconductor) for the measurement of heat change.

The major advantages of this type of thermal detection are stability, increased sensitivity, and the possibility of miniaturization (Ahmad et al., 2010).

5.3 METHODS OF IMMOBILIZING BIOSENSORS

The published literature reveals that immobilization of the biocomponent is one of the most important steps involved in biosensor design. The choice of the technique used for connecting the biological component to the transducer is crucial because the stability, the longevity, and the sensitivity largely depend on the configuration of the biocomponent layer. Biosensors based on immobilized biomolecules were developed to solve several problems related to the stability and operational life of the biosensor, as well as to reduce the time until the response and offer disposable devices that can be easily used in stationary or in flow systems.

Major methods of immobilization can be classified as covalent binding, cross-linking, adsorption, entrapment, and membrane system.

Covalent binding is the retention of biomolecules on support surfaces by covalent bond formation. Enzyme molecules bind to support material via certain functional groups, such as amino, carboxyl, hydroxyl, and sulfhydryl groups. These functional groups must not be in the active site. One common trick is to block the active site by flooding the enzyme solution with a competitive inhibitor before covalent binding. Functional groups on support material are usually activated by using chemical reagents, such as cyanogen bromide, carbodiimide, and glutaraldehyde.

Cross-linking of molecules with each other using agents such as glutaraldehyde, bis-diazobenzidine, and 2,2-disulfonic acid is another method of immobilization. Cross-linking can be achieved in several different ways: molecules can be cross-linked with glutaraldehyde to form an insoluble aggregate, adsorbed molecules may be cross-linked, or cross-linking may take place following the impregnation of porous support material with molecules.

Adsorption is the simplest method of immobilization. It involves attachment on the surfaces of support particles by weak physical forces, such as Van der Waals or dispersion forces. Adsorption is a mild method. The active site of the adsorbed molecule is usually unaffected, and nearly full activity is retained upon adsorption. Desorption is a common problem, however, especially in the presence of strong hydrodynamic force or drastic change in pH or ionic strength from those used at the time of adsorption, since binding forces are weak. Adsorption of enzymes may be stabilized by cross-linking with glutaraldehyde. Glutaraldehyde treatment can denature some proteins.

Support materials used for adsorption can be inorganic material, such as alumina, silica, porous glass, ceramics, diatomaceous earth clay, and bentonite, as well as organic materials such as cellulose (carboxymethyl cellulose), starch, activated carbon, and ion-exchange resins, such as Amberlite, Sephadex, and Dowex. The surfaces of the support materials may need to be pretreated chemically or physically for effective immobilization.

Entrapment is the physical enclosure of molecules in a small space. Matrix entrapment and microencapsulation are the two major methods of entrapment. Matrices used for enzyme immobilization are usually polymeric materials such as calcium-alginate, agar, potassium-carrageenin, polyacrylamide, and collagen. However, some solid matrices such as activated carbon, porous ceramic, and diatomaceous earth can also be used for this purpose.

The matrix can be a particle, a membrane, or a fiber. When immobilizing in a polymer matrix, enzyme solution is mixed with polymer solution before polymerization takes place. Polymerized gel is either extruded or a template is used to shape the particles from a liquid polymer-enzyme mixture. Entrapment and surface attachment may also be used in combination in some cases.

A special form of entrapment is *microencapsulation*. In this technique, microscopic hollow spheres are formed. The spheres contain the molecule in a solution, while the sphere is enclosed within a porous membrane. The membrane can be polymeric or an enriched interfacial phase formed around a microdrop.

Membrane entrapment of molecules can also be used to allow selective analyte transmission toward the transducer. Membranes of nylon, cellulose, polysulfone, and polyacrylate are commonly used. Configurations other than hollow fibers are possible, but in all cases a semipermeable membrane is used to retain high-molecular-weight compounds (enzyme or biomolecule), while allowing small-molecular-weight compounds (substrate or products) access to the enzyme or biomolecule. An advantage of this method is that it does not affect the activity of the enzyme or biomolecule.

5.4 PRACTICAL APPLICATIONS

Biosensors are used in food analysis for detecting the presence of both biological (microbial pathogens) and chemical contaminants. Practical applications of biosensors depend on their type. Uses of transducer-based biosensors are more common in food industries, and some of them are discussed in the following sections.

5.4.1 ELECTROCHEMICAL BIOSENSOR

Electrochemical-based biosensors have been used for identification and quantification of different components in food.

5.4.1.1 Microbial Pathogens

The food industry is primarily concerned about the presence of pathogenic bacteria as failure to detect certain pathogens that may lead to disease outbreaks with horrible consequences. Although conventional culturing techniques enable the detection of food-borne pathogens, the methodology is time-consuming and it often gives false-negative results in the identification of certain types of pathogens because of the high background level of competing microorganisms (Baeumner et al., 2004). Another method for detecting a pathogen through antibodies is less sensitive and requires the purification of bacteria. Therefore, methods based on biosensing are useful for determining food-borne pathogens because of its quick speed, convenience, and high sensitivity. *Escherichia coli* contamination in food and water represents a major threat for human health; *E. coli* is an infectious agent causing diarrhea, urinary tract infections, inflammation, and peritonitis in immunocompromised patients such as children and elderly people (Gfeller et al., 2005). Ercole et al. (2003) used a potentiometric biosensor for the detection of *E. coli* present on vegetables. The bacteria was recovered in a liquid medium by washing the vegetables with peptone water (pH 6.8) and then blending it in a sonicator. Analysis of the liquid phase by biosensors showed 10 cells/mL within 1.5 h. Similarly,

Hnaiein et al. (2008) reported the detection of *E. coli* using a conductometric biosensor, which involves the immobilization of anti−*E. coli* and biotinylated antibodies on magnetite nanoparticles modified with streptavidin through the interaction of biotin−streptavidin. The biosensor showed good selectivity, with a potential to detect 500 colony-forming units (CFUs)/L. In a recent study conducted by Majumdar et al. (2013), an amperometric immunosensor was developed for the detection of *Staphylococcus aureus* in food samples. The antibody was immobilized onto the platinum electrode surface through cross-linkage via glutaraldehyde precoated with a polyethyleneimine layer, which led to a change in response. The obtained response showed good linearity between the increasing concentrations of test bacteria and current output, along with detection limit of 10 CFUs/mL.

5.4.1.2 Viruses

Virus represents a major concern of human health; therefore, a tool for the rapid detection of viruses is required. Miodek et al. (2014) developed an electrochemical biosensor based on a immunodetection system for quantification of the PB1-F2 protein, which is a small proapoptotic protein that contributes to the virulence of influenza A virus. A specific anti-PB1-F2 monoclonal antibody was immobilized on the co-polypyrrole layer via a biotin−streptavidin system. The study demonstrated that the electrochemical system could sensitively detect purified recombinant PB1-F2 over a wide range of concentrations, from 5 nM to 1.5 mM. Another study investigated human hepatitis B and papillomaviruses using electrochemical impedance spectroscopy. The sensor was fabricated by depositing gold nanoparticles on the in situ−prepared single-walled carbon nanotube arrays and could detect lower than 1 amol complimentary hepatitis ssDNA, corresponding to 600 ssDNA molecules in a 1.0-mL sample, with a detection limit of 0.1 pmol for a 1-base mismatched hepatitis B ssDNA. When being applied to detect 24-base papillomavirus ssDNA, the experimentally determined low detection limit is 1 amol. In addition to the low detection limit, the single-walled carbon nanotube/gold/ssDNA sensor also showed great stability (Wang et al., 2013).

5.4.1.3 Toxins

Toxins represent a broad class of molecules that is responsible for a variety of human diseases and can be used as a biological warfare agent. Staphylococcal enterotoxin produced by *Staphylococcus aureus* is a common food contaminant that is mainly responsible for gastroenteritis, suppurative infections such as boils, abscesses, and wound infections. Pimenta-Martins et al. (2012) detected Staphylococcal enterotoxin A in contaminated and noncontaminated cheese samples using an amperometric immunosensor. The sensor consists of a gold electrode on which self-protein A and an assembly monolayer were immobilized. The immunosensor showed excellent response, with high sensitivity and specificity. Aflatoxins are another class of toxins that are produced by different species of fungi, such as *Aspergillus flavus*, *Aspergillus nomius*, and *Aspergillus parasiticus*. These toxins generally affect crops during transportation, storage, and processing. Therefore, Soldatkin et al. (2013) used an acetylcholinesterase-based conductometric biosensor for detection of aflatoxin B1, and results showed high signal reproducibility with a detection limit of 0.05 µg/mL. A recent study reported detection of other type of mycotoxin named citrinin. Citrinin has carcinogenic, toxic, and mutagenic properties and is found mainly in vegetables, crops, and fruits. In rice samples, citrinin mycotoxin content was detected using a horseradish peroxidase−based amperometric biosensor. In this method, carbon paste electrodes with multiwalled carbon nanotubes were embedded in horseradish peroxidase, mineral oil, and ferrocene (as a redox mediator). A linear response was observed in a concentration ranging from

1 to 11.6 nM, along with a detection limit and quantification limit of 0.25 and 0.75 nM, respectively (Zachetti et al., 2013).

5.4.1.4 Pesticides

Another application of electrochemical biosensors could be determining pesticides and herbicides. An increasing concern has developed about water and food contamination due to the exhaustive use of pesticides in agriculture and their potential toxicity to human health. A potentiometric sensor was used for the detection of atrazine by implementing the imprinted polymer membrane without macropores. Detection was carried out in acidic solution with a pH less than 1.8, which caused the protonation of atrazine. The membrane potential was increased with atrazine concentration in the range of 3×10^{-5} to 1×10^{-3} M (Agostino et al., 2006). A study also reported impedimetric label-free immunodetection of phenylurea herbicides (Bhalla et al., 2012). Gold nanoparticles were used as signal enhancers cum immobilization matrix and electrodeposited on carbon screen-printed electrodes. These gold nanoparticles were further functionalized with specific antidiuron antibodies for the development of the biointerface. The results showed a linear response in the useful analytic range of 1−1000 ng/mL, with a limit of detection equal to 5.46 ng/mL. Carbofurans are another group of pesticides that are toxic and cause immunological respiratory diseases and nerve disorders in humans. An amperometric biosensor based on alkaline phosphatase inhibition was designed for the detection of carbofurans. Alkaline phosphatase was immobilized on a carbon nanopowder paste electrode and the measurements were performed through chronoamperometric monitoring of the inhibition of enzyme activity. The optimum performance was observed at pH 8.5 and at an operation potential of +0.75 V versus Ag/AgCl. The dynamic range of 10−97 µg/L carbofuran with a detection limit at 10 µg/L was obtained with this method (Samphao et al., 2013).

5.4.1.5 Heavy Metals

Electrochemical biosensors also find applications in the detection of heavy metals. A potentiometric biosensor was developed for the determination of mercury ions using self-assembled gold nanoparticles. Gold nanoparticles were adsorbed on a $PVC-NH_2$ matrix membrane at the surface of pH electrodes containing N, N-didecylaminomethylbenzene; then the urease was immobilized on the gold nanoparticles. Mercury ions were detected in the range of 0.09−1.99 µmol/L, with a detection limit of 0.05 µmol/L (Yang et al., 2006). Soldatkin et al. (2012) designed a conductometric biosensor based on a three-enzyme system (invertase, mutarotase, glucose oxidase) for selective determination of heavy metal ions. The proposed biosensor showed the best sensitivity toward Hg^{2+} and Ag^+ ions. An ultrasensitive conductometric sensor based on the inhibition of alkaline phosphatase activity from *Arthrospira platensis* was recently developed for the detection of heavy metals (Tekaya et al., 2013). Cells of *A. platensis* were immobilized on the ceramic part of gold interdigitated transducers via physical adsorption. The heavy metals inhibited this activity, causing a variation of the local conductivity, which was measured after addition of the substrate. It was observed that cadmium and mercury inhibited alkaline phosphatase activity, with a detection limit of 10−20 M. The half-maximal inhibitory concentration was determined at 10−19 M for Cd^{2+} and 10−17 M for Hg^{2+}.

5.4.1.6 Sugars

Carbohydrates are a source of carbon and energy for life activities, and therefore the analysis of sugar in food is very important. Traditional methods to detect sugars, such as spectroscopic and chromatographic methods, are either less sensitive, expensive, and/or time-consuming. In comparison, biosensor techniques are rapid, cost-effective, and highly sensitive. An amperometric glucose biosensor based on a screen-printed electrode and Os-complex mediator for flow injection analysis was constructed (Liu et al., 2011). The biosensor displayed a linear response for glucose in the range of 0.5−30 mM, with a correlation coefficient up to 0.99. A high concentration of glucose is constantly present in patients with diabetes because they are unable to regulate it. Therefore, patients need to monitor their blood glucose regularly to manage the concentrations. Li et al. (2012) designed a selective and sensitive D-xylose electrochemical biosensor based on xylose dehydrogenase displayed on the surface of a bacteria and multiwalled carbon nanotube−modified electrode. The xylose dehydrogenase bacteria was prepared from *Pseudomonas borealis* DL7 using a newly identified ice nucleation protein that catalyzed the oxidization of xylose to xylonolactone in the presence of coenzyme NAD^+. The results demonstrated good analytical performance along with a wide dynamic range of 0.6−100 μM and a low detection limit of 0.5 μM D-xylose. Xylose and xylose-related products such as xylan and xylo-oligosaccharides are well known for their functional value in food as an ideal sweetener, as well as nutritional and therapeutic agents (Broekaert et al., 2011). A study made use of a light addressable potentiometric sensor (LAPS) for the simultaneous determination of three saccharides: sucrose, maltose, and glucose. The LAPS system is an integrated microbiosensor technique that is able to detect microsurface potentials only at a light-illuminated part of a silicon chip. It is a highly sensitive means of monitoring reactions in microscopic areas. The integrated LAPS biosensor was constructed using the conjugated bienzyme system catalyzed by glucokinase and oligosaccharide hydrolases such as invertase and ß-D-glucosidase in order to make use of the selectivity of biorecognition. Clear distinctions of the response patterns among the three saccharide samples were obtained (Aoki et al., 2002).

5.4.1.7 Amines

Biogenic amines are formed in certain food products during storage and processing, causing serious health problems, especially to sensitive persons. Histamine is a biogenic amine that accumulates in seafood upon bacterial spoilage and causes poisoning without altering the normal appearance or odor of the fish. It exerts its effects by binding to receptors present on cellular membranes of the cardiovascular, respiratory, gastrointestinal, and immunological systems. It is considered to be an indicator of earlier microbial decomposition of seafoods, and a guidance level of 50 ppm is considered to be the chemical index for fish spoilage. Therefore, a histamine biosensor has been developed for rapid monitoring of histamine concentrations (Keow et al., 2007). The biosensor responded in less than 1 min, and 7.4 was an optimum pH of operation, with reproducibility and repeatability of 4.87% and 5.26%, respectively, relative standard deviations. Another study carried out detection of tyramine in sauerkraut samples using an amperometric biosensor (Apetrei and Apetrei, 2013). An electrochemical tyrosinase electrode was developed via the cross-linking immobilization method over a phosphate-doped polypyrrole film; the amine was detected by the direct reduction of biocatalytically formed dopaquinone at −0.250 V. Under the optimized conditions, the current response had a linear relationship with the concentration of tyramine in the range of $4−80 \times 10^{-6}$ M, with a sensitivity of 0.1069 A/M and a limit of detection of 5.7×10^{-7} M. The selective detection is related to the specific action of tyrosinase toward the phenolic group from tyramine (1-hydroxy-4-ethylaminobenzene).

5.4.1.8 Enantiomers

The determination of the optical purity of various compounds used in the food industry is especially important in cases where they are not of natural origin or have been formed by transforming one enantiomer into another. The selective metabolic use of enantiomers is of primary importance for human beings because it can affect vital functions of the organism. There are many examples of chiral compounds whose one enantiomer has a beneficial role but the other is detrimental to the organism (Kaniewska et al., 2008). Guo et al. (2013) constructed an electrochemical chiral biosensor for tryptophan (Trp) enantiomers with the assistance of Cu(II). The results demonstrated the larger electrochemical response from L-Trp when Cu(II) was present, signifying that this strategy could be used to enantioselectively recognize Trp enantiomers. The chiral biosensor showed a good linear response to Trp enantiomers in the range of the concentration of [Cu(II)(Trp)2] from 5.0×10^{-4} to 2.5 mM, with a low limit of detection of 0.17 µM. Basozabal et al. (2011) determined the enantiomers of the pesticide dinoseb using an electrochemical method. They optimized the methodologies for both an R- and S-specific sensor to get the highest chiral discrimination ability. S-camphorsulfonic acid was used as a chiral inducing agent and R- and S-specific sensors were synthesized potentiostatically at 0.5 and 0.6 V, respectively. The proposed methodology allows the generation of sensors that are able to bind the different enantiomers of the pesticide dinoseb with good sensitivity and repeatability (residual standard deviation $< 10\%$).

5.4.1.9 Other Analytes

Electrochemical biosensors are also used in the detection of analytes such as urea, ethanol, and hydrogen peroxide (H_2O_2). An amperometric biosensor has been proposed for the measurement of urea (Bozgeyik et al., 2011). The biosensor was constructed through immobilization of urease enzyme onto a poly(N-glycidylpyrrole-co-pyrrole)-conducting film. The resulting biosensor exhibited a linear current response to urea concentration ranging from 0.1 to 0.7 mM, with a sensitivity of 4.5 µA/mM and a response time of 4 s. A study reported the determination of melamine in milk using a potentiometric sensor based on a molecularly imprinted polymer (Liang et al., 2009). The molecularly imprinted polymer for selective recognition was synthesized using melamine as a template molecule, methacrylic acid as a functional monomer, and ethylene glycol dimethacrylate as a cross-linking agent. The membrane electrode exhibited a near-Nernstian response (54 mV/decade) to the protonated melamine over the concentration range of 5.0×10^{-6} to 1.0×10^{-2} mol /L, with a response time of ~ 16 s. Ajay and Srivastava (2007) detected ethanol by exploiting a microtubular conductometric sensor. The microtubules of polyaniline were used as transducer cum immobilization matrix along with alcohol dehydrogenase enzyme and its coenzyme NAD^+ for improving the selectivity of the sensor. The sensor was able to detect ethanol as low as 0.02% (v/v) and showed good results at a physiological pH. Similarly, Shamsipur et al. (2012) used an impedimetric catalase nanobiosensor for determination of H_2O_2 and detected ultratraces of H_2O_2 in the range of 5−1700 nM. A study compared the amperometric and potentiometric biosensors for determination of phosphate (Lawal and Adeloju, 2013); a minimum detectable concentration of 20.0 mM phosphates and a linear concentration range of 20−200 mM were obtained with the potentiometric biosensor, whereas the amperometric biosensor attained a minimum detectable concentration of 10 mM and a linear concentration range of 0.1−1 mM.

5.4.2 PIEZOELECTRIC BIOSENSORS

Piezoelectric biosensors are suitable for very sensitive detection, in which transduction is based on small changes in mass.

5.4.2.1 Pathogens

Bacterial contamination alone is responsible for about 91% of all food-borne diseases and has been reported in various food-based samples such as processed foods, raw agricultural food products, and even nonagricultural raw materials. The main pathogens responsible for this contamination include *E. coli*, *Salmonella*, *Staphylococcus*, *Listeria*, and *Campylobacter* (Michino et al., 1999). In the past 10 years, polymerase chain reaction has been widely used for the detection of bacteria and viruses because of its high sensitivity, but the technique is unable to distinguish between viable and nonviable cells (de Viedma, 2003). Moreover, polymerase chain reaction is laborious and expensive compared with biosensor systems. Kim and Park (2003) developed a flow-type immunosensor system using a broad-spectrum anti−*E. coli* antibody and QCM; they observed a linear sensor response for the microbial suspensions ranging from 1.7×10^5 to 8.7×10^7 CFUs/mL within 20−30 min after a treatment. Similarly, *Salmonella typhimurium* has been studied in chicken meat samples using a QCM biosensor with simultaneous measurements of resonant frequency and motional resistance. In direct detection, resonant frequency and motional resistance were proportional to the cell concentration in the range of 10^5−10^8 and 10^6−10^8 cells/mL, respectively. The detection limit was reduced to 10^2 cells/mL using anti-*Salmonella* magnetic beads as a separator/concentrator for sample pretreatment. Li et al. (2010) demonstrated phage-based magnetoelastic biosensors with a magnetoelastic resonator platform coated with filamentous E2 phage for detection of *S. typhimurium* on fresh tomato surfaces, and they concluded that the biosensor was able to detect 5×10^2 CFUs /mL in 30 min. Phage-based detection is based on the principle that the phage infection causes bacterial lysis that gives rise to the development of visible plaques on a solid medium. On the other hand, in a liquid growth medium, the lysis results in the release of cell contents such as enzymes into the surrounding medium, which could be measured and quantified rapidly by sensors. This provides a powerful means for highly specific detection of a given bacterial strain.

5.4.2.2 Viruses

A report studied three piezoelectric sensors modified with anti-HIV-1 virion infectivity factor (Vif) single-fragment antibodies, single domains, and camelized single domains for detecting HIV-1 Vif. Dithio-bis-succinimidyl undecanoate and 11-hydroxy-1-undecanethiol-mixed self-assembled monolayers were generated at the sensor surface, onto which the antibodies were immobilized. All sensors detected specifically the target HIV-1 Vif antigen in solution, and no unspecific binding was monitored (Encarnacao et al., 2007).

5.4.2.3 Toxins

The contamination of food by toxins has become a subject of huge concern because they are responsible for many diseases (Richard, 2007). In a recent study by Chalupniak et al. (2014), low-frequency piezoelectric quartz tuning forks were used for measuring the concentrations of endotoxins and bacterial cells (*E. coli* O157:H19) in the range of 0.001−5 endotoxin units/mL and 10^2−10^7 CFUs/mL. In the same way, Yu et al. (2011) developed a novel piezoelectric biosensor based on a lipid layer with GM1 (monosialoganglioside) for the detection of cholera toxin. The frequency responses of the developed piezoelectric biosensors were linearly correlated to toxin concentration in the range of 0.25−1.0 µg/mL, with a detection limit of 95 ng/mL.

5.4.2.4 Pesticides

Organophosphate pesticides are toxic compounds and are commonly found as contaminants in agricultural products. Therefore, it is necessary to monitor the pesticide for its effective removal. Zimmermann et al. (2001) used a love-wave gas sensor based on a quartz piezoelectric substrate and a silicon dioxide guiding layer coated with a specific polysiloxane polymer for sensing organophosphorus compounds. A report showing the detection of parathion pesticide using hybrid organic/inorganic coating on SAW devices recently appeared in the literature (Mensah-Brown et al., 2014). Benzo[a]pyrene (BaP), a carcinogenic contaminant originating from the incomplete combustion of organic compounds, is often present in foodstuffs and is difficult to detect because of its low molecular weight and poor solubility in water. Boujday et al. (2010) designed piezoelectric immunosensors for the determination of BaP using QCM with dissipation measurements as transduction techniques and measured the BaP concentration with a sensitivity limit of 5 µM. A recent study reported a cost-effective protocol to fabricate acoustic microimmunosensors for the detection of atrazine herbicide and carbofuran insecticide using a single microstructured gold QCM (Jia et al., 2013). The proposed methodology allowed sequential specific detection of 4.5 µM carbofuran and 4.6 µM atrazine, respectively.

5.4.2.5 Heavy Metals

Heavy metals pollution is a main health issue and can cause damage or reduce mental and central nervous function. The toxicity is related to an excessive buildup of metals in the body. Conventional methods that have been used for heavy metal analysis in food are expensive and time-consuming. In comparison, sensor systems offer several advantages such as a high sensitivity and a real-time response. Gammoudi et al. (2014) designed a highly sensitive love-wave sensor coated with mesoporous titanium dioxide for detection of heavy metal in liquid medium using *E. coli* cells as bioreceptors because of their viscoelastic properties, which are affected by toxic heavy metals. The addition of a thin titania layer dip-coated onto the acoustic path of the sensor increased the specific surface area as well as improved the mass effect sensitivity of the acoustic device. In another study conducted by Sun et al. (2012), a highly sensitive coating containing 3-mercaptopropionic acid (MPA) monolayer−modified electrospun polystyrene (PS) membranes was constructed on the electrode of QCM through electrospinning for Cu^{2+} detection. The PS-MPA sensors displayed a quick response (2−3 s), with increased sensitivity to Cu^{2+} and a detection limit down to 100 ppb. The response of the PS-MPA sensors with a PS membrane was 40 times larger than that without membranes when exposed to 1 ppm of Cu^{2+}. In addition, the sensors also showed chelation selectivity for other transition metal ions (in descending order: $Cu^{2+} > Ni^{2+} > Zn^{2+} > Fe^{2+}$) at concentrations between 1 and 5 ppm. Similarly, a novel copolymer P (MBTVBC-co-VIM) was designed for fabrication of copolymer-coated QCM sensors to detect heavy metal ions in aqueous solution (Cao et al., 2011). The copolymer P made complexes with heavy metal ions using nitrogen and sulfur atoms present in the side groups as electron donors. The QCM results revealed that the sensor exhibited high sensitivity, stability, and selectivity for the detection of Cu^{2+} in aqueous solution, with the lowest detection limit of 10 ppm, which resulted in a frequency shift of 3.0 Hz.

5.4.2.6 Odors

The demand for tools that are able to detect odorant molecules is increasing rapidly with potential applications in the assessment of various foodstuffs and beverages. A SAW biosensor for odor detection in the food industry was also developed (Palla-Papavlu et al., 2014). The SAW biosensor was coated with wild-type bovine odorant binding protein solutions containing 20% and 50% glycerol through laser-induced forward transfer, which showed a high sensitivity (5 Hz/ppm) and a detection limit in parts per million. Escuderos et al. (2011) investigated the potential of eight QCM sensor arrays to differentiate the quality of olive oil samples based on their aromatic profiles. Five gas chromatographic stationary phases (OV-17, OV-275, PEG, Span 80, and Vaseline) were used as sensing films of QCM sensors, and the discrimination properties of each array were measured using principal component analysis. The results of the study provided promising perspectives for the use of a low-cost, easy-to-use, and rapid system for the differentiation of virgin and extra virgin olive oils from lampante ones.

5.4.2.7 Other Analytes

Olfactory sensing of specific volatile organic compounds released by bacterial pathogens is one of the unique ways to detect contamination in food products. Therefore, Sankaran et al. (2011) developed a biomimetic piezoelectric biosensor based on olfactory receptors to detect the specific gases (alcohols) released at low concentrations by bacterial pathogens. A computational simulation was used to determine the biomimetic peptide-based sensing material to be deposited on the QCM sensor. The developed QCM sensors were sensitive to 1-hexanol as well as 1-pentanol, with detection limits of $2-3$ and $3-5$ ppm, respectively. Bello et al. (2007) proposed a modified QCM sensor for quality control of ethanol concentrations during bakery production, ranging from 0.26% to 1.70% (w/w) in whole-meal bread and from 0.68% to 2.06% (w/w) in durum-wheat bread.

5.4.3 CALORIMETRIC BIOSENSORS

Calorimetric methods are also used in the food industry for the detection of various analytes.

5.4.3.1 Microbes

Enumeration of viable bacteria in liquid samples is important in environmental, food processing, and other applications. Since bacterial cells generate heat as a result of metabolic activities associated with cellular functions, thermal sensing techniques such as calorimetry have been used for detecting viable cells in biological samples. Ma et al. (2007) applied microcalorimetry to study the metabolic activity of microbes (*E. coli* and *Saccharomyces cerevisiae*) in microencapsulated cultures as well as in a free, nonencapsulated culture. It was concluded that microorganisms in the free, nonencapsulated culture have the highest metabolic rate because of the high heat output. Maskow et al. (2013) recently monitored the bacterial contamination of drinking water using a highly sensitive calorimetric biosensor and found that even minor initial bacterial contamination ($1-100$ cells/mL) gave rise to calorimetric signals after $5-17$ h of growth.

5.4.3.2 Pesticides

Calorimetric biosensors are also used for the determination of organophosphorus pesticides. The residue of these pesticides in fruits and vegetables is a serious problem in food safety because they are highly toxic and often cause respiratory paralysis and even death. A flow injection calorimetric biosensor equipped with a chicken liver esterase enzyme as the biorecognition element and acetyl-1-naphthol as the substrate was designed for the detection of dichlorvos pesticide (Zheng et al., 2006). The enzyme was immobilized on an ionic exchanging resin and packed in the enzyme reaction cell. At the same time, the same batch of the resin containing completely inactivated enzyme filled in the reference cell. As a result, temperature difference was generated at the outlets of the two cells because the enzymatic reaction occurred in the enzyme reaction cell but not in the reference cell. Calorimetric biosensing may be a good technique for rapid onsite detection of pesticide residue because it is insensitive to the electrochemical, optical, and other intrinsic properties of the real sample and does not need frequent recalibration, which makes the system easy to use.

5.4.3.3 Heavy Metals
Another application of calorimetric biosensors for determination of submillimolar concentrations of heavy metals has been proposed by Satoh (1991). He used apoenzyme thermistors to determine trace amounts of copper(II) in human blood sera using immobilized ascorbate oxidase and cobalt(II) and zinc(II) using immobilized alkaline phosphatase.

5.4.3.4 Enzymes
Enzymes constitute one of the most important groups of industrial products and are used in a wide range of processes in the food industry. A study generated thermokinetic data for the production of proteases from *Pseudomonas aeruginosa* to scale up, monitor, and control the process by following the heat signals. The study showed comparable profiles of protease production and heat generation, and oxygen uptake rate was observed to correlate well with both biomass growth and heat production rate at all lag, exponential, and stationary growth phases. The results revealed the potential of a biological reaction calorimeter for monitoring the secretion of protease by *P. aeruginosa* (Sivaprakasam et al., 2008). Zhou et al. (2013) established a fast, sensitive, and convenient method based on an enzyme thermistor (ET) for examination of β-lactamase enzyme, which is added illegally to milk to lower the concentration of antibiotic since there are regulatory restrictions on antibiotic residues in dairy products. An aliquot of the milk sample containing β-lactamase and its specific substrate, penicillin G, was directly injected into the ET system, which generated a temperature change corresponding to the remaining penicillin G. The β-lactamase present in the sample was quantified by measuring the penicillin G consumed during incubation. The linear range and the detection limit obtained with this method were 1.1–20 and 1.1 U/mL, respectively.

5.4.3.5 Sugars
Other applications of calorimetric biosensing include the detection of sugars such as glucose, xylose, and fructose. Bhand et al. (2010) developed a fructose-selective calorimetric biosensor based on an ET that was used to monitor the heat liberated during the conversion of fructose-6-phosphate to fructose-1,6-biphosphate by fructose-6-phosphate-kinase. Reproducible linearity of 0.5–6.0 mM, with a detection limit of 0.12 mM, was obtained with this method. The method was rapid, inexpensive, and selective for fructose analysis in food samples. Another study developed a microfluidic calorimeter to measure the dynamic temperature changes occurring during glucose oxidation in the presence of layer-by-layer self-assembled glucose oxidase (Tangutooru et al., 2012). To the lower channel wall of the microfluidic calorimeter, an antimony–bismuth thin-film thermopile was attached for the detection of glucose concentration by measuring the heat released (−79 kJ/mol) during the oxidation of glucose in the presence of glucose oxidase. The lowest detection limit of the device was 1 mg/dL for glucose.

5.4.3.6 Biochemical Process

Calorimetry is a fundamental and effective investigative tool for analyzing biochemical reactions. Many studies demonstrate the suitability of the heat production rate for examining and controlling aerobic growth, but it is difficult to monitor anaerobic bioprocesses because of the low sensitivity of the applied sensors. Paufler et al. (2013) recently described that the sensitivity of a standard heat flux reaction calorimeter can be improved by a factor of 10 (1.886 ± 0.012 K/W) by simply adding an additional internal thermal shield. Signals of about 5 mW/L could be detected this way, and the steady-state response time was about 10 min. The process was successfully demonstrated with acetone–butanol–ethanol fermentation using *Clostridium acetobutylicum*. Zhang and Tadigadapa (2004) presented a microthermopile device with a glass microfluidic reaction chamber for measuring real-time enthalpy changes of biochemical reactions and the thermal properties of biological fluids. The device was used to monitor the heat released from the catalytic action of glucose oxidase, catalase, and urease on glucose, H_2O_2, and urea, respectively. Sensitivities of 53.5 V/M for glucose, 26.5 V/M for H_2O_2, and 17 V/M for urea were obtained.

5.4.3.7 Other Analytes

H_2O_2 vapor is primarily used as an antimicrobial agent for inactivation of microorganisms during sterilization processes in the aseptic food industry. A novel thin-film calorimetric gas sensor with two platinum resistances as temperature-sensing elements, passivated with spin-coated perfluoralkoxy and manganese oxide particles as a catalytically active material, was developed for the detection of H_2O_2 in aseptic filling processes (Kirchner et al., 2011). The device showed the highest sensitivity of 0.57 °C/% (v/v) toward H_2O_2. Verhaegen et al. (2000) developed the MiDiCal array calorimeter using a thermopile-based sensor to analyze the concentration of ascorbic acid in food through the measurement of heat evolved by ascorbate oxidase. A similar device was used in another study for the determination of urea in adulterated milk, with a lower detection limit of 0.1 mM and a dynamic range of detection of 1–200 mM (Mishra et al., 2010). The technique was easy to perform, economical, highly stable, and sensitive and thus confirmed its suitability for routine analysis in the dairy industry. Another study presented a calorimetric gas sensor for the detection of vapor-phase H_2O_2 at elevated temperatures during sterilization processes in the aseptic food industry. The sensor setup consists of two temperature-sensitive platinum, thin-film resistances passivated by a layer of SU-8 photoresist and is catalytically activated by manganese(IV) oxide. Instead of an active heating structure, the sensor uses the elevated temperature of evaporated H_2O_2 aerosol. The sensor has shown a sensitivity of 4.78 °C/(%, v/v) in an H_2O_2 concentration range of 0–8%, and retained the same signal even at varied medium temperatures of the gas stream (between 210 and 270 °C) (Kirchner et al., 2012).

5.4.4 OPTICAL BIOSENSORS

A variety of optical biosensors for the rapid detection of microbes, toxins, and contaminants in food were developed (Narsaiah et al., 2012b), but their practical usage still is in scarce; a few of them are described briefly below.

5.4.4.1 Microbial Pathogens

Ohk and Bhunia (2013) developed and optimized a multiplex fiber optic sensor for the simultaneous detection of the three most common food-borne bacterial pathogen (*E. coli* O157:H7, *Listeria monocytogenes*, and *Salmonella enterica*). The experiment was conducted by inoculating each pathogen with ready-to-eat beef, chicken, and turkey meats in a multipathogen selective enrichment broth for 18 h. It tested for the presence of pathogens using the proposed optical biosensor. The biosensor eliminated the use of multiple single-pathogen detection platforms because it successfully detected each pathogen with a very little cross-reactivity and a detection limit of $\sim 10^3$ CFUs/mL. Narsaiah et al. (2012a) reported the potential of adenosine triphosphate bioluminescence for the estimation of total bacterial load on a mango surface.

5.4.4.2 Toxins
Ochratoxin A is a mycotoxin that widely occurs in food as a contaminant. A novel, rapid-detection, plasmonic-based optical biosensor has been developed using the metal-enhanced fluorescence phenomenon for detection of this toxin. Ochratoxin A on dried milk, in juices, and in a wheat mix was detected with this biosensor, which was able to sense the mycotoxin at concentrations lower than the EU specification of 0.5 μg/kg (Todescato et al., 2014).

5.4.4.3 Pesticides
Marcos et al. (2014) recently developed a mathematical pesticide model to design an optical sensor that uses horseradish peroxidase immobilized in a polyacrylamide gel for the continuous determination of tetrachlorvinphos pesticide. The sensor responded linearly to tetrachlorvinphos concentrations above the range of 4.0×10^{-7} to 4.0×10^{-6} mol/L.

5.4.4.4 Metals
A recent study presented a gold nanoparticle–based optical microfluidic system for the determination of Hg(II) ions. The system consisted of a newly synthesized thiourea-based ionophore attached to the gold nanoparticles, which led to a change on the gold surface plasmon resonance band. The sensor recognized the heavy metals specifically with a lower detection limit of 11 ppb (Pedro et al., 2014).

5.4.4.5 Sugar
A study reported an optical sensor based on the mechanism of phosphorescence for monitoring the concentration of glucose in beverage samples and human serum (Ho et al., 2014). The sensing material was coordination polymers having crystalline iridium(III) doped in a sol–gel matrix and built up under glucose oxidase encapsulated in hydrogel and then immobilized on an egg membrane. Upon exposure of the glucose solution, the oxygen content depleted with increase in the phosphorescence of the coordination polymers in a good linear relationship from 0.05 to 5.0 mM, with a detection limit of 0.01 mM. Extensive research and development are required to bring the optical biosensor into practical use for the detection of harmful adulterants, contaminants, microbes, and toxins.

REFERENCES

Agostino, G.D., Alberti, G., Biesuz, R., Pesavento, M., 2006. Potentiometric sensor for atrazine based on a molecular imprinted membrane. Biosens. Bioelectron. 22, 145–152.

Ahmad, L.M., Towe, B., Wolf, A., Mertens, F., Lerchner, J., 2010. Binding event measurement using a chip calorimeter coupled to magnetic beads. Sens. Actuators, B 145, 239–245.

Ajay, A.K., Srivastava, D.N., 2007. Microtubular conductometric biosensor for ethanol detection. Biosens. Bioelectron. 23, 281–284.

Aoki, K., Uchida, H., Katsube, T., Ishimaru, Y., Iida, T., 2002. Integration of bienzymatic disaccharide sensors for simultaneous determination of disaccharides by means of light addressable potentiometric sensor. Anal. Chim. Acta 471, 3–12.

Apetrei, I.M., Apetrei, C., 2013. Amperometric biosensor based on polypyrrole and tyrosinase for the detection of tyramine in food samples. Sens. Actuators, B 178, 40–46.

Baeumner, A.J., Leonard, B., McElwee, J., Montagna, R.A., 2004. A rapid biosensor for viable *B. anthracis* spores. Anal. Bioanal. Chem. 380, 15–23.

Barsoukov, E., Macdonald, J., 2005. Impedance Spectroscopy: Theory, Experiment, and Applications. Wiley, Hoboken.

Basozabal, I., Gómez-Caballero, A., Unceta, N., Goicolea, M.A., Barrio, R.J., 2011. Voltammetric sensors with chiral recognition capability: the use of a chiral inducing agent in polyaniline electrochemical synthesis for the specific recognition of the enantiomers of the pesticide dinoseb. Electrochim. Acta 58, 729–735.

Bello, A., Bianchi, F., Careri, M., Giannetto, M., Mastria, V., Mori, G., Musci, M., 2007. Potentialities of a modified QCM sensor for the detection of analytes interacting via H-bonding and application to the determination of ethanol in bread. Sens. Actuators, B 125, 321–325.

Bhalla, V., Sharma, P., Pandey, S.K., Suri, C.R., 2012. Impedimetric label-free immunodetection of phenylurea class of herbicides. Sens. Actuators, B 171–172, 1231–1237.

Bhand, S.G., Soundararajan, S., Surugiu-Wärnmark, I., Milea, J.S., Dey, E.S., Yakovleva, M., Danielsson, B., 2010. Fructose-selective calorimetric biosensor in flow injection analysis. Anal. Chim. Acta 668, 13–18.

Bhunia, A.K., 2008. Biosensors and bio-based methods for the separation and detection of foodborne pathogens. Adv. Food Nutr. Res. 54, 1–44.

Boujday, S., Nasri, S., Salmain, M., Pradier, C.M., 2010. Surface IR immunosensors for label-free detection of benzo [a] pyrene. Biosens. Bioelectron. 26, 1750–1754.

Bozgeyik, I., Senel, M., Cevik, E., Abasıyanık, M.F., 2011. A novel thin film amperometric urea biosensor based on urease-immobilized on poly(N-glycidylpyrrole-co-pyrrole). Curr. Appl. Phys. 11, 1083–1088.

Broekaert, W.F., Courtin, C.M., Verbeke, K., Van de Wiele, T., Verstraete, W., Delcour, J.A., 2011. Prebiotic and other health-related effects of cereal-derived arabinoxylans, arabinoxylan-oligosaccharides, and xylooligosaccharides Crit. Rev. Food Sci. Nutr. 51, 178–194.

Cao, Z., Guo, J., Fan, X., Xu, J., Fan, Z., Du, B., 2011. Detection of heavy metal ions in aqueous solution by P(MBTVBC-co-VIM)-coated QCM sensor. Sens. Actuators, B 157, 34–41.

Chałupniak, A., Waszczuk, K., Hałubek-Głuchowska, K., Piasecki, T., Gotszalk, T., Rybka, J., 2014. Application of quartz tuning forks for detection of endotoxins and Gram-negative bacterial cells by monitoring of Limulus Amebocyte Lysate coagulation. Biosens. Bioelectron 58, 132–137.

Daly, C.J., McGrath, J.C., 2003. Fluorescent ligands, antibodies, and proteins for the study of receptors. Pharmacol. Ther. 100, 101–118.

de Viedma, D.G., 2003. Rapid detection of resistance in *Mycobacterium tuberculosis*: a review discussing molecular approaches. Clin. Microbiol. Infect. 9, 349–359.

Encarnacao, J.M., Rosa, L., Rodrigues, R., Pedro, L., da Silva, F.A., Goncalves, J., Ferreira, G.N.M., 2007. Piezoelectric biosensors for biorecognition analysis: application to the kinetic study of HIV-1 Vif protein binding to recombinant antibodies. J. Biotechnol. 132, 142–148.

Ercole, C., Gallo, M.D., Mosiello, L., Baccella, S., Lepidi, A., 2003. *Escherichia coli* detection in vegetable food by a potentiometric biosensor. Sens. Actuators, B 91, 163–168.

Escuderos, M.E., Sanchez, S., Jimenez, A., 2011. Quartz Crystal Microbalance (QCM) sensor arrays selection for olive oil sensory evaluation. Food Chem. 124, 857–862.

Gammoudi, I., Blanc, L., Morote, F., Grauby-Heywang, C., Boissiere, C., Kalfat, R., Rebiere, D., Cohen-Bouhacin, T., Dejous, C., 2014. High sensitive mesoporous TiO_2-coated love wave device for heavy metal detection. Biosens. Bioelectron. 57, 162–170.

Gfeller, K.Y., Nugaeva, N., Hegner, M., 2005. Micromechanical oscillators as rapid biosensor for the detection of active growth of *Escherichia coli*. Biosens. Bioelectron. 21, 528−533.

Grate, J.W., Frye, G.C., 1996. Acoustic wave sensors. In: Baltes, H., Göpel, W., Hesse, J. (Eds.), Sensors Update, vol. 2. Wiley-VCH, Weinheim.

Grieshaber, D., MacKenzie, R., Voros, J., Reimhult, E., 2008. Electrochemical biosensors − sensor principles and architectures. Sensors 8, 1400−1458.

Guo, L., Zhang, Q., Huang, Y., Han, Q., Wang, Y., Fu, Y., 2013. The application of thionine−graphene nano-composite in chiral sensing for Tryptophan enantiomers. Bioelectrochemistry 94, 87−93.

Hnaiein, M., Hassen, W.M., Abdelghani, A., Fournier-Wirth, C., Coste, J., Bessueille, F., Leonard, D., Jaffrezic-Renault, N., 2008. A conductometric immunosensor based on functionalized magnetite nano-particles for *E. coli* detection. Electrochem. Commun. 10, 1152−1154.

Ho, M.L., Wang, J.C., Wang, T.Y., Lin, C.Y., Zhu, J.F., Chen, Y.A., Chen, T.C., 2014. The construction of glucose biosensor based on crystalline iridium(III)-containing coordination polymers with fiber-optic detection. Sens. Actuators, B 190, 479−485.

Jia, K., Adam, P.M., Ionescu, R.E., 2013. Sequential acoustic detection of atrazine herbicide and carbofuran insecticide using a single micro-structured gold quartz crystal microbalance. Sens. Actuators, B 188, 400−404.

Kaniewska, M., Sikora, T., Kataky, R., Trojanowicz, M., 2008. Enantioselectivity of potentiometric sensors with application of different mechanisms of chiral discrimination. J. Biochem. Biophys. Methods 70, 1261−1267.

Keow, C.M., Bakar, F.A., Salleh, A.B., Heng, L.Y., Wagiran, R., Bean, L.S., 2007. An amperometric biosensor for the rapid assessment of histamine level in tiger prawn (*Penaeus monodon*) spoilage. Food Chem. 105, 1636−1641.

Kim, N., Park, I.S., 2003. Application of a flow-type antibody sensor to the detection of *Escherichia coli* in various foods. Biosens. Bioelectron. 18, 1101−1107.

Kirchner, P., Li, B., Spelthahn, H., Henkel, H., Schneider, A., Friedrich, P., Kolstad, J., Keusgen, M., Schöning, M.J., 2011. Thin-film calorimetric H_2O_2 gas sensor for the validation of germicidal effectivity in aseptic filling processes. Sens. Actuators, B 154, 257−263.

Kirchner, P., Oberländer, J., Friedrich, P., Berger, J., Rysstad, G., Keusgen, M., Schöning, M.J., 2012. Realisation of a calorimetric gas sensor on polyimide foil for applications in aseptic food industry. Sens. Actuators, B 170, 60−66.

Lawal, A.T., Adeloju, S.B., 2013. Polypyrrole based amperometric and potentiometric phosphate biosensors: a comparative study B. Biosens. Bioelectron. 40, 377−384.

Li, L., Liang, B., Shi, J., Li, F., Mascini, M., Liu, A., 2012. A selective and sensitive d-xylose electrochemical biosensor based on xylose dehydrogenase displayed on the surface of bacteria and multi-walled carbon nanotubes modified electrode. Biosens. Bioelectron. 33, 100−105.

Li, S., Li, Y., Chen, H., Horikawa, S., Shen, W., Simonian, A., Chin, B.A., 2010. Direct detection of Salmonella typhimurium on fresh produce using phage-based magnetoelastic biosensors. Biosens. Bioelectron. 26, 1313−1319.

Liang, R., Zhang, R., Qin, W., 2009. Potentiometric sensor based on molecularly imprinted polymer for deter-mination of melamine in milk. Sens. Actuators, B 141, 544−550.

Liu, J., Sun, S., Liu, C., Wei, S., 2011. An amperometric glucose biosensor based on a screen-printed electrode and Os-complex mediator for flow injection analysis. Measurement 44, 1878−1883.

Ma, J., Qi, W.T., Yang, L.N., Yu, W.T., Xie, Y.B., Wang, W., Ma, X.J., Xu, F., Sun, L.X., 2007. Microcalorimetric study on the growth and metabolism of microencapsulated microbial cell culture. Journal of Microbiological Methods 68, 172−177.

Majumdar, T., Chakraborty, R., Raychaudhurin, U., 2013. Development of PEI-GA modified antibody based sensor for the detection of *S. aureus* in food samples. Food Biosci. 4, 38−45.

Marcos, S.D., Callizo, E., Mateos, E., Galbán, J., 2014. An optical sensor for pesticide determination based on the autoindicating optical properties of peroxidise. Talanta 122, 251−256.

Maskow, T., Wolf, K., Kunze, W., Enders, S., Harms, H., 2013. Rapid analysis of bacterial contamination of tap water using isothermal calorimetry. Thermochim. Acta 543, 273–280.

Mensah-Brown, A.K., Wenzel, M.J., Bender, F., Josse, F., 2014. Analysis of the absorption kinetics for the detection of parathion using hybrid organic/inorganic coating on SH-SAW devices in aqueous solution. Sens. Actuators, B 196, 504–510.

Michino, H., Araki, K., Minami, S., Takaya, S., Sakai, N., Miyazaki, M., Ono, A., Yanagawa, H., 1999. Massive outbreak of *Escherichia coli* O157: H7 infection in school children in Sakai City, Japan, associated with consumption of white radish sprouts. Am. J. Epidemiol 150 (8), 787–796.

Miodek, A., Sauriat-Dorizon, H., Chevalier, C., Delmas, B., Vidic, J., Korri-Youssoufi, H., 2014. Direct electrochemical detection of PB1-F2 protein of influenza A virus in infected cells. Biosens. Bioelectron. 59, 6–13.

Mishra, G.K., Mishra, R.K., Bhand, S., 2010. Flow injection analysis biosensor for urea analysis in adulterated milk using enzyme thermistor. Biosens. Bioelectron. 26, 1560–1564.

Mol, N.J.D., Fischer, M.J.E., 2010. Surface plasmon resonance: a general introduction. In: Mol, N.J.D., Fischer, M.J.E. (Eds.), Surface Plasmon Resonance Methods and Protocols. Springer, New York, pp. 1–14.

Monk, D., Walt, D.R., 2004. Optical fiber-based biosensors. Anal Bioanal. Chem. 379, 931–945.

Narsaiah, K., Jha, S.N., Jaiswal, P., Singh, A.K., Gupta, M., Bharadwaj, R., 2012a. Estimation of total bacteria on mango surface by using ATP bioluminescence. Sci. Hortic. 146, 159–163.

Narsaiah, K., Jha, S.N., Bharadwaj, R., Sharma, R., Kumar, R., 2012b. Optical biosensors for food quality and safety assurance – a review. J. Food Sci. Technol. 49 (4), 383–406.

Ohk, S.H., Bhunia, A.K., 2013. Multiplex fiber optic biosensor for detection of *Listeria monocytogenes, Escherichia coli* O157:H7 and *Salmonella enterica* from ready-to-eat meat samples. Food Microbiol. 33, 166–171.

Palla-Papavlu, A., Patrascioiu, A., Pietrantonio, F.D., Fernandez-Pradas, J.M., Cannata, D., Benetti, M., D'Auria, S., Verona, E., Serra, P., 2014. Preparation of surface acoustic wave odor sensors by laser-induced forward transfer. Sens. Actuators, B 192, 369–377.

Paufler, S., Weichler, M.T., Harms, H., Maskow, T., 2013. Simple improvement of the sensitivity of a heat flux reaction calorimeter to monitor bioprocesses with weak heat production. Thermochim. Acta 569, 71–77.

Pedro, S.G.D., Lopes, D., Miltsov, S., Izquierdo, D., Alonso-Chamarro, J., Puyol, M., 2014. Optical microfluidic system based on ionophore modified gold nanoparticles for the continuous monitoring of mercuric ion. Sens. Actuators, B 194, 19–26.

Perumal, V., Hashim, U., 2014. Advances in biosensors: principle, architecture and applications. J. Appl. Biomed. 12, 1–15.

Pimenta-Martins, M.G.R., Furtado, R.F., Heneine, L.G.D., Dias, R.S., Borges, M.D.F., Alves, C.R., 2012. Development of an amperometric immunosensor for detection of staphylococcal enterotoxin type A in cheese. J. Microbiol. Methods 91, 138–143.

Richard, J.L., 2007. Some major mycotoxins and their mycotoxicoses—an overview. Int. J. Food Microbiol. 119, 3–10.

Samphao, A., Suebsanoh, P., Wongsa1, Y., Pekec, B., Jitchareon1, J., Kalcher, K., 2013. Alkaline phosphatase inhibition-based amperometric biosensor for the detection of carbofuran. Int. J. Electrochem. Sci. 8, 3254–3264.

Sankaran, S., Panigrahi, S., Mallik, S., 2011. Olfactory receptor based piezoelectric biosensors for detection of alcohols related to food safety applications. Sens. Actuators, B 155, 8–18.

Satoh, I., 1991. An apoenzyme thermistor microanalysis for Zinc (II) ions with use of an immobilized alkaline phosphatase reactor in a flow system. Biosens. Bioelectron. 6, 375.

Shamsipur, M., Asgari, M., Maragheh, M.G., Moosavi-Movahedi, A.A., 2012. A novel impedimetric nanobiosensor for low level determination of hydrogen peroxide based on biocatalysis of catalase. Bioelectrochemistry 83, 31–37.

Sivaprakasam, S., Mahadevan, S., Sekar, S., 2008. Calorimetric on-line monitoring of proteolytic activity of *P. aeruginosa* cultivated in a bench-scale biocalorimeter. Biochem. Eng. J. 39, 149–156.

Soldatkin, O.O., Burdak, O.S., Sergeyeva, T.A., Arkhypova, V.M., Dzyadevych, S.V., Soldatkin, A.P., 2013. Acetylcholinesterase-based conductometric biosensor for determination of aflatoxin B1. Sens. Actuators, B 188, 999–1003.

Soldatkin, O.O., Kucherenko, I.S., Pyeshkova, V.M., Kukla, A.L., Jaffrezic-Renault, N., Elskaya, A.V., Dzyadevych, S.V., Soldatkin, A.P., 2012. Novel conductometric biosensor based on three-enzyme system for selective determination of heavy metal ions. Bioelectrochemistry 83, 25–30.

Sun, M., Ding, B., Yu, J., Hsieh, Y.L., Sun, G., 2012. Self-assembled monolayer of 3-mercaptopropionic acid on electrospun polystyrene membranes for Cu^{2+} detection. Sens. Actuators, B 161, 322–328.

Tangutooru, S.M., Kopparthy, V.L., Nestorova, G.G., Guilbeau, E.J., 2012. Dynamic thermoelectric glucose sensing with layer-by-layer glucose oxidase immobilization. Sens. Actuators, B 166–167, 637–641.

Tekaya, N., Saiapina, O., Ouada, H.B., Lagarde, F., Ouada, H.B., Jaffrezic-Renault, N., 2013. Ultra-sensitive conductometric detection of heavy metals based on inhibition of alkaline phosphatase activity from *Arthrospira platensis*. Bioelectrochemistry 90, 24–29.

Todescato, F., Antognoli, A., Meneghello, A., Cretaio, E., Signorini, R., Bozio, R., 2014. Sensitive detection of Ochratoxin A in food and drinks using metal-enhanced fluorescence. Biosens. Bioelectron. 57, 125–132.

Velusamy, V., Arshak, K., Korostynska, O., Oliwa, K., Adley, C., 2010. An overview of foodborne pathogen detection: in the perspective of biosensors. Biotechnol. Adv. 28, 232–254.

Verhaegen, K., Baert, K., Simaels, J., Van, D.W., 2000. A high throughput silicon microphysiometer. Sens. Actuators, B 82, 186–190.

Wang, S., Li, L., Jin, H., Yang, T., Bao, W., Huang, S., Wang, J., 2013. Electrochemical detection of hepatitis B and papilloma virus DNAs using SWCNT array coated with gold nanoparticles. Biosens. Bioelectron. 41, 205–210.

Wang, Y., Xu, H., Zhang, J., Li, G., 2008. Electrochemical sensors for clinic analysis. Sensors 8, 2043–2081.

Yang, Y., Wang, Z., Yang, M., Guo, M., Wu, Z., Shen, G., Yu, R., 2006. Inhibitive determination of mercury ion using a renewable urea biosensor based on self-assembled gold nanoparticles. Sens. Actuators, B 114, 1–8.

Yu, H.W., Wang, Y.S., Li, Y., Shen, G.L., Wu, H.L., Yu, R.Q., 2011. One step highly sensitive piezoelectric agglutination method for cholera toxin detection using GM1 incorporated liposome. Procedia Environ. Sci. 8, 248–256.

Zachetti, V.G.L., Granero, A.M., Robledo, S.N., Zon, M.A., Fernandez, H., 2013. Development of an amperometric biosensor based on peroxidases to quantify citrinin in rice samples. Bioelectrochemistry 91, 37–43.

Zhang, Y., Tadigadapa, S., 2004. Calorimetric biosensors with integrated microfluidic channels. Biosens. Bioelectron. 19, 1733–1743.

Zhang, Z., Zhang, S., Zhang, X., 2005. Recent developments and applications of chemiluminescence sensors. Anal. Chim. Acta 541, 37–46.

Zheng, Y.H., Hua, T.C., Sun, D.W., Xiao, J.J., Xu, F., Wang, F.F., 2006. Detection of dichlorvos residue by flow injection calorimetric biosensor based on immobilized chicken liver esterase. J. Food Eng. 74, 24–29.

Zhou, S., Zhao, Y., Mecklenburg, M., Yang, D., Xie, B., 2013. A novel thermometric biosensor for fast surveillance of β-lactamase activity in milk. Biosens. Bioelectron. 49, 99–104.

Zimmermann, C., Rebiere, D., Dejous, C., Pistre, J., Chastaing, E., Planade, R., 2001. A love-wave gas sensor coated with functionalized polysiloxane for sensing organophosphorus compounds. Sens. Actuators, B 76, 86–94.

SPECTROSCOPY AND CHEMOMETRICS

6

Spectroscopy together with chemometrics is a powerful tool for detecting and quantitatively predicting the chemical composition of a material. Studying properties of matter through their interaction with different-frequency components of the electromagnetic spectrum is called "spectroscopy." In Latin *spectron* means "ghost" or "spirit," and in Greek, σκοπειν means "to see." With light, you are not looking directly at the molecule—the matter—but its "ghost." You observe the light's interaction with different degrees of freedom of the molecule. Each type of spectroscopy with a different light frequency gives a different picture, which is called a spectrum. Spectroscopy is a general methodology that can be adapted in many ways to extract the needed information (energies of electronic, vibrational, rotational states; structure and symmetry of molecules; dynamic information). Extraction of information is possible using different techniques, and one of them is chemometrics, which means performing calculations using measurements of chemical data. This can be anything from calculating pH from a measurement of hydrogen ion activity to computing a Fourier transform (FT) interpolation of a spectrum. Chemometrics, therefore, is the science of extracting information from chemical systems by data analyses using various multivariate statistical techniques. This chapter deals with different types of infrared (IR) spectroscopy, data analysis techniques, and their practical applications.

6.1 ULTRAVIOLET AND VISUAL SPECTROSCOPY

The ultraviolet (UV) (about 200–400 nm) and visible (about 400–700 nm) ranges of electromagnetic spectrum (Figures 6.1 and 6.2) are of shorter wavelengths with higher-energy radiation that causes many organic molecules to undergo electronic transitions; interaction with infrared light, on the other hand, causes molecules to undergo vibrational transitions. When the energy from UV or visible light is absorbed by a molecule, one of its electrons jumps from a lower-energy to a higher-energy molecular orbital, creating a spectrum of energy absorbed for analysis, and the same can be correlated with food attributes.

Consider an example of simple molecular hydrogen, H_2. The molecular orbital (MO) picture for the hydrogen molecule consists of one bonding σ MO and a higher-energy antibonding σ* MO. When the molecule is in the ground state, both electrons are paired in the lower-energy bonding orbital—this is the highest occupied molecular orbital (HOMO). The antibonding σ* orbital, in turn, is the lowest unoccupied molecular orbital (LUMO), as shown in Figure 6.3. If the molecule is exposed to light of a wavelength with energy equal to ΔE, the HOMO—LUMO energy gap, this wavelength is absorbed and the energy is used to bump one of the electrons from the HOMO to the LUMO—in other words, from the σ to the σ* orbital. This is referred to as *σ–σ* transition*. ΔE for this electronic transition is 258 kcal/mol, corresponding to light with a wavelength of 111 nm. When a double-bonded molecule

Rapid Detection of Food Adulterants and Contaminants. http://dx.doi.org/10.1016/B978-0-12-420084-5.00006-8
Copyright © 2016 Elsevier Inc. All rights reserved.

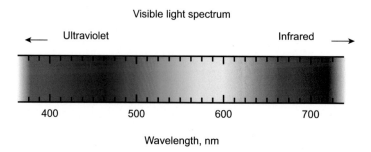

FIGURE 6.1

Color spectrum of electromagnetic waves in the ultraviolet and visible range of wavelength.

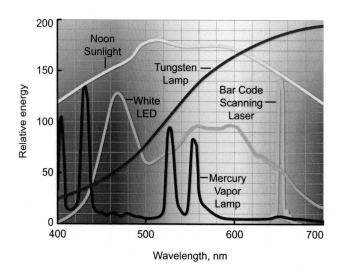

FIGURE 6.2

Visible electromagnetic wave spectra of common light sources. LED, light-emitting diode.

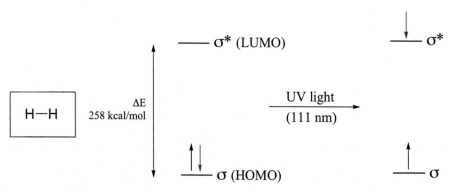

FIGURE 6.3

The molecular orbital (MO) of a hydrogen molecule. HOMO, highest occupied molecular orbital; LUMO, lowest unoccupied molecular orbital; UV, ultraviolet.

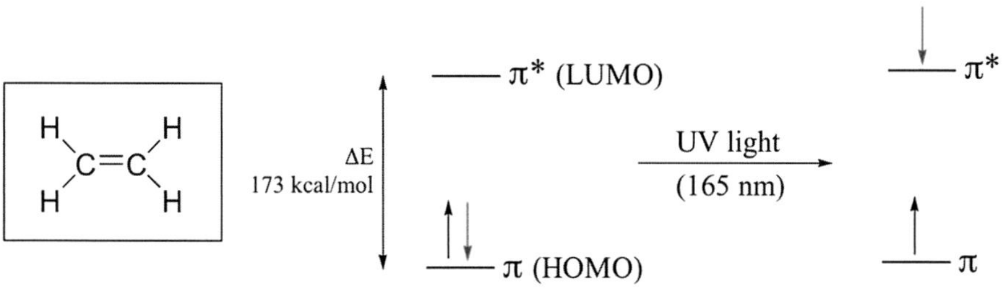

FIGURE 6.4

Depiction of absorption of light by ethylene. HOMO, highest occupied molecular orbital; LUMO, lowest unoccupied molecular orbital; UV, ultraviolet.

such as ethylene absorbs light, it undergoes a $\pi-\pi^*$ transition. Because $\pi-\pi^*$ energy gaps are narrower than $\sigma-\sigma^*$ gaps, ethylene absorbs light at 165 nm, a longer wavelength than molecular hydrogen (Figure 6.4).

The electronic transitions of both molecular hydrogen and ethylene are too energetic to be accurately recorded by standard UV spectrophotometers, which generally have a range of 220–700 nm. UV–visible spectroscopy, therefore, becomes useful to most organic and biological chemists in the study of molecules with conjugated *pi* systems. In these groups, the energy gap for $\pi-\pi^*$ transitions is smaller than for isolated double bonds, and thus the wavelength absorbed is longer. Molecules or parts of molecules that absorb light strongly in the UV–visible region are called *chromophores*.

Let us revisit the MO picture for 1,3-butadiene, the simplest conjugated system. Recall that we can draw a diagram showing the four *pi* MOs that result from combining the four $2p_z$ atomic orbitals (Figure 6.5). The lower two orbitals are bonding, while the upper two are antibonding. Comparing this image of 1,3-bitadiene's MO with that of ethylene, our isolated *pi*-bond example, we see that the HOMO–LUMO energy gap is indeed smaller for the conjugated system. 1,3-Butadiene absorbs UV light with a wavelength of 217 nm. As conjugated *pi* systems become larger, the energy gap for a $\pi-\pi^*$ transition becomes increasingly narrower, and the wavelength of light absorbed correspondingly becomes longer. The absorbance resulting the $\pi-\pi^*$ transition in 1,3,5-hexatriene, for example, occurs at 258 nm, corresponding to a ΔE of 111 kcal/mol (Figure 6.6).

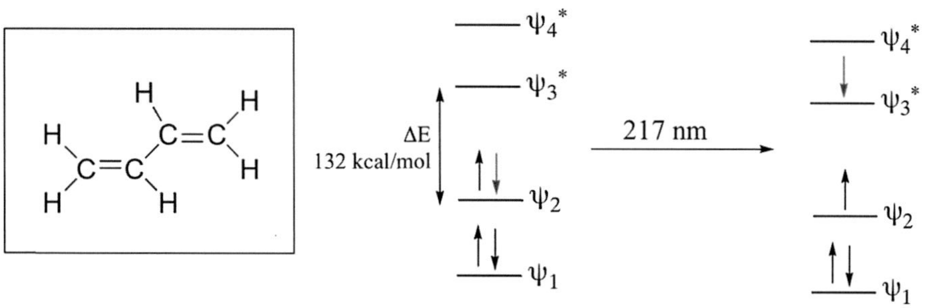

FIGURE 6.5

Molecular orbital for 1,3-butadiene.

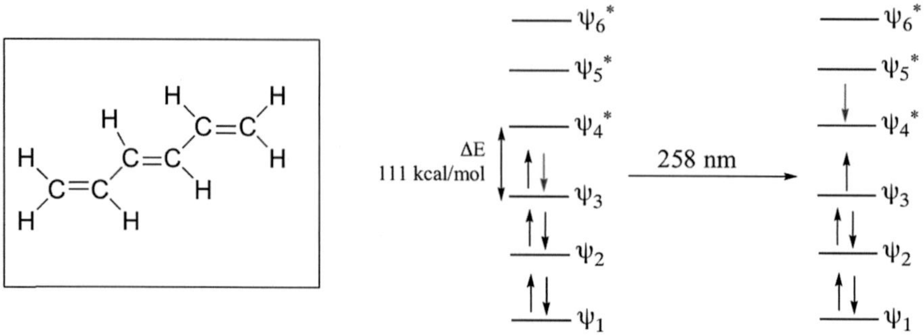

FIGURE 6.6

Absorbance caused by the $\pi-\pi^*$ transition in 1,3,5-hexatriene.

In molecules with extended *pi* systems, the HOMO—LUMO energy gap becomes so small that absorption occurs in the visible rather than the UV region of the electromagnetic spectrum. Beta-carotene, with its system of 11 conjugated double bonds (Figure 6.7), absorbs light with wavelengths in the blue region of the visible spectrum, while allowing other visible wavelengths— mainly those in the red-yellow region—to be transmitted. This is why carrots are orange.

The conjugated *pi* system in 4-methyl-3-penten-2-one gives rise to a strong UV absorbance at 236 nm because of a $\pi-\pi^*$ transition. However, this molecule also absorbs at 314 nm. This second absorbance is the result of the transition of a nonbonding (lone pair) electron on the oxygen up to a π^* antibonding MO (Figure 6.8). This is referred to as an n—π^* *transition*. The nonbonding (n) MOs have higher energy than the highest bonding p orbitals, so the energy gap for an n—π^* transition is smaller than that of a $\pi-\pi^*$ transition; thus the n—π^* peak is at a longer wavelength. In general, n—π^* transitions are weaker (less light absorbed) than $\pi-\pi^*$ transitions.

6.1.1 UV—VISIBLE SPECTRA

We have been discussing in general terms how molecules absorb UV and visible light. Now let's look at some actual examples of data from a UV—visible absorbance spectrophotometer. The basic setup is the same as for IR spectroscopy (IRS), which is described in next section: Radiation with a range of

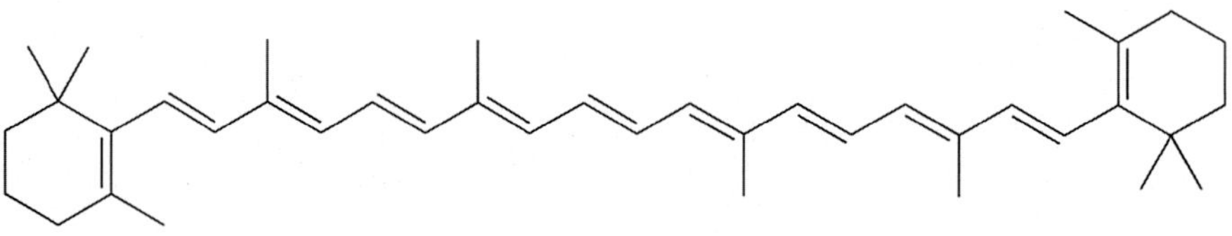

β-carotene

FIGURE 6.7

Beta-carotene, with its system of 11 conjugated double bonds.

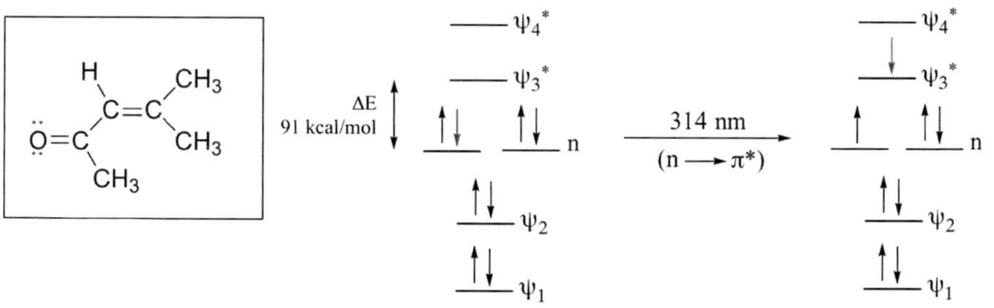

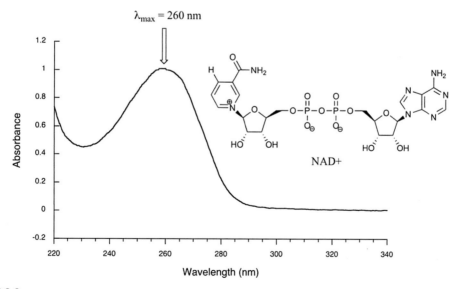

FIGURE 6.8

The conjugated *pi* system in 4-methyl-3-penten-2-one.

wavelengths is directed through a sample of interest, and a detector records which wavelengths were absorbed and to what extent the absorption occurs. Figure 6.9 is the absorbance spectrum of an important biological molecule called nicotinamide adenine dinucleotide, abbreviated NAD^+. This compound absorbs light in the UV range because of the presence of conjugated *pi*-bonding systems.

You will notice that this UV spectrum is much simpler than the IR spectra. This one has only one peak, although many molecules have more than one. Notice also that the convention in UV–visible spectroscopy is to show the baseline at the bottom of the graph, with the peaks pointing up. Wavelength values on the *x*-axis are generally measured in nanometers rather than in centimeters as is the convention in IRS.

Peaks in UV spectra tend to be broad, often spanning well over 20 nm at half-maximal height.

FIGURE 6.9

Absorbance spectrum of the biological molecule nicotinamide adenine dinucleotide.

6.1.2 DATA ACQUISITION AND ANALYSES

Typically, there are two things that we look for and record from a UV—visible spectrum. The first is λ_{max}, which is the wavelength at maximal light absorbance. As you can see in Figure 6.9, NAD^+ has $\lambda_{max} = 260$ nm. We also want to record how much light is absorbed at λ_{max}. Here we use a unitless number called absorbance, abbreviated "A." This contains the same information as the percentage transmittance value used in IRS, just expressed in slightly different terms. We also analyze the entire spectrum to get useful information or correlate biochemical compounds with one or a group of wavelengths for their prediction, which is similar to that of analysis of IR/near-IR (NIR) spectra. Details about different laws governing absorbance, transmittance, and reflectance, and data acquisition systems and their analyses, are discussed in subsequent sections.

6.2 NIR SPECTROSCOPY

Nowadays NIR spectroscopy (NIRS) is common in developed countries for rapid and often nonde-structive measurement of the composition of biological as well as food materials. This method is now no longer new, even in the field of food; it started in early 1970 in Japan, just after some reports from America. Even an official method to determine the protein content of wheat and many others are available. The National Food Research Institute, Tsukuba, Japan, has since become a leading institute in NIRS research in Japan, and the Central Institute of Post-harvest Engineering and Technology, Ludhiana, India, is considered a pioneer of research in the field of quality determination of food using NIRS in India. The major advantages of NIRS are that almost no sample preparation is required; it is often nondestructive, and therefore no wasted materials must to be analyzed; it is very fast and almost as accurate as the wet chemistry method; samples can be measured even in glass containers (test tubes, beakers, etc.); samples with high moisture/water content can easily be measured in NIR systems; it is easy to operate and has online applicability, and therefore has higher throughput, robustness, and portability in field operations.

6.2.1 THEORY OF NIRS

NIRS is a spectroscopic method that uses the near-IR region of the electromagnetic spectrum (from about 700 to 2500 nm). It is based on molecular overtones and combination vibrations. As a result, the molar absorptivity in the NIR region is usually small, but one advantage is that NIR can typically penetrate much deeper into a sample than mid-IR radiation. Therefore NIRS not only is a sensitive technique but also is very useful in probing bulk material with little or no sample preparation. The molecular overtone and combination bands seen in NIR light are typically very broad, leading to complex spectra. It therefore can be difficult to assign specific features to particular chemical com-ponents. Multivariate (multiple wavelengths) calibration techniques (e.g., principal components analysis or partial least squares, which are described later in this chapter) are often used to extract the desired chemical information from absorbance, transmittance, or reflectance data. Careful develop-ment of a set of calibration samples and application of multivariate calibration techniques are essential for spectroscopic analytical methods.

6.2.1.1 *Properties of Electromagnetic Radiation*

IR is a part of electromagnetic radiation, which is considered to be a simple harmonic wave. The electric and magnetic properties of these waves are interconnected and interact with matter to give rise to a spectrum. A simple harmonic motion has the property of a sine wave defined by Eqn (6.1):

$$y = A \sin \theta \tag{6.1}$$

where y is the displacement with a maximum value of A, and θ is an angle varying between zero and 2π radians.

Consider a point P traveling with uniform angular velocity ω/rads along a circular path of radius A (Figure 6.10); P describes an angle $\theta = \omega t$ radians after t seconds of passing Q, and its vertical displacement is written as

$$y = A \sin \omega t \tag{6.2}$$

The right-hand side of Figure 6.10 presents the graphical form of Eqn (6.2). P returns to Q after $2\pi/\omega$ seconds and completes one cycle in 1 s; the same is repeated $\omega/2\pi$ times, which is called the frequency (ν) of the wave. The basic equation for the same is Eqn (6.3):

$$y = A \sin 2\pi \nu t \tag{6.3}$$

The wavelength (λ), the distance traveled in a complete cycle, is another property of the wave. Equation (6.3) must be expressed in terms of variation of displacement with distance instead of with time. This is obtained by substituting $t = l/c$, where l is the distance covered by the wave in time t at velocity c. The velocity c is known as the velocity of light in a vacuum and is the universal constant. The wavelength may therefore be defined as:

$$\lambda = c/\nu \tag{6.4}$$

Wavenumber ($\bar{\nu}$), defined as the reciprocal of wavelength in centimeters, is another way of expressing the character of electromagnetic radiation. The wavenumber, therefore, is considered as the number of waves or cycles per centimeter of radiation. Spectroscopists describe the position of an IR

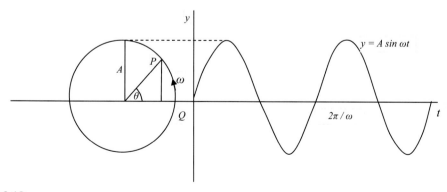

FIGURE 6.10

Sine wave representation of electromagnetic radiation.

absorption band in terms of wavenumber. It is directly proportional to frequency ($v = c\bar{v}$) and is related to the energy changes involved in transitions between different vibrational states.

The wave model is used to explain many properties of electromagnetic radiation, but it fails to describe phenomena associated with the absorption or emission of energy. It is therefore necessary to view electromagnetic radiation as a stream of discrete particles called photons, with an energy proportional to the frequency of the radiation (Eqn (6.14)).

6.2.1.2 Properties of Vibrating Molecules
6.2.1.2.1 Harmonic Oscillators

Mechanical Model: Consider a mass m at one end of a spring that is fixed at the other end for understanding the harmonic oscillation of a vibrating molecule. The force of gravity is constant, and therefore the only influence is the equilibrium point, not the motion of the mass about that point. The disturbance of the mass along the axis of the spring results in motion, which can be described by Hooke's law. This states that the restoring force F exerted by the spring is proportional to distance y that it has traveled from the equilibrium position and can be expressed as Eqn (6.5):

$$F = -ky \tag{6.5}$$

where k is a constant called the force constant. The acceleration a of the mass from equilibrium is

$$a = \frac{d^2y}{dr^2} \tag{6.6}$$

and by applying Newton's second law of motion, $F = ma$,

$$\frac{md^2y}{dr^2} = -ky \tag{6.7}$$

A solution of the differential Eqn (6.7) is

$$y = A \sin \alpha t \tag{6.8}$$

where α is the positive square root of k/m. After one period of motion, y returns to its initial value and the *sine* wave repeats each time; αt is increased by $2\pi v$. From this, we may derive Eqn (6.9):

$$v = \frac{1}{2\pi} \sqrt{\frac{k}{m}} \tag{6.9}$$

Putting $\alpha = \sqrt{k/m} = 2\pi v$ into Eqn (6.8), we get Eqn (6.3). Therefore, the electromagnetic waves and mechanical oscillators, by first approximation, may be described in the same terms. The significance of spectroscopic measurements lies in the association between the frequency of radiant energy and the frequencies of molecular motions. Equation (6.9) can be modified to describe the behavior of a system consisting of two masses m_1 and m_2 connected by a spring by substituting the reduced mass $\mu = (m_1 m_2)/(m_1 + m_2)$ for m:

$$v = \frac{1}{2\pi} \sqrt{\frac{k}{\mu}} \tag{6.10}$$

Any system containing more than two masses follows similar or more complex equations. The vibration of a chemical bond may therefore be considered analogous to the behavior of the spring when

m_1 and m_2 become the masses of two atoms and k is the force constant for the chemical bond. Using this simple mechanical model, it is possible to explain many spectral observations in the IR. For example, a compound containing a carbonyl group C=O has been found experimentally to have an IR band in the region $\bar{v} = 1500$ to 1900 cm^{-1} (Osborne et al., 1983).

Putting $v = c\bar{v}$ and approximate values of 1×10^3 N/m for the force constant of a double bond, 2×10^{-26} and 2.7×10^{-26} kg for the masses of carbon and oxygen, respectively, and 3×10^8 m/s for the velocity of light into Eqn (6.10), we get

$$\bar{v} = \frac{1}{2 \times 3.14 \times 3 \times 10^8} \sqrt{\frac{1 \times 10^3 (2 + 2.7) \times 10^{-26}}{2 \times 2.7 \times 10^{-52}}}$$
$$= 1.565 \times 10^5 \text{ m}^{-1}, \text{ or } 1565 \text{ cm}^{-1}$$

Considering the mass and spring model, it is obvious that the energy of the system undergoes cyclic conversion from potential energy to kinetic energy. The potential energy diagram for a harmonic oscillator is presented as the dotted curve in Figure 6.11. At the equilibrium position, potential energy may be considered to be zero, but as the spring is compressed or stretched by a small amount (dE),

$$dE = -F\mathrm{d}y \tag{6.11}$$

Combining Eqns (6.5) and (6.11), we get:

$$dE = ky\mathrm{d}y \tag{6.12}$$

Integrating from $y = 0$ to y,

$$\int_0^E dE = k \int_0^y y\mathrm{d}y$$

$$E = k\frac{y^2}{2} \tag{6.13}$$

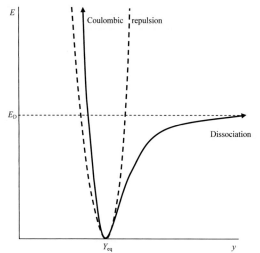

FIGURE 6.11

Energy of a diatomic molecule under anharmonic vibration (solid line) and simple harmonic (dotted line).

The kinetic energy is zero and the total energy is potential energy at the turning point of the motion corresponding to maximum amplitude A. At the equilibrium position, the spring is compressed or stretched, and the energy reverts to kinetic and decreases parabolically to zero.

Quantum Mechanical Model: Max Planck proposed in 1900 that the energy of an oscillator is discontinuous; changes in its content occur only by means of transition between two discrete energy states, and these changes are brought about by the absorption or emission of discrete packets of energy called "quanta." This idea was known as the quantum theory, and the energy levels are identified by integers called "quantum numbers."

When an appropriate amount of energy ($\Delta E = E_2 - E_1$) is either absorbed or emitted by the system, transitions occur between energy levels E_1 and E_2 (Figure 6.12). Planck further proposed that this energy takes the form of electromagnetic radiation, and the frequency of that radiation is related to the energy change ΔE, shown by Eqn (6.14):

$$\Delta E = h\nu \tag{6.14}$$

where h is the universal Planck's constant. Equation (6.14) signifies that if a radiation beam containing a wide range of frequencies is directed onto a molecule in energy state E_1, energy will be absorbed from the beam and a transition to energy state E_2 will occur. A detector placed to collect the radiation after its interaction with the molecule shows that the intensity of the radiation decreases at frequency $\nu = \Delta E/h$, and all other frequencies remain undiminished and produce the absorption (Figure 6.12). In practice, the number of energy levels for a molecule is infinite; there are many possible transitions, and therefore a spectrum even from the simplest molecule would be very complex.

A molecule in space possesses many forms of energy, such as vibrational energy, resulting from the periodic displacement of its atoms from their equilibrium position, and rotational energy, caused by the molecule's rotation about its center of gravity. Absorption of IR radiation is largely confined to molecular species for which energy differences exist between different vibrational and rotational states. The energy required to cause a change in rotational states is, however, much smaller than that required for vibrational states, and rotational absorption bands are absorbed only in the case of gases.

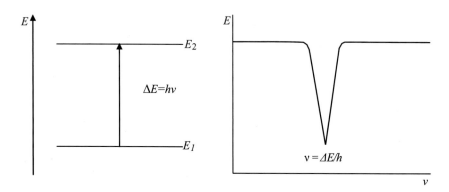

FIGURE 6.12

A typical depiction of the quantum theory of absorption and radiation.

Therefore, for the study of IR spectra of solid and liquid samples, only vibration motion needs to be considered.

Vibrational energies, like other molecular energies, are quantized, and the allowed vibrational energies for any particular system are found by solving a series of partial differential equations known as "quantum mechanical wave equations." Solution of these equations, assuming a simple harmonic oscillator, is found for energy levels, as in Eqn (6.15):

$$E = (v + 0.5)hv \tag{6.15}$$

where v is the vibrational quantum number $(0, 1, 2,...n)$. From Eqn (6.15), it can be seen that the lowest vibrational energy E_0 at $v = 0$ is $hv/2$. Therefore, a molecule can never have zero vibrational energy because an atom can never be completely at rest relative to another atom. E depends only on the strength of the chemical bond and the atoms' atomic masses. Prediction of E is the basic difference between wave mechanical and classical approaches to molecular vibrations. Promotion to the first excited state $(v = 1)$ thus requires absorption of radiation of energy, $(3hv/2) - (hv/2) = hv$; the frequency v of radiation that will bring about this change is identical to the vibration frequency of the bond defined by the Eqn (6.10). Therefore, as Eqn (6.16) we get the following:

$$\Delta E = \frac{h}{2\pi} \sqrt{\frac{k}{\mu}} \tag{6.16}$$

From Eqn (6.15) it can be seen that ΔE (given by Eqn (6.16)) is the energy associated with the transition between any pair of adjacent levels.

Selection rules for spectral bands:

1. According to quantum theory, the allowed vibrational transitions are those in which v changes by one $(\Delta v = \pm 1)$.
2. Spectral bands are observed if the vibration interacts with the radiation. Vibration caused by radiation is therefore electromagnetic in origin, and such interaction depends on the existence of an electric moment across the vibrating bond. Thus, homonuclear diatomic molecules, for example, do not exhibit vibrational absorption bands. It is sufficient, however, for polar bonds to present molecular vibration that induces a temporary dipole moment. To understand the anharmonicity selection rules, Holland et al. (1990) at Reading University (England) used a high-resolution spectrophotometer to study the NIR spectra of small molecules in the vapor phase.

6.2.1.2.2 Anharmonic Oscillators

The observed IR absorption bands caused by the fundamental modes of molecular vibration are explained by quantum-mechanical treatment of a harmonic oscillator, but this does not explain the presence of overtone bands in the NIR spectra. These bands arise from transitions when Δv is ± 2, ± 3, and so on, and therefore they are forbidden by selection rule 1 above. This anomaly is because real molecules do not obey exactly the laws of simple harmonic motion, and real bonds, although elastic, do not obey Hooke's law exactly. As two atoms approach one another, Coulombic repulsion between the two nuclei causes the potential energy to increase more quickly than the harmonic approximation predicts, and when the interatomic distance approaches that at which dissociation occurs, potential energy levels off (Figure 6.11). It may be seen from the dotted curve in Figure 6.11 that the success of the harmonic model stems from the fact that the two curves are almost identical at low potential energies. An empirical form of Morse function fits the solid curve in Figure 6.11 to a good approximation, as in Eqn (6.17):

$$E = E_d\left(1 - e^{-\alpha y}\right)^2 \tag{6.17}$$

where α is a constant for a particular molecule and E_d is dissociation energy. Equation (6.17) is used to solve the wave mechanical equation. The solution for the so-called anharmonic oscillator becomes that in Eqn (6.18):

$$E = (v + 0.5)h\nu - (v + 0.5)^2 h\nu x - (v + 0.5)^3 h\nu x' - \tag{6.18}$$

where x, x' are small and positive anharmonicity constants that decrease in magnitude. For small values of v, the third term and beyond in Eqn (6.18) may be ignored; then we get Eqn (6.19):

$$E \approx (v + 0.5)h\nu - (v + 0.5)^2 h\nu x \approx h\nu[1 - x(v + 0.5)](v + 0.5) \tag{6.19}$$

Equation (6.15) gives the same energy level as Eqn (6.19) if ν is replaced by ν'

$$\nu' = \nu[1 - x(v + 0.5)] \tag{6.20}$$

The anharmonic oscillator thus behaves like a harmonic oscillator, with an oscillation frequency that decreases steadily with increasing v. E_0 is now $(h\nu/2)(1 - 0.5x)$, and the energy associated with a transition from v to $v + \Delta v$ can be expressed as Eqn (6.21):

$$\Delta E = h\nu[1 - (2v + \Delta v + 1)x] \tag{6.21}$$

According to selection rules, $\Delta v = \pm 1, \pm 2, \pm 3, \dots n$. Changes in vibrational quantum number are therefore same as for the harmonic oscillator but with the addition of the possibility of larger jumps. In practice these are of rapidly diminishing probability, and normally only bands resulting from $\Delta v = \pm 1$, ± 2, and ± 3 at the most have observable intensity. Furthermore, according to the Maxwell–Boltzmann law, almost all molecules in a particular sample at room temperature remain at the lowest energy level. According to this law, the proportionality of molecules in an excited state n_1/n_2, where n_1 is the number of molecules in the excited state and n_2 is the number of molecules in the ground state, is in the form of an exponential function, such as Eqn (6.22):

$$\frac{n_1}{n_2} = e^{-\Delta E/kT} \tag{6.22}$$

where k is the Boltzmann constant and T is the absolute temperature. $\Delta E = h\nu \gg kT$ at room temperature.

Based on the above discussion, the three most important transitions in IRS are the following:

1. $v = 0 \rightarrow v = 1$; $\Delta v = +1$
   $\Delta E = h\nu (1 - 2x)$
2. $v = 0 \rightarrow v = 2$; $\Delta v = +2$
   $\Delta E = 2h\nu (1 - 3x)$
3. $v = 0 \rightarrow v = 3$; $\Delta v = +3$
   $\Delta E = 3h\nu (1 - 4x)$

At $x \approx 0.01$ the three bands approximately lie very close to v, $2v$, and $3v$. The line near v is called the fundamental absorption, whereas those near $2v$ and $3v$ are called the first and second overtones, respectively. The highest wavenumber at which absorption of radiation at fundamental vibration frequencies occurs is about 4000 cm^{-1}, and the region between 4000 and 14,300 cm^{-1} is considered the NIR spectrum in which absorption at overtones frequencies occurs. In addition to overtone bands, combination and difference bands are possible theoretically if two or more different vibrations interact to create bands with frequencies that are the sums or differences of multiples of their fundamental frequencies, as in Eqn (6.23):

$$v_{comb} = n_1 v_1 \pm n_2 v_2 \pm n_3 v_3 \pm \cdots \tag{6.23}$$

where n_1, n_2,... are positive integers. Combination bands are of very low probability unless they arise from no more than two vibrations involving either bonds that are connected through a common item or multiple bonds. Difference bands, which are caused by absorption by molecules residing in excited vibrational states, are of very low probability at room temperature as a consequence of Eqn (6.22).

The NIR region of the electromagnetic spectrum is from about 700 to 2500 nm, and the entire IR range can be divided both instrumentally and functionally into near, middle, and far IR (Table 6.1). The far IR region is that in which rotation absorptions occur; it is not discussed further in this chapter.

6.2.1.2.3 Chemical Assignments of NIR Bands

Modes of Vibration: It is necessary to define the number of momentum coordinates required to describe the system for calculating the number of possible modes of vibration for a polyatomic molecule. A space is defined using three coordinates. To define n points in space, $3n$ coordinates are required. Three of these momentum coordinates define the translational motion of the entire molecule, and another three define the rotational motion of the same. This gives $3n$—6 coordinates to describe interatomic vibrations. A linear molecule requires only two coordinates to describe the rotational motion because rotation about the bond axis is not possible. Therefore there are $3n$—5 modes are possible in this case. Each of these possible vibrations is represented by a separate potential energy curve and is subjected to the selection rules described above. For a very simple molecule it is possible to describe the nature as well as the number of the vibrational modes, which are shown in Figure 6.13, with reference to a triatomic molecule or group AX_2.

Vibration is categorized either as stretching or bending. If there is a continuous change in the interatomic distance along the axis of the bond between the two atoms, vibration is called "stretching," which may occur symmetrically in a triatomic group of atoms AX_2, whereas the two A—X bonds vibrate in and out together, or asymmetrically when they vibrate in opposite directions. Vibration

Table 6.1 Approximate Ranges of the Infrared Regions

Region	Characteristic Transitions	Wavelength Range (nm)	Wavenumber (cm^{-1})
Near infrared	Overtone combinations	700–2500	14300–4000
Middle infrared	Fundamental vibrations	$2500–5 \times 10^4$	4000–200
Far infrared	Rotations	$5 \times 10^4–10^6$	200–10

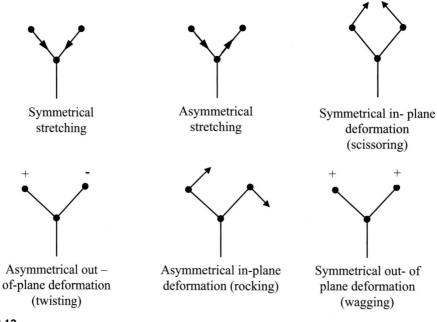

FIGURE 6.13

Modes of vibration of a group AX_2.

involving a change in bond angle is known as bending and is further classified into four types: scissoring, rocking, wagging, and twisting. These are also known as symmetrical in-plane deformation, asymmetrical in-plane deformation, symmetrical out-of-plane deformation, and asymmetrical out-of-plane deformation, respectively (Figure 6.13). Each of these vibration modes gives rise to overtone combinations that are observable in the NIR spectra. The intensity of such bands, however, depends on the degree of anharmonicity. Bonds involving hydrogen, the lightest of atoms, vibrate with a large amplitude when undergoing stretching, and therefore motion deviates appreciably from the harmonic. Consequently, almost all the absorption bands observed arise from overtones of hydrogenic stretching vibrations involving AH_y functional groups, or combinations thereof involving the stretching and bending modes of vibration of such groups in the NIR range.

6.2.1.2.4 Interpretation of NIR Spectra

It is evident from the above discussion that most NIR spectra may be explained by assigning the band of overtones and the combination of fundamental vibrations involving hydrogenic stretching modes. The position of fundamental bonds in the IR spectra is well documented (Bellamy, 1975); as a starting point, it is possible to use them to predict the positions of the corresponding overtone bands.

Carbon-Hydrogen Overtone Bands: The carbon—hydrogen (CH) bond is the most important in organic molecules, and its fundamental stretching bands lie in the region $\bar{v} = 2972-2843$ cm^{-1} for alkanes, 3010—3095 cm^{-1} for alkenes, and, in aromatic systems, 3310—3320 cm^{-1} for alkynes.

The first overtone bands in alkanes would therefore be expected to be found around $2\bar{v}$ cm^{-1} or $10^7/2\bar{v}$ nm $= 1700$ nm. The position may, however, be predicted more accurately by applying Eqn (6.24):

$$\bar{v} = n\bar{v_0}(1 - nx) \tag{6.24}$$

where $\bar{v}$ is the wavenumber of the overtone band, x is the anharmonicity constant (0.01 as a first approximation), and n is an integer that has the same value as Δv (i.e., 2 for the first overtone). Using Eqn (6.24), it is possible to construct a table of predicted wavelengths and compare those with observed values. Tosi and Pinto (1972) studied NIR spectra of 50 alkanes from 1,690 to 1,770 nm and identified five bands at 1693, 1710, 1724, 1757, and 1770 nm, whereas Wheeler (1959) reported bands at 1695 and 1705 nm for the methyl group and 1725 and 1765 nm for the methylene group. Rose (1938) reported methylene bands at 1730 and 1764 nm, and methyl bands at 1703, 1707, and 1724 nm. Many researchers have since identified various bands corresponding to various chemical constituents and combinations thereof.

A similar analysis may be applied to the weaker second overtone band, which is expected to be around 1150 nm. The observed position for absorption by methyl and methylene, respectively, are 1190 and 1210 nm (Liddel and Kasper, 1933; Rose, 1938) or 1195 and 1215 nm (Wheeler, 1959). In polymers, methylene bands were observed at 1214 nm in polyethene and 1206 nm in nylon (Glatt and Ellis, 1951). The third overtone exhibits extremely weak absorption in the region of 880 nm. In the case of chloroform, the relative absorptivities of the first, second, and third CH stretch overtones are a ratio of 484:22:1 (Lauer and Rosenbaum, 1952). The first and second overtones of the aromatic CH group are at 1685 and 1143 nm, respectively (Wheeler, 1959); the second overtone lies in the region of 1136−1149 nm (Liddel and Kasper, 1933); and the first and second overtones at 1695 and 1143 nm, respectively (Rose, 1938). The first overtone bands of terminal methylene lie in the region of 1611−1636 nm (Goddu, 1960), and for terminal methyne as in pent-1-yne they lie at 1533 nm. The presence of polar substituent may considerably change the wavelength at which fundamental CH vibration and overtones occur.

More complex molecules follow the similar CH stretch overtone bands, though they have more complex patterns because of the effect of C−C coupling of the chain units. Polythene exhibits bands at 1730 and 1763 nm because of the CH$_2$ first overtone and 1214 nm because of the second overtone (Glatt and Ellis, 1951). Poly(2-methylpropane) exhibits bands at 1760 and 1210 nm because of methylene and 1690 and 1200 nm because of methyl; polypropene shows bands at 1750, 1700, and 1200 nm; and for polyphenylethene typical aromatic bands occur at 1650 and 1150 nm (Foster et al., 1964). The CH bands in nylon (Glatt and Ellis, 1951) and proteins (Hecht and Wood, 1956) occur in the same region as in similar compounds without the peptide group, for example, 1190, 1700, 1730, and 1760 nm in wheat protein (Law and Tkachuk, 1977). Starch and pentosans have bands at 1200, 1700, 1720, and 1780 nm, whereas wheat lipid has bands at 1170, 1210, 1720, 1760 nm (Law and Tkachuk, 1977). Fatty acids have CH$_2$ first overtone bands at 1740 and 1770 nm and a $=$CH first overtone at 1680 nm (Holman and Edmondson, 1956).

Oxygen−Hydrogen Overtone Bands: The NIR bands assignable to first to third overtones of oxygen−hydrogen (OH) bonds stretch; in the case of water, these are at 1450, 970, and 760 nm (Curcio and Petty, 1951). The position of three bands with all OH bonds is dependent on temperature and hydrogen bonding environment, although the first overtone of OH in water bound to protein was observed at 1450 nm. The OH first overtone bands in alcohol and phenols occur in the

Table 6.2 Bands Caused By O−H Stretch in the First Overtone Band (Trott et al., 1973)			
	Free OH (nm.pl)	**H-bond (Intramolecular, nm.pl)**	**H-bond (Intermolecular, nm.pl)**
Cyclohexanol (DMSO)	1417	1520	1573
α-D-glucose (DMSO)	1443	1484	1581
Glycogen (DMSO)	1445	1481	1581
(solid)	1440	1470	1575
Starch (solid)	1440	1528	1588
	1450	1540	−

DMSO, dimethyl sulfoxide.

region of 1405−1425 nm (Goddu, 1960), with the second overtone between 945 and 985 nm. Iwamoto et al. (1987) postulated that the band in the 1450-nm region is a composite of three bands of water molecules with no hydrogen bond, one hydrogen bond, and two hydrogen bonds at 1412, 1466, and 1510 nm, respectively. These bands shift as a consequence of water-solute interactions and hence may be useful in the study of the state of water in foods.

The effect of hydrogen bonding in carbohydrates and cyclic alcohols was studied, and three bands were observed in spectra of cyclohexanol (Trott et al., 1973), α-D-glucose, and glycogen (Table 6.2). The first band, at about 1440 nm in the carbohydrates, was attributed to the first overtone of the stretching vibration of a free OH group, whereas those about 1490 and 1580 nm were the result of intra- and intermolecular hydrogen-bonded OH groups, respectively. Osborne and Douglas (1981) reported a similar effect in the case of wheat starch, and the first two bands were observed at 1450 and 1540 nm by Law and Tkachuk (1977).

Nitrogen−Hydrogen Overtone Bands: Overtone bands of primary amines are expected to be at about 1500 and 1530 nm because of the nitrogen−hydrogen (NH) stretch first overtone and a band at about 1000 nm resulting from the second overtone. Secondary amines should have a single band at about 1520−1540 nm and 1030 nm, and aromatic primary amines at 1460, 1500, and 1000 nm. No band caused by the amine group would be expected in the case of tertiary amines. Methylamine and dimethylamine have bands at 1520 and 1528 nm, respectively. NH bands caused by amine are not temperature sensitive, though they are displaced by hydrogen bonding. In the spectrum of nylon, for example (Glatt and Ellis, 1951; Foster et al., 1964), the free NH band at 1500 nm is shifted to 1559 nm by hydrogen bonding. The origin of amide absorption bands was proposed by Krikorian and Mahpour (1973). Ethanamide has asymmetrical and symmetrical NH stretch first overtone bands at 1430 and 1490 nm, respectively, whereas N-methyl ethanamide has a single band at 1475 nm. Protein molecules contain peptide (−CONH−) linkage and, in some cases, free amine or amide side groups. Hecht and Wood (1956) reported an NH band at 1550 nm, whereas Law and Tkachuk (1977) observed this band's first overtone at 1500 and 1570 nm due to NH in wheat protein.

Miscellaneous Overtone Bands: A few bands other than those caused by CH, OH, or NH, which may be of importance in the NIR spectra of foods, also exist. The carbonyl group has a strong fundamental band at about 1700 cm^{-1} and would therefore expected to have first to fifth overtones at 2900, 1950, 1450, 1160, and 970 nm. The second overtone has been observed at 1960, 1900−1950,

1920, and 1900 nm in aldehydes and ketones, esters, peptides, and carboxylic acids, respectively (Wheeler, 1959). These bands are too weak and too close to the water band at 1940 nm. Overtones of hydrogenic groups (other than those discussed above) should, in theory, be observable in simple compounds. It is possible, for example, to observe a very weak band at 1892 nm caused by PH in some organophosphorous compounds (Wheeler, 1959). Terminal epoxides have absorption properties very similar to those of terminal alkenes, and their first overtone band may be seen at about 1640−1650 nm. The observed frequency of NIR bands depends on the masses of the vibrating atoms; exchange of hydrogen for deuterium causes a shift, which is a useful tool for the study of NIR spectra. It is possible to assign NIR bands to a combination of frequencies using Eqn (6.23). In practice, however, there are an enormous number of possible combinations that can be accounted for by a very simple compound. Kaye (1954) assigned all 62 possible combination bands in the case of haloforms, which have simple symmetry and only a simple CH group.

CH Combination Band: Bands in stretching and various deformation modes involving a CH group are the most important combination bands, which occur between 2000 and 2500 nm. Much weaker combination bands have also been observed between 1300 and 1450 nm and 1000−1100 nm. Thus, bands observed at 1390, 2290, 2300, 2340, and 2740 nm in wheat gluten, 1370, 2290, 2320, and 490 nm in wheat starch, 1390, 2310, and 2340 nm in wheat lipid (Law and Tkachuk, 1977); 2314, 2354, and 2371 nm in polyethene, 2303, 2349, and 2370 nm in nylon (Glatt and Ellis, 1951), 1400, 2260, 2310, 2350, and 2470 nm in poly(2-methylpropane), and 1400, 2270, 2310, 2400, and 2470 nm in polypropane (Foster et al., 1964) are all CH combination bands. Holman and Edmondson (1956) observed bands at 2140 and 2190 nm in fatty acids due to *cis*-unsaturation. These bands are undoubtedly caused by combination arising from =CH or CH_2 and C=C vibrations. For example, =CH stretching ($3020 \ cm^{-1}$) plus C=C stretching ($1660 \ cm^{-1}$) gives 2137 nm, whereas CH_2 asymmetrical stretching ($2915 \ cm^{-1}$) plus C=C stretching ($1600 \ cm^{-1}$) gives 2186 nm. The CH bond involving the carbonyl carbon atom of an aldehyde has a pair of characteristic fundamental vibration bands at 2820 and $2720 \ cm^{-1}$. A combination of the $2820 \ cm^{-1}$ band with the C=O band at $1735 \ cm^{-1}$ was observed near 2200 nm in simple saturated aldehydes.

OH Combination Bands: Bands at 1940 nm in the spectrum of liquid water (Curcio and Petty, 1951) and water bonded with protein at 1945 nm were observed (Hecht and Wood, 1956). These bands are considered to be the most important absorption bands in NIRS from an analytical point of view. An OH stretching/OH deformation combination band occurs in all hydroxyl compounds, for example, at 2080 nm in ethanol (OH stretching $3500 \ cm^{-1}$ plus OH deformation $1300 \ cm^{-1}$). The OH combination bands shifted, as in the case of overtone bands, with hydrogen bonding combination bands may also be observed between OH stretching and C−O or C−C stretching (Osborne and Douglas, 1981).

NH Combination Bands: Krikorian and Mahpour (1973) reported bands at 1960, 2000, 2050, 2100, and 2150 nm, which they assigned to NH asymmetrical stretching plus amide II, NH symmetrical stretching plus amide II, NH asymmetrical stretching plus amide III, NH symmetrical stretching plus amide III, and twice amide I plus amide III, respectively. Primary amines would be expected to have an NH stretching/NH deformation combination band at about 2000 nm.

Proteins have three prominent bands in the NH combination region, at 1980, 2050, and 2180 nm (Law and Tkachuk, 1977). Bands at 2058 and 2174 nm were noted by Elliott et al. (1954) in the spectra of α-polypeptides. Hecht and Wood (1956) assigned the 2060-nm band to NH stretching H-bonded ($3280 \ cm^{-1}$) plus NH deformed amide II ($1550 \ cm^{-1}$) and the 2180-nm band to twice C=O stretching, amide I (1650 cm^{-1}) plus amide III (1250 cm^{-1}). Three bands of this type were reported for secondary amides: 2000 nm due to NH stretching plus amide II, 2100 nm due to NH stretching plus amide III, and 2160 nm due to twice amide I plus amide III. An NH stretching/NH deformation band at 2050 nm was observed in nylon (Glatt and Ellis, 1951).

Murray (1987, 1988) traced the characteristic absorption patterns for a number of functional groups by studying the NIR spectra of a homologous series of organic compounds. Using this work, wavelengths in the spectra of agricultural products have been assigned, and a concept of food composition in terms of CH, OH, and NH structures was proposed.

6.2.1.2.5 Summary of Chemical Assignments

Chemical assignments for NIR bands available are summarized in Figure 6.14. Though these summaries represent only some of the important NIR bands, the degree of complexity of spectra is obvious. It is of interest to note also that the spectral information is repeated through the successive overtones and combination regions; since the bands involved become weaker by an order of magnitude each time, this represents a useful built-in dilution series. A library of 328 NIR spectra of pure chemicals and agricultural and food products is contained in the monograph by Williams and Norris (1987). The NIRS library "NICODOM NIR Pharmaceuticals" is recent and contains more than 700 NIR spectra (4.200−11.000 cm^{-1}) of active substances and excipients used in pharmaceutical industry. The sources of the samples were local pharmaceutical companies. A basic version of this library is also available in the form of a book, which contains more than 385 NIR spectra.

6.2.2 NIR IMAGING

In addition to chemical assignments of different NIR bands for analysis of foods and other materials, NIR imaging is also being used nowadays for chemical imaging of samples. Chemical imaging is the analytical capability (as quantitative mapping) to create a visual image from the simultaneous measurement of spectra (as quantitative chemical), spatial, and time information. The technique is most often applied to either solid or gel samples, and it has applications in chemistry, biology, medicine, pharmacy, food science, biotechnology, agriculture, and industry. NIR imaging is also referred to as hyperspectral, spectroscopic, spectral, or multispectral imaging. Chemical imaging techniques can be used to analyze samples of all sizes, from a single molecule to the cellular level in biology and medicine, to images of planetary systems in astronomy; however, different instrumentation is used for making observations of such widely different systems, which is beyond the scope of this book. Some work, however, is reported in Chapter 7.

6.2.3 NIRS INSTRUMENTATION

In previous sections we have learned what NIRS is and how it works in determining quality. The user usually faces an increasingly difficult task in determining the most appropriate instrument for their application, be it in the laboratory, in the field, or in online applications.

The generalized instrumentation used in nondestructive spectrometry for measurements of absorption and/or reflectance consists of a light source, a wavelength selector or isolator holder, a

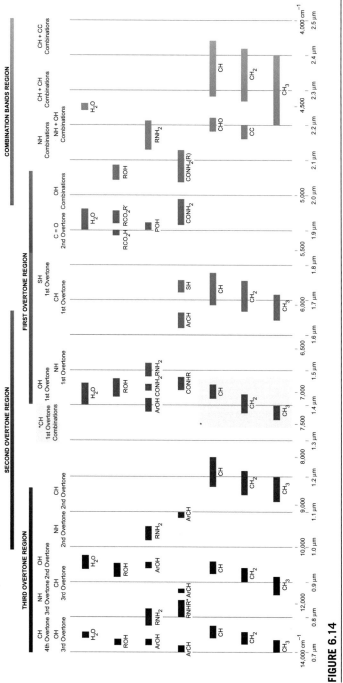

FIGURE 6.14

Near-infrared bands versus spectra structure. (Borrowed from Central Queensland University/Camo India Limited.)

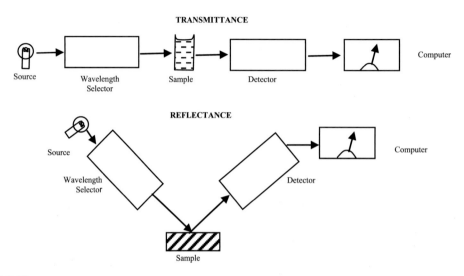

FIGURE 6.15

Schematic instrument layout for transmittance and reflectance measurements.

detector, and a computer (Figure 6.15). Nowadays these components (except the computer, which is not considered a part of spectrometry, though it is an essential requirement) are connected through a fiber optics cable or a single thread. Wavelength isolators and detectors are combined in a spectrometer. Selections of these components of NIRS are of paramount importance and are discussed briefly in the following sections.

6.2.3.1 Light Source

Common incandescent or quartz halogen light bulbs are most often used as broadband sources of NIR radiation for analytical applications. The quartz halogen lamp is the most popular source of NIR energy because of its widespread use for household lighting. In the case of NIRS, the use of the word *energy* is more suitable than *light* because NIR energy is not visible to the human eye and therefore is not light. Light-emitting diodes (LEDs) are the most popular source of NIR energy today. They have a long lifetime, good spectral stability, and low power requirements.

The method by which light is spectrally modulated or selected defines the optical operating principle of an NIR spectrometer. All such instruments are grouped into three categories: dispersive, interferometric, and nonthermal. Devices in the first two groups generally use broadband, thermal radiation produced by an incandescent filament. The third group consists of nonthermal or "cold" sources wherein wavelength selection is inherent in the source's spectrally narrow emission range. The resolving power of an instrument (its ability to distinguish between two close spectral elements) and its transmission (the amount of light that can be admitted) are both fundamental features of the optical operating principle. The resolving power (resolution) R is defined as $\lambda/\delta\lambda$, where $\delta\lambda$ is the resolution (usually expressed in microns). The same expression can, of course, be defined in wavenumber (per centimeters). The solid angle of light (luminosity) admitted by the instrument, Ω, is related to P, such that

$$R\Omega = \text{constant} \qquad (6.25)$$

The resolution luminosity product is constant. This criterion, a figure of merit for spectrometers, has been fully developed by Jaquinot (1958). In a given instrument, resolving power has to be traded for luminosity. The higher the value of $R\Omega$, the better the instrument is.

NIR Spectrometers: NIR spectrometers are classified according to wavelength isolation techniques. Wavelength isolators broadly come under two groups: (1) discrete-value and (2) full-spectrum devices. Full-spectrum spectrometers, also known as "scanning instruments," produce spectra with equally spaced data, that is, data at fixed intervals across the full range, from 700 to 2500 nm. An interval may be any one number such as 1.5, 2, 2.2 nm, and so on. Discrete-value spectrometers can be further categorized by the technology used to produce narrow wavelength bands. Table 6.3 presents major wavelength-isolating technologies used in spectrometry.

6.2.3.2 Diodes

LED Arrays: Several LEDs are commercially available and are making possible the construction of multichannel or multiband NIR spectrometers, which are being used for numerous applications in the food industry. A schematic diagram of an LED array spectrometer is shown in Figure 6.16. Four LED arrays are systematically arranged in a matrix array. As per the requirements, the bandpasses of individual diodes are narrowed by adding narrow-band interference filters. A diffuser is interposed between the sample and the diode matrix to provide uniform illumination from all diodes simultaneously.

Photodiode Detector Arrays: Nowadays, most NIR spectrometers have not a single moving part, which is useful for many applications. Photodiode detector arrays (PDAs) are used in such spectrometers. A basic setup of a PDA spectrometer is shown in Figure 6.17, which is configured with a fiber-optic bundle for conducting diffusely reflected energy from the sample to the fixed grating monochromator. This instrument uses a linear diode array for measuring the reflected energy. These arrays cover the range of 200−1000 nm, but the signal becomes more noisy below 450 nm and above 900 nm.

Greensill and Walsh (2002) developed procedures for standardizing miniature Zeiss MMSI PDA spectrometers, and Clancy (2002) demonstrated a simple linear slope and bias correction for effective normalization of instruments, thus allowing a single master calibration to be used on all instruments. Morimoto (2002) developed two field-operable PDA instruments for determining fruit quality based on PDA technology.

Laser Diodes: The advantages of laser diode spectrometers are that the bandwidths of the lasers are very narrow and the output intensity can be very high compared with that of other spectrometers. A schematic of a laser diode spectrometer is shown in Figure 6.18. The cost of laser diodes has decreased much in the past 40 years.

6.2.3.3 Filters

Fixed Filter: A typical fixed-filter spectrometer (FFS) includes filters that are useful for calibrations in high-volume food applications, such as determining moisture, protein, and fat. The first commercial grain NIR analyzer was made using fixed filters. Today several companies produce multiple fixed-filter instruments using fast FT for noise reduction. A schematic of an FFS is shown in Figure 6.19. Since an FFS is capable of producing only a limited number of bands, it cannot do everything for everyone. Yet, because the NIR absorption of food and food products is broad and overlapping, it is surprising how many things an FFS can do. Morimoto et al. (2001) demonstrated that derivative calibrations can be developed with filters, which further increased the interest in FFS spectrometers.

Table 6.3 Near-Infrared Spectrometer Classification on a Wavelength-Isolating Basis

Wavelength Isolation Method	Status
Diodes	
LEDAs	LEDA is a nonscanning instrument. Using these emitters, making very compact spectrometers for specialized applications requiring only a few wavelengths is possible. Different kinds of LEDs are available in the market.
DAD	DAD spectrometers are full-spectrum instruments. A DAD is used in conjunction with a fixed grating for making compact spectrometers with resolution limited only by the number of receptors in the array.
Laser diodes	Laser-based instruments are wavelength limited and therefore are not full-spectrum instruments. Production of many wavelengths from laser diodes is still limited.
Filters	
Fixed	The bandpass of these filters may be as narrow as 1 nm. Fixing the plane of a narrow-band interference filter normal to an NIR beam provides a single wavelength band. Fixed-filter spectrometers are not capable of producing a full spectrum.
Wedge	A WIF consists of two quartz plates spaced by a dielectric wedge. Moving the wedge in front of a slit produces incremental wavelength bands. Spectrometers that incorporate a WIF are full-spectrum instruments.
Tilting	Tilting results in an increased bandwidth and diminished transmission. Tilting an NBIF in a beam of parallel NIR energy produces incremental wavelength bands. Tilting-filter spectrometers, though limited to a narrow range of wavelengths, are full-spectrum spectrometers.
AOTF	An AOTF is a specialized optic whose bandpass is determined by the radiofrequency applied across the optic. AOTF spectrometers are full-spectrum instruments.
LCTF	LCTF spectrometers may be designed to operate in the visible, NIR, mid-IR, and far IR range. However, the switching speed is much slower than that of an AOTF (maximum of 1 ms).
Prism	Prisms, used in the early days of spectrometry, produce nonlinear dispersion of the NIR spectrum, making it difficult to coordinate prism position with wavelength. Prism-based spectrometers are full-spectrum instruments.
Grating	Grating produces a near-linear dispersion with wavelength. The remaining nonlinearity can be removed with software. Spectrometers incorporating grating are said to be full-spectrum instruments.
FT-NIR	FT-NIR spectrometers produce reflection spectra by moving mirrors. Once plagued by noise, modern FT-NIR spectrometers boast noise levels equivalent to grating-based instruments. FT-NIR spectrometers are full-spectrum instruments.
Hadamard	This initially was made using a complex arrangement of shutters. Hadamard technology has never competed with dispersion-type instruments. This technology is capable of producing a full spectrum.

AOTF, acousto-optical tunable filter; DAD, diode array detector; FT, Fourier transform; LCTF, liquid crystal tunable filter; LED, light-emitting diode; LEDA, light-emitting diode array; NBIF, narrow-band interference filter; NIR, near infrared; WIF, wedge interference filter.

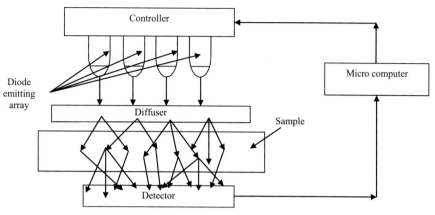

FIGURE 6.16

Schematic of diffuser and narrow-band interference filters with near infrared energy-emitting diodes.

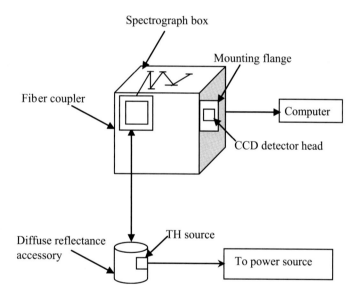

FIGURE 6.17

Schematic of the basic setup of a photodiode detector array spectrometer. CCD, charge-coupled device; TH, tungsten halogen.

Wedge-Interference Filters: A wedge-interference filter (WIF) is similar to an FF with a single difference: the optical dielectric between the plates is wedge-shaped. That is, the dielectric at one end is thicker than that at the other end, producing longer to shorter wavelengths, respectively. A slit between the source and the sample allows a narrow band of wavelengths to pass, with the band changing as the wedge is moved from one end to the other. WIFs are also available in a circular form in

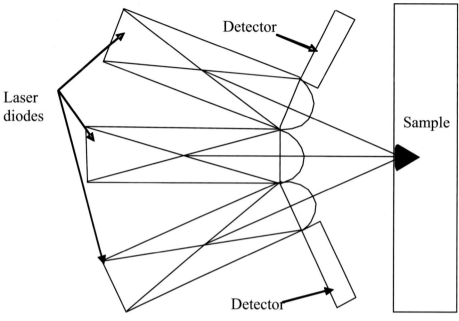

FIGURE 6.18

Schematic of a laser-diode array spectrometer for reflection measurements.

FIGURE 6.19

Schematic of a fixed-filter spectrometer.
NIR, near infrared.

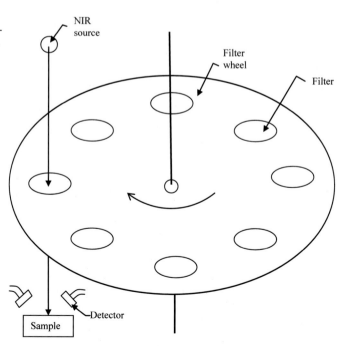

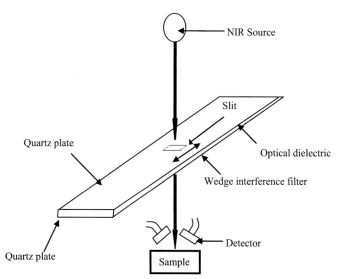

FIGURE 6.20

Schematic of a wedge-interference filter spectrometer. NIR, near infrared.

which the thickness of the optical dielectric varies with filter rotation. Figure 6.20 illustrates construction of a WIF spectrometer.

A recent development is a variable filter array (VFA) spectrometer. It enables the acquisition of NIR spectra of a variety of materials wherever they occur, whether in a production plant or in the field. It consists of an attenuated total reflectance (ATR) sample plate with an elongated, pulseable source mounted close to one end and a linear variable filter attached to a detector array mounted close to the other. Use of this technology has resulted a very compact spectrometer with no moving parts and no optical path exposed to air, and that is able to produce NIR spectra of powders, films, liquids, slurries, semisolids, and surfaces. Sample cups are not required in this type of instrument, and sample loading involves simply loading the ATR with a suitable thickness of material.

Tilting Filter: This filter has definite advantages over wavelength-isolation techniques. First, narrow-band NIR interference filters can be produced for any wavelength in the NIR region. Second, NIR filter characteristics can be reproduced, making it much easier to duplicate spectrometer characteristics. Third, the bandpass of a filter may be increased or decreased (increasing or decreasing the energy falling on the sample, respectively) depending on the application. Spectrometers implementing narrow-band interference filters compete well in online and field (hand-held) applications where the objective is to measure a limited number of parameters.

The following are major disadvantages of tilting-filter spectrometers (Figure 6.21): the relationship between the angle and wavelength is nonlinear, the bandpass of the filter increases as the angle from the normal energy beam increases (clockwise or anticlockwise), and the peak transmission of the filter decreases as the angle of the filter to the source beam increases (clockwise or counterclockwise from normal), in addition to being expensive. The first three disadvantages make it difficult to reproduce specifications from one instrument to another. Because of these disadvantages this kind of spectrometer is now rarely produced.

Acousto-optical Tunable Filter: An acousto-optical tunable filter (AOTF) is a solid-state electronically tunable spectral bandpass filter (Anon, 2010). It operates on the principle of acousto-optic

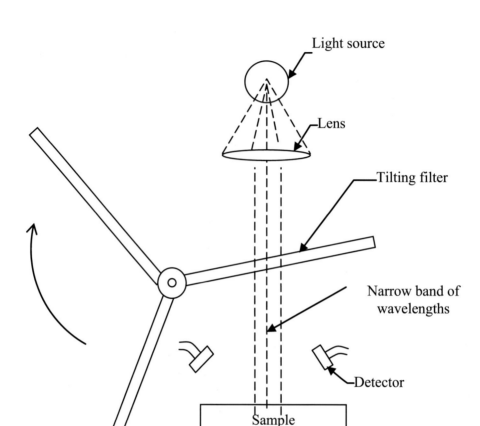

FIGURE 6.21

Schematic of a tilting-filter spectrometer.

interaction in an anisotropic medium. The AOTF has reached technological maturity, moving from the research laboratory to the commercial environment (Figure 6.22). It utilizes an anisotropic, birefringent medium for its operation. It has a relatively long acoustic interaction length, which is required to achieve a narrow spectral bandwidth, which can be achieved only for a certain momentum-matching configuration, that is, the group velocity for the extraordinary wave should be collinear with the ordinary wave. This is shown in the Figure 6.23, where the momentum-matching vectors represent the phase velocities of the incident light (k), diffracted light (k_d), and acoustic waves (k_a). In this geometry, the tangents for the incident and diffracted light beams are parallel to each other. Note that the two optical beams do not separate until the light propagates inside the acousto-optic medium. This long

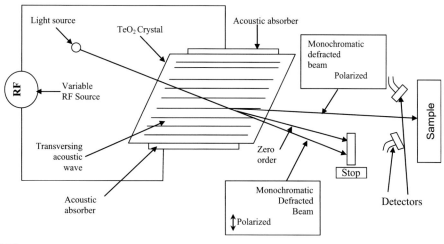

FIGURE 6.22

Schematic of an acousto-optical tunable filter spectrometer's arrangements. RF, radiofrequency.

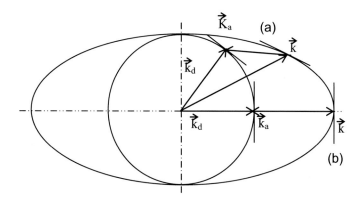

FIGURE 6.23

Phase matching condition for (a) a noncollinear acousto-optical tunable filter (AOTF) and (b) a collinear AOTF. k, wave vector of entrance radiation; k_a, wave vector of acoustic wave; k_d, wave vector of diffracted radiation.

coherent buildup of the diffracted beam is only partially beneficial to the AOTF. To a first order, any momentum mismatch, due to the light arriving at some angle from this ideal condition, is compensated for a change in the birefringence for this orientation, and thus the AOTF displays its most important characteristic: a large angle of view, which is unique among acousto-optical devices. The tuning dependence of an AOTF is calculated using $F = c \times D_n/\lambda$, where F is the frequency of the radio-frequency signal; $D_n = n_e - n_o$, where n_e, n_o are refraction indices for extraordinary and ordinary rays; c is the constant; and λ is the electromagnetic wavelength.

Liquid Crystal Tunable Filters: A liquid crystal tunable filter (LCTF) is a stack of sensitive plates polarizing electromagnetic wave or light. The NIR spectrum is fixed specific regions by the number and thicknesses of plates, usually made of crystal quartz. Switching speeds from one wavelength band to another is dependent on the relaxation time of the crystal and can be as high as 50 ms. Although special crystals with a switching time of 5 ms have been made, this time far exceeds that of grating and AOTF technology and is restricted to a short sequence of wavelength. The spectral resolution is on the order of 10−20 nm, although special crystals can reduce the bandpass to 5−10 nm, which nowadays is considered not good enough. It therefore has limited success in NIRS; however, it can perform considerably well in the visual region. Tilotta et al. (1987) did further development in this field and utilized liquid crystals to modulate radiation in a spectrometer with no moving parts called a Hadamard spectrometer.

Hadamard Spectrometer: The basis of Hadamard-transform NIRS (HT-NIRS) is combination of multiplexing and dispersive spectrometers in which the choice of transparent or opaque elements of a multislit mask provides information that may be transformed into the conventional NIR spectrum using Hadamard mathematics (Hammaker et al., 1986; Tilotta et al., 1987). A tungsten halogen lamp is used as a light source. The energy is first dispersed into spectral elements (wavelengths) and then collected and focused onto a focal plane. Unlike purely dispersive systems, where there is only one exit slit, the focal plane of a Hadamard system implements a multislit array. Signals from this multiple-slit arrangement are collected by a single element detector. HT-NIRS is still not commonly used because of various practical issues.

Prisms: There are three types of prisms: (1) polarizing, (2) dispersing, and (3) reflecting. Polarizing prisms are made of birefringent materials that determine which polarization emerges. Dispersing (or transmission) prisms were most popular in the early development of NIR technology. Reflecting prisms are designed to change the orientation or direction (or both) of an NIR beam. Prism-based spectrometers (Figure 6.24) initially were used to acquire automatically absorption NIR spectra in the range of 210−2700 nm (Kaye et al., 1951) by replacing the photomultiplier tube with a lead sulfide (PbS) cell and including a chopper and an electronic recorder. Absence of a digitizer for the photometric signal was a major drawback of this kind of instrument.

Gratings: The first commercial gratings used were grooved by machine using a diamond ruling tool. Nowadays holographic gratings are usually used in this kind of spectrometer. They are made by a deposing photosensitive material onto a flat glass plate. Lasers are used to etch grooves in the material, and aluminum is vacuum-deposited on the surface of the grooves to make them reflective. Holographic gratings are a little less efficient than the original replica gratings, but the precision of the grooves reduces scattered light. This kind of spectrometer is capable of much higher resolution than prisms, and its use makes it much easier to implement volumes (Figure 6.25).

Fourier Transform-NIR: No dispersion is involved with an FT-NIR spectrometer. These are made using an entirely different method for producing spectra. Energy patterns set up by an interaction with a sample and reference moving mirrors (or other optical components) produce sample and reference interferograms that are decoded using a well-known mathematical technique called Fourier transformation, with the help of a microcomputer in the spectrometer, which produces the desired spectral information to users for interpretation and further manipulations. There are two distinct advantages of FT-NIR spectrometers that make them more attractive. The first is the throughput advantage. In the absence of dispersion, the energy at the output of an FT-NIR interferometer (similar to a monochromator) can be many times greater than that obtained from a grating monochromator.

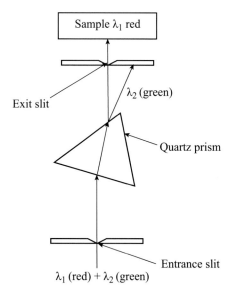

FIGURE 6.24

Schematic arrangement of a simple prism spectrometer.

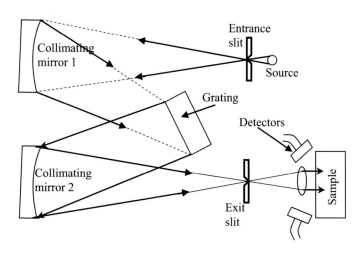

FIGURE 6.25

Schematic of a simple grating spectrometer.

Second, FT-NIR improves spectra reproducibility and wavenumber precision, which may give better accuracy in prediction using chemometrics.

IR Detectors: There are two kinds of infrared detectors: photon detectors, comprising photo-conductors and photodiodes, and thermal detectors. The latter respond to total radiated energy, irre-spective of wavelength, and are not efficient enough for serious use in the NIR region. The response characteristics of some detectors useful in the NIR are shown in Table 6.4 and Figure 6.26.

Responsivity (R), noise equivalent power (NEP), and specific detectivity (D^*) are terms that are commonly used to define detectors. R is expressed as:

$$R = VW^{-1} \tag{6.26}$$

where V is the root mean square (rms) output voltage and W is the rms power input.

The NEP is defined as the power input required to give an output noise voltage equal to the rms noise voltage, whereas the inverse of the NEP is the detectivity (D). When D is expressed in a normalized form for a detector with a 1-cm^2 area and 1-Hz bandwidth, it is called the "specific detectivity" and expressed as D^*. The unit of D^* is cm Hz$^{1/2}$ W^{-1}. Silicon and germanium photodi-odes cover the very NIR part of the spectrum. These cells consist of reverse-based p–n junctions. Sufficiently energetic photons create an electron–hole pair across the boundary, producing a small potential $\approx \mu V$. Silicon detectors respond through the visible part of the spectrum and up to 1 μm, peaking at 0.85 μm. Germanium detectors peak at about 1.5 μm (Figure 6.26).

Compound lead–salt semiconductors are the most widely used detectors in the NIR range and operate in a photoconductive mode. PbS is used over the range of 1–3.6 μm and lead selenide (PbSe) is useful from 1.5 to 5.8 μm (Table 6.4). PbSe is much less sensitive than PbS (Figure 6.26). Both are available individually as squares of material from 1 × 1 mm to 10 × 10 mm and can be operated at room temperature or in a cooled environment. Cooling shifts the sensitivity of the cell to longer wavelengths but increases the response time and signal-to-noise ratio.

The absorption of photons causes the creation of electron–hole pairs across the n–p junctions, and electrons are excited from valence bands to conduction bands. When electrons are in this state, they can conduct electricity, and the resistance of the cell is reduced. The decay time back to the valence bands dictates the response time of the cell, which is typically 100–200 μs for PbS and less than 1 μs for PbSe. Lead–salt detectors are formed by chemically depositing the precipitation of the salt. PbS is an n-type semiconductor, rendered photoconducting by oxygen treatment. Oxide layers some 200 Å thick are grown into the semiamorphous crystals. In effect, a two-dimensional array of n–p–n junctions is produced. Variability of crystal size and p-layer thickness means that the sensitivity and wavelength response of the cell vary over its surface.

Another detector with an exciting feature is epitaxially grown indium gallium arsenide. This detector operates over the range 0.7–1.7 μm, peaking at 1.7 μm. These detectors are available with a 1- to 6-mm diameter and are a few times more sensitive than PbS. Response times are less than 1 μs. Lattice mismatch occurs at indium levels required to give a peak response at longer wavelengths. Although a progressive matching technique can produce such detectors, lattice strain and defects render them noisier, and production costs also are higher. Performance of any detector varies with the environmental conditions in which it is used. Indium gallium arsenide detectors are very sensitive to temperature and humidity.

Fiber Optics: Fiber optics nowadays are common for transmitting energy without much loss. The use of fiber optics as a means of delivery and transfer of NIR energy and information is of paramount

Table 6.4 Type of Infrared Detectors and Their Characteristics

Types		Detectors	Spectral Response (μm)	Operating Temperature (K)	D^* (cm Hz$^{1/2}$/W)
Thermal	Thermocouple, thermopile	—	Depends on window material	300	$D^*(\lambda, 10, 1) = 6 \times 10^8$
	Bolometer	—		300	$D^*(\lambda, 10, 1) = 1 \times 10^8$
	Pneumatic cell	Golay cell, condenser-microphone		300	$D^*(\lambda, 10, 1) = 1 \times 10^9$
	Pyroelectric detector	PZT, TGS, LiTaO^3		300	$D^*(\lambda, 10, 1) = 2 \times 10^8$
Photon detector	Intrinsic Photoconductive	Silicon	0.85–1.0	295	$D^* \approx 4 \times 1012$
		Lead sulfide	1–3.6	300	$D^*(500, 600, 1) = 1 \times 10^9$
		PbSe	1.5–5.8	300	$D^*(500, 600, 1) = 1 \times 10^8$
		InSb	2–6	213	$D^*(500, 1200, 1) = 2 \times 10^9$
		HgCdTe	2–16	77	$D^*(500, 1000, 1) = 2 \times 10^{10}$
	Photovoltaic	Ge	0.8–1.8	300	$D^*(\lambda_p) = 1 \times 10^{11}$
		InGaAs	0.7–1.7	300	$D^*(\lambda_p) = 5 \times 10^{12}$
		Ex. InGaAs	1.2–2.55	253	$D^*(\lambda_p) = 2 \times 10^{11}$
		InAs	1–3.1	77	$D^*(500, 1200, 1) = 1 \times 10^{10}$
		InSb	1–5.5	77	$D^*(500, 1200, 1) = 2 \times 10^{10}$
		HgCdTe	2–16	77	$D^*(500, 1200, 1) = 1 \times 10^{10}$
	Extrinsic	Ge:Au	1–10	77	$D^*(500, 900, 1) = 1 \times 10^{11}$
		Ge:Hg	2–14	4.2	$D^*(500, 900, 1) = 8 \times 10^9$
		Ge:Cu	2–30	4.2	$D^*(500, 900, 1) = 5 \times 10^9$
		Ge:Zn	2–40	4.2	$D^*(500, 900, 1) = 5 \times 10^9$
		Si:Ga	1–17	4.2	$D^*(500, 900, 1) = 5 \times 10^9$
		Si:As	1–23	4.2	$D^*(500, 900, 1) = 5 \times 10^9$

D^* normally is expressed in the format D^*(a, b, c), where a is the temperature (Kelvin) or wavelength (micrometers) of radiant energy, b is the chopping frequency, and c is the bandwidth. Subscript p denotes the peak wavelength.

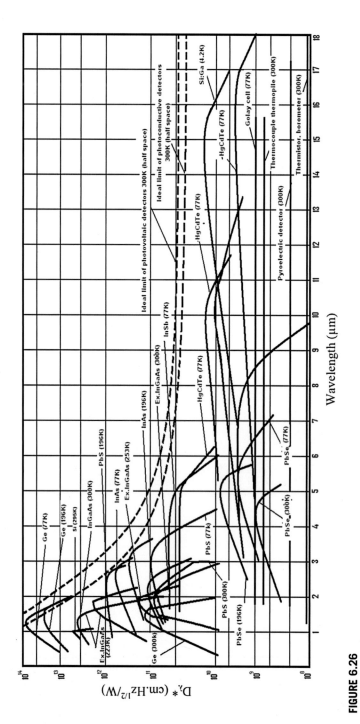

FIGURE 6.26

Approximate response characteristics of different infrared detectors.

importance. There are many situations, particularly on the production line, where the operating environment is unsuitable for sensitive equipment. This may be because of temperature, vibration, space, or explosion hazards. In these situations, the ability to deliver and collect NIR energy via optical fiber is advantageous. The total internal reflection principle governs the science of fiber optics. It covers all aspects of transmission of light energy along optically transparent fibers by means of the phenomenon of total internal reflection. The transparent medium may be silica glass, plastic, or any other similar materials.

Unlike normal reflections from metallic or dielectric surfaces, which may be up to 99% efficient, total internal reflection is highly efficient, typically better than 99.999% at each reflection. Even with several hundred or several thousand reflections per meter, significant amounts of energy can be transmitted over useful distances. The attenuation of fiber is measured in dB/km or dB/m. Figure 6.27 shows the attenuation for the three most suitable fibers for NIR.

Chalcogenide fibers best cover the wavelength region from 4 to 11 μm, but transmission is limited to distances of a few meters. Silica glass is ideal below the 2-μm region. Zirconium fluoride is still very expensive. It offers very low attenuations (<1 dB/km).

6.2.3.3.1 Optical Concepts

The basic structure of a fiber is presented in Figure 6.28. The cladding is a low-index material and the outer is optional and for protective purpose only. If the refractive indices of the core and cladding are n_1 and n_2, where $n_1 > n_2$, an internal ray of light, incident on the boundary at an angle ϕ_1, will be partly reflected at the same angle and partly refracted into the cladding at an angle ϕ_2, according to Snell's law:

$$\frac{\sin \phi_1}{\sin \phi_2} = \frac{n_2}{n_1} \tag{6.27}$$

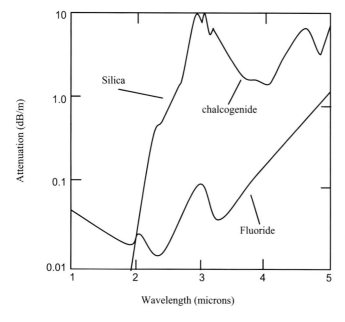

FIGURE 6.27

Approximate attenuation of near infrared transmission in optical fibers.

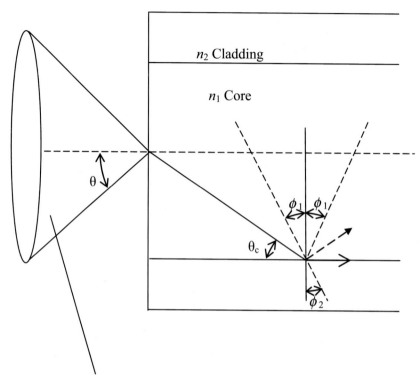

Numerical Aperture defined by acceptance cone ...... half angle θ

FIGURE 6.28

Schematic structure of fiber showing light rays.

When ϕ_1 increases, a point is reached where ϕ_2 becomes $90°$ and no light escapes the core—it is totally internally reflected. The complementary angle at which this occurs is known as the critical angle (θ_{critical}). From Snell's law, this critical angle is defined in terms of the refractive indices as Eqn (6.28):

$$\theta_{\text{critical}} = \cos^{-1} n_2/n_1 \tag{6.28}$$

Rays with angles less than the critical angle are totally internally reflected with high efficiency. A similar arrangement is used to calculate the maximum angle to the normal, at which light is admitted to the fiber and transmitted by total internal reflection. The sine of this is defined as the numerical aperture (NA), as in Eqn (6.29):

$$NA = \sin\theta = \left(n_{\text{core}}^2 - n_{\text{clad}}^2\right)^{1/2/n_{\text{air}}} \tag{6.29}$$

The angle θ defines an acceptance cone for the fiber. The square of the numerical aperture defines the light-gathering capacity of the fiber. There is obviously no point in attempting to feed light onto the fiber at greater angles than this. In theory, light entering a fiber at a given angle should emerge at the other end at the same angle. However, the azimuthal angle varies rapidly with input angle, and rays generally emerge on an annular cone, centered on the input angle. Scattering and inhomogeneities also cause the degradation of images.

6.2.3.3.2 Monofilaments and Bundles

Fibers are available in form of clad, single fibers (monofilament) or as bundles of filaments. Bundles are randomized, branched, or structured as spatially coherent arrays for optical uniformity and mixing, and for multiple input/output applications; or they can be structured as spatially coherent arrays for imaging purposes. Core diameters normally range from 0.02 to 2 mm. Bundles are available with diameters from less than 1−8 mm. Monofilament fibers are less expensive and can be easier to use than bundles, provided that the energy required can be imaged through the end of the fiber. Large-diameter fibers are attractive for these reasons, although the minimum bending radius must be taken into account. For a 1-mm silica glass fiber, this is approximately 150 mm. Repeated bending may cause microfractures (silica is bad for this), resulting in a loss of transmission by scattering into the cladding.

Thermal and mechanical stressors affect transmission characteristics and are disastrous for accurate NIR transmission measurements. For example, a 1-mm monofilament fiber stressed by 10% bending can change its relative transmission characteristics of, say, a 1.8- to 1.9-μm ratio by as much as 0.5%. This problem can be obviated with a double-beam optical design. Fiber bundles are normally made from much smaller core sizes (30−100 μm), giving much better flexibility. The minimum bend is defined by the protective jacket. However, bundles are less efficient than monofilaments because a proportion of the incident light is lost on the cladding and interstices between fibers. Packing fractions of 35−45% are typical, but this means that more than 50% of the available energy is lost every time light is coupled into a bundle.

6.2.3.4 Sample Preparation and Presentation

Two main areas of measurement, the laboratory and the production line, can be subdivided into solids and liquids. The former are more common and the latter usually require less sample preparation.

6.2.3.4.1 Solids Online

1. There may be wide variations in dimensions of some products. Generally, a diffuse reflection sensor tolerates $\pm 25-50$ mm of random product height fluctuation at maximum sensitivity. If the height variation exceeds the observed performance of the sensor, then it is necessary to plane the surface immediately before it passes by the target area.
2. NIR penetration is generally superficial and may range from 10 μm to a few millimeters. The penetration can be much greater at shorter wavelengths (in the region of 0.8−1.1 μm), but currently the very NIR part of the spectrum is not widely used in online sensing. Particularly if moisture is being measured, it is important to ensure that the surface is typical of the bulk material. Surface drying in that case may be a problem. Mixing, turning, or ploughing may solve these problems to some extent for granular or leaf materials.
3. The sensor reading depends, to a small degree, on the mean orientation of the product in relation to the sensor. Generally, leaf or laminar products are more problematic than granular materials and powders, and thus instantaneous sensor readings may be erroneous. So, it is useful if orientation is averaged. As the product passes under the sensor, the readings fluctuate around the correct value for the mean orientation on the production line. These fluctuations can be called

"presentation noise" and must be integrated with an appropriate time constant. Averaging a few second signals may typically be required.

4. Product coming out from driers may contain steam and produce erroneous results, so proper care is needed to avoid such a situation.
5. It is essential that a sample patch should be filled with product and that the sensor not be able to "see" the transport mechanism. If gaps in production flow are unavoidable, gauging biscuits, for example, then it is essential that the sensor has a fast and efficient "gating" facility.
6. An extreme case, as stated above, is that of vacuum-transported powders such as flour. Here, the instantaneous density is too low for a successful NIR diffuse reflection measurement; that type of situation must be avoided.

All of these product presentation problems can be minimized by an intelligent location of the sensor. In practice, excellent results can be obtained.

A limited number of applications are used to present samples to an off-line or laboratory sensor. In the case of measurement of fruits and vegetables in a tree, or manual sorting in a laboratory or in a pack house on a small scale, this arrangement seems to be more appropriate to avoid mechanical complexity of the continuous sorting system and to fit to requirements of the analyzer.

6.2.3.4.2 Liquids in General

The measurement of liquids is generally easier than that of solids. The Beer—Lambert absorption law is obeyed unless the liquid is full of particulate matter or extremely turbid. The width of the cuvette of a flow-through cell can be adjusted to optimize the absorption sensitivity. Finally, it is relatively easy to adapt existing equipment to make the measurement. A simple test tube may also be used in place of a cuvette for holding the liquid sample.

There is a choice between diverting a portion of the product to an off-line instrument or installing a cell or an insertion probe directly in the production flow in online conditions. Tables 6.5 and 6.6 summarize the advantages and disadvantages of working in a laboratory environment and online, respectively. The profiles of some absorption features are significantly temperature dependent. Some form of correction, such as calibrated temperature correction, a chemometrically applied correction, or calculating a ratio against a reference cell, may be required. Two cells with the same working gap are maintained at the same temperature by the process stream. In this way, automatic temperature correction is achieved.

6.2.3.4.3 Cells

Most instrument and accessory manufacturers produce a range of glass or quartz cells. These can be of fixed-gap or adjustable types. The increasing use of fiber optic coupling has led to the development of more

Table 6.5 Advantages and Disadvantages of Off-line Measurements

Advantages	Disadvantages
Preparation and filtering of samples are possible	There may be some difficulty in controlling the flow of samples
Sampling from various points is possible	Condition of the sample may change in transit
An instrument may be used	May not be suitable for the factory environment, and only a limited quantity of product can be measured
Existing laboratory spectrometers can be used	There is the possibility of transit delay
There is the possibility of stabilized temperature and pressure during the process	"Clean in place" systems are required; otherwise there is a possibility of wastage of samples

versatile arrangements of open-path cells and ATR attachments that are equally suitable for immersion into laboratory beakers or into process streams, vats, and reaction vessels via insertion mechanisms. Some of these devices use a reflective surface, and it is useful to note the distinction between direct transmission and retroreflection. Figure 6.29 shows two cells with the same effective absorption path. When liquid in

Table 6.6 Advantages and Disadvantages of Online Measurements	
Advantages	**Disadvantages**
Fast, accurate, and direct measurement	Multiple point measurement is not possible
No wastage of samples	Some form of temperature compensation might be required since control of temperatures and pressures may not be possible
A significant quantity of samples can be tested	Cleaning, maintenance, and standardization of the probe are difficult
Product usually remains undisturbed	Some desirable parts may left out from measurements

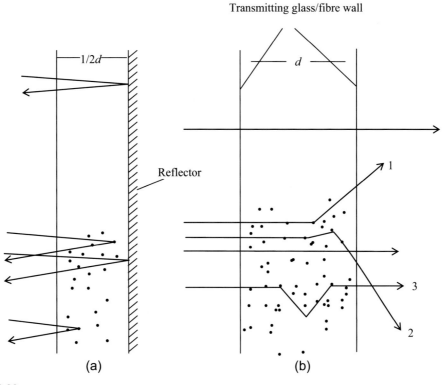

FIGURE 6.29

Scattering of light by suspended particles in a (a) reflecting and (b) transmitting sample holder of diameter *d* (Jha, 2010).

the cell is clear and the cell is clean, there is no significant distinction between the performance of the two arrangements. If the liquid is cloudy, however, the reflection cell will suffer from an apparent gap reduction as a proportion of the light is scattered back from particles or turbidity. The cloudier the contents, the shorter the effective path. A similar problem occurs if the cell wall is dirty. With a true transmission cell this problem does not arise. Provided that the illumination and collection optics use quasi-parallel light, the cell will have a high tolerance to particle scattering. Most of the singly and doubly scattered light will not be collected. Light must be at least triply scattered to be returned to the collection optics after significant deviation, and a high level of scattering can be tolerated before this occurs. For a situation where scattering dominates, a path-to-extension ratio that relates to the effective path length through the cell can be defined. Figure 6.30 shows what is known as a transflectance cell, originally developed by Technicon to fit the sample drawer of an InfraAnalyzer (Fearn, 1982). Rather than using a mirror, the light is scattered back from a ceramic surface to be compatible with the diffuse reflection characteristics of the instrument. It can be noted that the term *transflectance* also has another meaning in NIR applications. Referring to the measurement of scattering films or sheets, it describes the simultaneous collection of backscattered and transmitted energy produced by the instrument. Another simple, less costly holder is made of aluminum, which can hold liquid samples in an ordinary test tube at constant temperature (Jha and Matsuoka, 2004a,b; Jha et al., 2001).

6.2.3.4.4 Attenuated Total Reflectance

The phenomenon of ATR relates to what happens at the boundary between high- and low-index materials. It can be shown from Maxwell's equations (Born and Wolf, 1977) that an electromagnetic disturbance exists in the low-index medium beyond the totally reflecting interface. This energy exhibits the same frequency as the incoming wave, but it is evanescent, and the amplitude of the electric field decreases exponentially away from the boundary:

$$E = E_0 \exp(-d/D) \tag{6.30}$$

E_0 is the field strength at the boundary, D is the length of the attenuation distance or penetration depth, and d is the actual distance. The penetration depth is proportional to the wavelength and also depends on the ratio of the refractive indices and the angle of incidence. The better the index matches, the greater the penetration depth is. Also, the penetration depth is infinite at the critical angle and rapidly

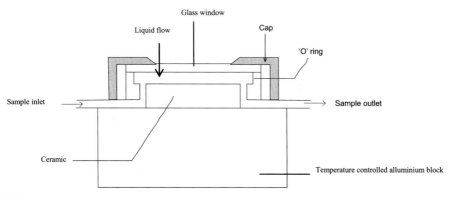

FIGURE 6.30

Flowing liquid samples for acquiring transflectance data (Jha, 2010).

decreases to about $\lambda/10$ at grazing incidence. ATR occurs when an absorbing medium is very close to, or in contact with, the reflecting surface. The internally reflected light is attenuated, the energy being dissipated as a tiny amount of heat. The reflection loss does not follow a simple law, and ATR spectra differ from classical absorption spectra.

Because the extent of an evanescent wave is typically only a few wavelengths, one reflection is equivalent to a very small transmission cell. For ATR to be useful, multiple reflections are required. There are numerous designs of rods and plates developed for specific applications; the basic concepts are shown in Figure 6.31. The single-pass arrangement is suitable for mounting across a pipe or in an insertion probe. The double-pass system is ideal for laboratory applications, where it can be lowered into a test tube or flask. The cylindrical rod is interesting because it corresponds to a portion of an optical fiber with the cladding removed. A simple probe incorporating many reflections can be produced this way. In practical instruments, it is useful to calculate a ratio against a reference plate or crystal or use dimensions similar to those of the absorption crystal. A clear advantage of the ATR technique is the avoidance of using a cell. If the equivalent cell needs to be narrow, for example, significantly less than a millimeter, ATR may be the only solution for viscous materials. A disadvantage is that the ATR crystal is vulnerable to the buildup of deposits on its surface and to abrasive substances. Currently, the ATR approach is more popular in the mid-IR than the NIR range, where absorptions are stronger and the penetration depth is greater.

6.2.3.4.5 Presentation of Solids

Solid materials may be measured in transmission or by diffuse reflectance. Measurement can be made with intact or grinded samples. Nowadays people usually prefer intact samples and nondestructive measurements, such as protein measurement of whole grain. A few hundred grams of grain are run past the sample area through a cell with a gap of typically 20 mm. There are various commercially produced sampling systems for achieving this, all deriving an average over many measurements. For wavelengths in the region of $0.8-1.1$ μm, transmittances can be obtained through a few centimeters of tissue, such as apple or potato, 20 mm of whole grain, 10 mm of homogenized meat or cream, and a few millimeters of powders such as flour (Norris, 1984). In the case of diffuse reflectance measurements, the 1- to 2.5-μm region is typically used. Most instruments use a similar construction. A circular quartz cover some

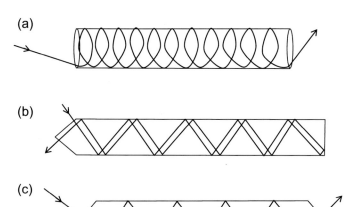

(a)

(b)

(c)

FIGURE 6.31

Configurations of an attenuated total reflectance crystal: (a) optical fiber without cladding, (b) double-pass reflection, (c) single-pass reflection (Jha, 2010).

FIGURE 6.32

The presentation of various types of samples: (a, b) fruits, (c) liquid, (d) grain.

3.5 cm in diameter is mounted in a black plastic cylinder to form a sample holder 1 cm deep. For most substances, this depth ensures negligible transmittance.

In many cases little preparation is required, but it is important that the sample should be thoroughly mixed or homogenized before analysis. Grains and seeds need to be ground before packing into the cell, if nondestructive measurement is not required. Most substances have a significant particle size distribution or are a heterogeneous mixture. It is vital to prevent stratification, which can cause either a partial separation of components in the sample or larger particles to come to the surface.

The sample should be scooped into the holder and then leveled off with minimum disturbance. The lid of the cup, a black plastic cover, fitted with lugs to engage the other half of the cup is then placed in position. A backing plate mounted on a rubber pad is included. This arrangement ensures constant and reproducible packing density. The window of the sample holder is brushed clean before insertion into the measuring instrument. Some materials, such as biscuits, dough, and bread, can be placed directly in the sample cup after trimming to size using a knife. Small samples and liquid samples can be supported on glass or fiber paper, which act as a diffusely scattering matrix. This method has been used by Meurens (1984) for determining sucrose in soft drinks. Nowadays, various grain analyzers using these principles are available. Numerous holders for solid samples, especially for individual fruits, have been developed and are in practical use in laboratories (Kawano et al., 1992; Jha and Matsuoka, 2000; Jha, 2007; Jha and Garg, 2010). Self-explanatory photographs of some of them are presented in Figure 6.32.

Commercial instruments usually offer their own software, but increasingly systems are becoming more open, with control and acquisition software running on any standard computers. Standard file formats allow the transfer of raw data to a range of proprietary software, such as The Unscrambler for chemometric analysis for spectral manipulation. Both microprocessors and their associated software are evolving rapidly. Electronics and information technologies now play an important role in the automatic control and monitoring of instrument and product quality parameters; potential users therefore are advised to consult the manufacturers and suppliers for desirable features in their instruments.

6.3 FOURIER-TRANSFORM IRS

The theory and principles of IRS is almost the same as those of NIRS; however, IR and NIR use different electromagnetic wavelengths. IR and mid-IR cover the spectral range of 2500—25,000 nm, with the fundamental absorption of molecular functional groups or fingerprints. NIR spectra are obtained from the overtone or combination absorption of the molecular functional groups and covers the wavelength range of 700—2500 nm (as explained earlier). Both techniques are nondestructive. NIR is usually more suitable for analyzing the bulk chemistry of a sample (versus contaminants), and it must be used for inhomogeneous samples or samples with various particle sizes. It is also used widely for in-line analysis during production with fiber optic probes. IR is used intensively in the laboratory environment and requires higher technical skill. The calibration development process of NIR requires a larger number of calibration standards than do IR calibrations. Selection of the right technology is dependent on the actual application, expertise on chemometrics and spectroscopy, and/or the compromise of these two technologies. However, FT IRS is now considered to be much better than simple NIRS or IRS. FT IRS is a technique that is used to obtain an IR spectrum of absorption, emission, photoconductivity, or Raman scattering of a solid, liquid, or gas sample.

The common technologies used to produce IR spectra are dispersive and FT. FT spectroscopy is a less intuitive way to get the same information—that is, rather than allowing only one wavelength at a time to pass through the detector, his technique lets through a beam containing many different wavelengths of light at once and measures the *total* beam intensity. Next, the beam is modified to contain a different combination of wavelengths, giving a second data point. This process is repeated many times. Afterward, a computer takes all these data and works backward to infer how much light there is at each wavelength. FT systems are much more advanced and have numerous performance advantages over dispersive instrumentation. FT is a mathematical calculation used to transform a frequency-domain spectrum into an IR spectrum. FT instruments scan the full spectral range faster. The high signal-to-noise ratio and high-resolution measurement of FT generate higher spectral quality and better sensitivity of small peaks (i.e., finer absorbance), which makes the details in a sample spectrum clearer and more distinguishable. More calibrations and lower detection limits can be achieved. An FT instrument uses a laser to calibrate the wavelength and makes the accuracy and precision of IR spectra much higher over other technologies. On the other hand, FT IR technology is rather complicated and requires higher technical skills to develop, manipulate, and optimize calibrations with robust performance; in addition, the costs of the instruments are manifold higher than those of simple NIR spectrometers. Data acquisition, preprocessing, and analysis and chemometrics modeling are almost the same as those of NIRS, as described in the next sections.

6.4 MULTIVARIATE ANALYSIS

NIR spectra contain a great deal of physical and chemical information about molecules of a material. The characteristics of NIR bands can be summarized as follows:

- NIR bands are much weaker and broad than IR bands.
- NIR bands strongly overlap with each other because of their origins in overtones and combinations of fundamentals, and thus yield severe multicollinearity.
- NIR bands are difficult to assign because of overlapping of bands, complicated combinations of vibrational modes, and possible Fermi resonance.
- The NIR region is dominated by bands created by functional groups containing a hydrogen atom (e.g., OH, CH, NH). This is partly because the anharmonic constant of an XH bond is large and partly because the fundamentals of XH stretching vibrations have a high frequency.
- Band shift in the NIR range is much larger for a particular band compared with that in the mid-IR spectrum because of hydrogen bonding.

From the above characteristics, one may consider that NIRS has properties that seem to be disadvantageous because NIR bands are weak and broad and overlap heavily. However, NIR spectra are still very rich in inherent information that can be extracted through rigorous multivariate analyses.

Multivariate analysis in spectroscopy is a process similar to churning cream to get better quality and a larger amount of butter or *ghee* from the same amount of milk. There is no limit on independent and dependent variables. It is like an ocean in which you have to dive to get some useful information as per your needs and level of satisfaction. A large number of variables are considered, and their effects on selected attributes are observed. To simplify the model, independent variables are reduced to a minimum possible number by following certain rules and techniques without scarifying the accuracy of prediction of attributes (Jha, 2010).

NIRS is a field where multivariate calibration, with its ability to embed unknown phenomena (interfering compounds, temperature variations, etc.) in the calibration model, is an efficient tool. Multivariate data analyses are used for a number of distinct, different purposes and have been described in detail in books about chemometrics. The objective here is to give to newcomers brief exposure to three main aspects:

- Data description (explorative data structure modeling)
- Regression and prediction
- Discrimination and classification

6.4.1 DATA DESCRIPTION

A large part of multivariate analysis is concerned with simply "looking" at data, characterizing it by useful summaries, and often displaying the intrinsic data structures visually using suitable graphic plots. As a case in point, the data in question can be stated parameter values monitored in an industrial process at several locations, or measured variables (temperature, refractive indices, reflux times, etc.) from a series of organic syntheses—in general, any p-dimensional characterization of n samples.

The objective of univariate and multivariate data description can be manifold: determination of simple means and standard deviations, as well as correlations and functional regression models. For example, in the case of organic synthesis, one may naturally be interested in seeing which variables most affect the product yield or the selectivity of the yield. The variables from the synthesis could also be used to answer questions such as, how correlated is temperature with yield? Is distillation time of importance for the refraction index? The principal component analysis (PCA) method is frequently used for data description and explorative data structure modeling of any generic (n, p)-dimensional data matrix.

There are different multivariate techniques: PCA, principal component regression (PCR), partial least squares (PLS), and multiple linear regression (MLR). PCA, PCR, and PLS are also known as "bilinear modeling." These methods denote a more geometric and mathematical approach. One may opt, for instance, to start with fundamental mathematics and statistics, which form the basis of all these methods.

6.4.1.1 Principal Component Analysis

PCA involves decomposing one data matrix, X, into a "structure" part and a "noise" part. There is no Y-matrix, no properties, at this stage. Representing the data as a matrix, the starting point is an X-matrix with n objects and p variables, namely an $n \times p$ matrix. This matrix is often called the "data matrix," the "data set," or simply "the data." The objects can be observations, samples, experiments, and so on, whereas the variables typically are "measurements" for each object. The important issue is that the p variables collectively characterize each, and all, of the n objects. The exact configuration of the X-matrix, such as which variables to use for which set of objects, is of course a strongly problem-dependent issue. The main up-front advantage of PCA—for any X-matrix—is that one is free to use practically any number of variables for the multivariable characterization. The purpose of all multivariate data analysis is to decompose the data in order to detect, and model, the "hidden phenomena." The concept of variance is important. It is a fundamental assumption in multivariate data analysis that the underlying "directions with maximum variance" are more or less directly related to these hidden phenomena.

6.4.2 REGRESSION AND PREDICTION

Regression and prediction is an approach for relating two sets of variables to each other. It corresponds to predicting one (or several) Y-variables on the basis of a well-chosen set of relevant X-variables, where X in general must consist of, say, more than three variables. Note that this is often related to indirect observations as discussed earlier. The indirect observation is X, and the property we are really interested in is Y. Prediction means determining Y-values for new X-objects, based on a previously estimated (calibrated) X–Y model, thus relying only on the new X-data. Though various types of analysis methods are available in statistical books, only certain ones that are important and directly usable in NIR spectral modeling/analysis—PLS, PCR, and MLR—are described briefly here.

6.4.2.1 PLS Regression

PLS regression is also known as a projection to latent structure, a method for relating the variations in one or several response variables (Y-variables) to the variations of several predictors (X-variables), with explanatory or predictive purposes. This method performs particularly well when the various X-variables express common information, that is, when there is a large amount of correlation or even colinearity.

PLS is a method of *bilinear modeling* whereby information in the original X-data is projected onto a small number of underlying ("latent") variables called PLS components. The Y-data are actively used in estimating the latent variables to ensure that the first components are those that are most relevant for predicting the Y-variables. Interpretation of the relationship between X-data and Y-data is then simplified as this relationship is concentrated on the smallest possible number of components.

By plotting the first PLS components, one can view main associations between X-variables and Y-variables, as well as interrelationships within X-data and within Y-data. There are two versions of the PLS algorithm: PLS1 deals with only one response variable at a time; PLS2 handles several responses simultaneously. The PLS procedure is depicted in Figure 6.33.

Bilinear Modeling: Bilinear modeling is one of several possible approaches to data compression. These methods are designed for situations where colinearity exists among the original variables. Common information in the original variables is used to build new variables that reflect the underlying (latent) structure. These variables are therefore called "latent variables." The latent variables are estimated as linear functions of both the original variables and the observations—hence the name "bilinear." PCR, PCA, and PLS are bilinear methods.

$$\boxed{\text{Observations}} = \boxed{\text{Data structure}} + \boxed{\text{Error}}$$

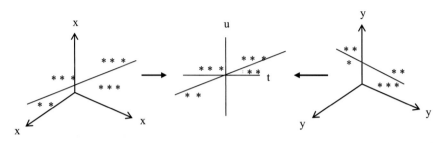

FIGURE 6.33

Depiction of the partial least squares procedure.

In these methods, each sample can be considered as a point in a multidimensional space. The model is built as a series of components onto which the samples, and the variables, can be projected. Sample projections are called "scores" and variable projections are called "loadings." The model approximation of data is equivalent to the orthogonal projection of the samples onto the model. The residual variance of each sample is the squared distance to its projection. Both the *X*- and *Y*-matrices are modeled simultaneously to find the latent variables in *X* that best predict the latent variables in *Y*. These PLS components are similar to principal components and are referred to as PCs.

6.4.2.1.1 Principles of Projection

The principle of PCA is to find the directions in space along which the distance between data points is the longest. This can be translated as finding the linear combinations of the initial variables that contribute most to making the samples different from each other. These directions, or combinations, are called PCs. They are computed iteratively in such a way that the first PC is the one that carries the most information (or, in statistical terms, the most explained variance). The second PC then carries the maximum share of the residual information (i.e., that not taken into account by the previous PC), and so on. Figure 6.34 describes PCs 1 and 2 in a multidimensional space. This process can go on until as many PCs have been computed as there are variables in the data table. At that point, all the variation between samples has been accounted for, and the PCs form a new set of coordinate axes, which has two advantages over the original set of axes (the original variables). First, the PCs are orthogonal to each other. Second, they are ranked so that each one carries more information than any of the subsequent ones. Thus, their interpretation can be prioritized. Start with the first ones, since it is known that they carry more information.

The way it was generated ensures that this new set of coordinate axes is the most suitable basis for a graphical representation of the data that allows easy interpretation of the data structure. Score plots with PCs are very good for getting the idea of different kinds of adulterants in a sample.

6.4.2.2 Principal Component Regression

PCR is a method that is suited to situations as PLS. It is a two-step method. First, a PCA of the *X*-variables is carried out. The PCs then are used as predictors in an MLR method. Figure 6.35 describes the PCR procedure.

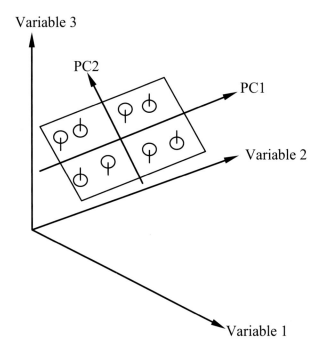

FIGURE 6.34

Description of principal components (PCs).

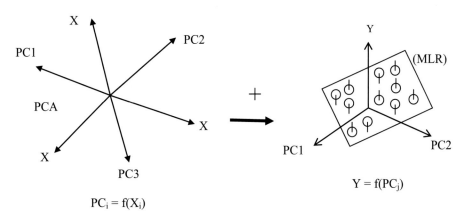

FIGURE 6.35

Description of the principal component (PC) regression procedure. MLR, multiple linear regression; PCA, principal component analysis.

6.4.2.3 Multiple Linear Regressions

MLR is a method for relating the variations in a response variable (Y-variable) to the variations of several X-variables with explanatory or predictive purposes. An important assumption for the method is that the X-variables are linearly independent; that is, no linear relationship exists between the X-variables. When the X-variables carry common information, problems can arise as a result of exact or approximate colinearity.

In MLR, all the X-variables are supposed to participate in the model independent of each other. Their covariations are not taken into account, so X-variance is not meaningful here. Thus the only relevant measure of how well the model performs is provided by the Y-variances.

6.4.3 SELECTION OF REGRESSION METHOD

Selection of regression method is of paramount importance in spectral modeling for nondestructive evaluation of food quality. One should know well that which type of analysis is useful for better prediction based on the data. For understanding the suitability of regression methods, knowledge of their characteristics is essential to save time. Otherwise one has to analyze the data using all techniques and compare their results for selection, which is tedious and time-consuming.

In MLR, the number of X-variables must be smaller than the number of samples. In the case of colinearity among X-variables, the b-coefficients are not reliable and may be unstable. MLR tends to overfit when noisy data are used.

As explained earlier, PCR and PLS, like PCA, are projection methods. Model components are extracted in such a way that the first PC conveys the largest amount of information, followed by the second PC, and so on. At a certain point, the variation modeled by any new PC is mostly noise. The optimal number of PCs—modeling useful information but avoiding overfitting—is determined with the help of residual variances. If the difference between the standard error of calibration and the standard error of prediction (SEP), as well as between biases of calibration and prediction sets of samples, is minimal, one may assume the model is stable (Jha, 2010). SEP is a variation in the precision of predictions over several samples and is computed as the standard deviation of the residuals; the standard deviation is computed as the square root of the mean square of deviations from the mean. Bias is the systematic difference between predicted and measured values and is computed as the average value of the residuals.

PCR uses MLR in the regression step; a PCR model using all PCs gives the same solution as MLR (and so does a PLS1 model using all PCs). If one runs MLR, PCR, and PLS1 on the same data, one can compare their performance by checking validation errors (predicted versus measured Y-values for validation samples, rms error of prediction). Also note that both MLR and PCR model only one Y-variable at a time.

The difference between PCR and PLS lies in the algorithm. PLS uses the information lying in both X and Y to fit the model, switching iteratively between X and Y to find the relevant PCs. Therefore PLS often needs fewer PCs to reach the optimal solution because the focus is on the prediction of the Y-variables, not on achieving the best projection of X, as in PCA.

If there is more than one Y-variable, PLS2 is usually the best method if the goal is to interpret all variables simultaneously. It is often argued that PLS1 or PCR has better prediction ability. This is usually true if there is strong nonlinearity in the data. On the other hand, if the Y-variables are somewhat noisy but strongly correlated, PLS2 is the best way to model all information and leave noise aside. The difference between PLS1 and PCR is usually small, but PLS1 usually gives results comparable to those of PCR using fewer components.

Formal tests of significance for the regression coefficients are well known and accepted for MLR. If one chooses PCR or PLS, he or she may check the stability of the results and the significance of the regression coefficients with the Marten uncertainty test. The final model must be properly validated, however, preferably using a test set (alternatively with cross-validation), but never with just leverage correction.

6.4.4 CLASSIFICATION ANALYSIS

Classification, which is sometimes called "discrimination of samples," deals with the separation of groups of data. Suppose that you have a large number of measurements of apples and, after the data analysis, it turns out that the measurements are clustered in two groups, perhaps corresponding to sweet and sour apples. There is now the possibility to derive a quantitative data model in order to classify these two groups. Similarly, if a food and drug inspector collects milk samples from different sources to determine which milk sample is adulterated and which not, the inspector must assign numerical values to these two categories to make a digital decision. Classification has a somewhat similar purpose, but a set of relevant groupings in the data set—that is, which groups are relevant to the model—typically is known before the analysis.

Classification thus requires a priori class description. Note that discrimination/classification deals with dividing a data matrix into two or more groups of samples. Classification thus can be seen as a predictive method, where the response is a category variable. The purpose of an analysis is to be able to predict which category a new sample belongs to. Interestingly, PCA can also be used here to great advantage, but there are many other competing multivariate classification methods.

6.4.4.1 Classification Methods

Any classification method uses a set of *features* or *parameters* to characterize each object; these features should be relevant to the task at hand. We consider here methods for *supervised* classification, meaning that a human expert both has determined into what classes a sample may be categorized and also has provided a set of sample objects with known classes, for example, adulterated and pure milk. This set of known objects is called the "training set" because it is used by classification programs to learn how to classify objects. There are two phases of constructing a classifier. In the training phase, the training set is used to decide how the parameters ought to be weighted and combined in order to separate the various classes of objects. In the application phase, the weights determined in the training set are applied to a set of objects to determine the likely classes they belong to. If a problem has only a few (two or three) important parameters, classification is usually an easy task. For example, with two parameters one can often simply make a scatter plot of the feature values and can determine graphically how to divide the plane into homogeneous regions where the objects are of the same classes. The classification problem becomes very difficult, though, when there are many parameters to consider. Not only is the resulting high-dimensional space difficult to visualize, but there are so many different combinations of parameters that techniques based on exhaustive searches of the parameter space rapidly become computationally infeasible. Practical methods for classification always involve a heuristic approach intended to find a "good-enough" solution to the optimization problem. There are many classification methods; a few are discussed briefly here.

6.4.4.1.1 Neural Networks

Among the classification methods, neural network methods are probably the most widely used. The biggest advantage of neural network methods is that they are general: They can handle problems with many parameters, and they are able to classify objects well even when the distribution of objects in the N-dimensional parameter space is complex. The disadvantage of neural networks is that they are very slow, not only in the training phase but also in the application phase. Another significant disadvantage of neural networks is that it is difficult to determine how the net is making its decision. Consequently, it is difficult to determine which of the features of the sample being used are important and useful for classification and which are worthless. The choice of the best features (discussed below) is an important part of developing a good classifier, and neural nets do not give much help in this process.

6.4.4.1.2 Nearest-Neighbor Classifiers

A simple classifier can be based on a "nearest-neighbor" approach. In this method, one simply finds in the N-dimensional feature space the object from the training set closest to an object being classified. Since the neighbor is nearby, it is likely to be similar to the object being classified and so is likely to be the same class as that object. Nearest-neighbor methods have the advantage of being easy to implement. They can also give good results if the features are chosen carefully. There are several serious disadvantages to nearest-neighbor methods. First, they (like neural networks) do not simplify the distribution of objects in parameter space to a comprehensible set of parameters. Instead, the training set is retained in its entity as a description of the object distribution. (Some thinning methods can be used on the training set, but the result still does not usually constitute a compact description of the object distribution.) This method is also slow if the training set has many examples. The most serious shortcoming of nearest-neighbor methods is that they are very sensitive to the presence of irrelevant parameters. Adding a single parameter that has a random value for all objects can cause these methods to fail miserably.

6.4.4.1.3 Decision Trees

Decision tree methods have also been used for solving many problems. In axis-parallel decision tree methods, a binary tree is constructed in which a single parameter is compared to some constant at each node. If the feature value is greater than the threshold, the right branch of the tree is taken; if the value is smaller, the left branch is followed. After a series of these tests, one reaches a leaf node of the tree where all the objects are labeled as belonging to a particular class. These are called "axis-parallel trees" because they correspond to partitioning the parameter space with a set of hyperplanes that are parallel to all of the feature axes except the one being tested.

Axis-parallel decision trees are usually much faster in the construction (training) phase than neural network methods, and they also tend to be faster during the application phase. Their disadvantage is that they are not as flexible at modeling parameter space distributions with complex distributions compared with either neural networks or the nearest-neighbor method. In fact, even simple shapes can cause difficulties for these methods. For example, consider a simple two-parameter, two-class distribution of points with parameters x, y that are all of type 1 when $x > y$ and are of type 2 when $x < y$. To classify these objects with an axis-parallel tree, it is necessary to approximate the straight diagonal line that separates the classes with a series of steps. If the density of points is high, many steps may be required. Consequently, for realistic problems axis-parallel trees tend to be rather elaborate, with many nodes.

6.4.4.1.4 Oblique Decision Trees

Oblique decision trees attempt to overcome the disadvantage of axis-parallel trees by allowing the hyperplanes at each node of the tree to have any orientation in parameter space. Mathematically, this means that at each node a linear combination of some or all of the parameters are computed (using a set of feature weights specific to that node), and the sum is compared with a constant. The subsequent branching until a leaf node is reached is just like that used for axis-parallel trees.

Oblique decision trees are considerably more difficult to construct than axis-parallel trees because there are so many possible planes to consider at each tree node. As a result, the training process is slower. However, they usually can be constructed much more quickly than neural networks. They have one major advantage over all the other methods: They often produce very simple structures that use only a few parameters to classify the objects. A straightforward, thorough examination of an oblique decision tree can determine which parameters were most important in helping to classify the objects and which were not used.

There are many other methods and software for classifying the samples in different categories. For example, The Unscrambler has a soft, independent modeling by class analogy (SIMCA) model using PCA and PLS regression for these purposes. SIMCA focuses on modeling similarities between members of the same class. A new sample is recognized as a member of the class if it is similar enough to the other members, or it is rejected. Readers are advised to consult books about software that deal exclusively in these topics.

6.4.4.2 Steps in Developing a Classifier

There are many choices of algorithm for classification. Two major hurdles must be faced before these methods can be used: A training set must be constructed for which the true classifications of the objects are known, and a set of object parameters that are powerful discriminators for classification must be chosen. Once a possible classifier has been identified, it is necessary to measure its accuracy using training and validation sets of data.

6.4.4.2.1 Training Set

A training set must contain a list of objects with known classifications. Ideally, the training set will contain many examples (typically thousands of samples) so that it includes both common and rare types. Creating a training set requires a source of true object classifications, which is usually difficult even for human experts to generate if it must rely on the same data being used by the classifier.

To construct a training set to identify, for example, adulterated samples of milk, one first has to identify all possible adulterants, including water, and identify the more sensitive factor and accordingly simulate them, giving proper weight. It is difficult, but to identify the adulterants it should be done. To determine simply whether samples are adulterated is easier because in this case there are only two classes, and they can be assigned specific values to make a separate group.

6.4.4.2.2 Feature Selection

Adding many irrelevant parameters makes classification harder for all methods, not just the nearest-neighbor method. Training classifiers is an optimization problem in a many-dimensional space. Increasing the dimensionality of the space by adding more parameters makes the optimization more difficult, and the difficulty increases exponentially with the number of parameters. It is always better to give the algorithm for only the necessary parameters rather expecting it to learn to ignore the irrelevant parameters.

One should not ask the classifier to rediscover everything that is already known about the data. Not only should irrelevant parameters be omitted, but highly correlated parameters should be combined when possible to produce a few powerful features. For example, if you expect the shapes of images of a particular class of object to be similar, include a brightness-independent shape parameter rather than simply giving the classifier raw pixel values and expecting it to figure out how to extract shape information from them. If the training process does not require too much computation, a useful approach to identify the best parameter is to train many times on subsets of the features. This method can be used in two ways: starting from the complete list of features and reducing it by removing parameters, and starting from a minimal list and augmenting it by adding parameters. Both methods have proven effective at pruning unnecessary parameters. This procedure can be very fast if axis-parallel trees are used for the exploration. Another useful approach with decision tree methods is to examine directly the weights assigned to the various features. Important features are given a high weight, whereas unimportant features may not be used at all.

6.4.5 DATA PREPROCESSING

Detection of outliers, groupings, clusters, trends, and so on is just as important in multivariate calibration as in PCA, and these tasks should, in general, always be first on the agenda. In this context one may use any validation method (discussed in subsequent section) in the initial screening of data analytical process because the actual number of dimensions of a multivariate regression model is of no real interest until the data set has passed this stage, that is, until it is cleaned up for outliers and is internally consistent. In general, removal of outlying objects or variables often significantly influences the model complexity; that is, the number of components often changes.

There are, however, still a large number of applications that use only two or three wavelengths in routine prediction. These applications have shown that the full PLS model is sometimes inferior to a model based on a relatively small number of variables found in various methods for variable selection. This is partly because of the redundancy and the large amount of noisy, irrelevant variables in NIR spectra. Recent results show that variable selection based on jack-knife estimates is a fast and reliable method with a low risk of overfitting.

6.4.5.1 Spectroscopic Transformation

Most spectroscopists prefer to work with absorbance data because they are more familiar with this type of data and feel more comfortable interpreting absorbance spectra. Many modern multichannel analytical instruments can provide spectra as absorbance readings, using some form of correction and transformation. If you do not know which formula is used in the instrument software, it may be wise to import the raw spectra instead and make the appropriate transformations yourself. In general, it is recommended that one start by analyzing the absorbance spectra. If this does not work, then try to transform the data. Transmission data are often nonlinear, so they are "always" transformed into, for example, absorbance data using a modified logarithmic transformation. Diffuse reflectance data are "always" transformed into Kubelka-Munk units, but exceptions may occur in more problem-specific cases. Multiplicative scatter correction is another useful transformation for spectroscopic data. Many new transformation methods have been introduced in spectral data analysis software; readers are advised to go through them before initiating an actual analysis.

6.4.5.2 Reflectance to Absorbance

The instrument readings R (reflectance) and T (transmittance) are henceforth expressed in fractions between 0 and 1. The readings may thereafter be transformed to apparent absorbance, which is also called the "optical density" of the sample.

6.4.5.3 Absorbance to Reflectance

An absorbance spectrum may be directly transformed to reflectance/transmittance using appropriate analysis software or even using the spectra acquisition software.

6.4.5.4 Absorbance to Kubelka-Munk

The apparent absorbance units may also be transformed into the pertinent Kubelka-Munk units by performing two steps: first transform absorbance units to reflectance units, then transform reflectance to Kubelka-Munk units.

6.4.5.5 Transmission to Absorbance and Back

It can be performed through a computing function using the expression $X = -\log(X)$ to transform X from transmission into absorbance data. $X = 10^{(-X)}$ gives transmission data again.

6.4.5.6 Multiplicative Scatter Correction

Spectroscopic measurements of powders, aggregates of grains of different particle sizes, slurries, and other particulate-laden solutions often display light-scattering effects. This especially applies to NIR data, but it is also relevant to other types of spectra; scatter effects in IR spectra may be caused by background effects, varying optical path lengths, temperature, and pressure variations. Raman spectra also often suffer from background scattering. In UV−visible spectra, varying path lengths and pressure may cause scatter. In general, these effects comprise both a so-called multiplicative effect as well as an additive effect. Other types of measurements may also suffer from similar multiplicative and/or additive effects, such as instrument baseline shift, drift, and interference effects in mixtures. Multiplicative scatter correction (MSC) is a transformation method that can be used to compensate for both multiplicative and additive effects. MSC was originally designed to deal specifically with light scattering. However, a number of analogous effects can also be successfully treated with MSC. In The Unscrambler software, MSC transformation can be done from the Modify−Transform menu. The idea behind MSC is that two undesired general effects, amplification (multiplicative) and offset (additive), should be removed from the raw spectral signals to prevent them from dominating over the chemical or other similar signals, which often are of lesser magnitude. Thus one or more PLS components may be saved in modeling of the relevant Y-phenomena if (most of) these effects are able to be eliminated before multivariate calibration. This, in general, enables more precise and accurate modeling to proceed, based on the cleaned-up spectra. MSC can be a very powerful general preprocessing tool.

6.4.5.7 Derivatives Computation

First or second derivatives are common transformations on continuous function data where noise is a problem, and these are often applied in spectroscopy. Some local information gets lost in the differentiation, but the "peakedness" is supposed to be amplified; this trade-off is often considered advantageous. It is always possible to "try out" differentiated spectra, since it is easy to see whether the model gets any better. As always, however, one should preferentially have a specific reason to choose a particular transformation. Again, this is not to be understood as a trial-and-error optional "supermarket"—it requires experience, reflection, and more experience! The first derivative is often used to correct the baseline shifts. The second derivative is used as an alternative to MSC for handling scatter effects.

6.4.5.8 Averaging

Averaging is used to reduce the number of variables or objects in a data set, to reduce uncertainty in measurements, to reduce the effect of noise, and so on. Data sets with many replicates of each sample can often be averaged over all sets of replicates to ease handling regarding validation and to facilitate interpretation. The result of averaging is a smoother data set. A typical situation in routine applications is fast instrumental measurements, for instance, spectroscopic X-measurements that replace time-consuming Y-reference methods. It is not unusual for several scans to be done for each sample. Should the scans be averaged and used to predict one Y-value for each sample, or should several predictions be made and averaged? Both give the same answer, which is why averaging can also be done on the calibration data and their reference values.

6.4.5.9 Normalization

In the above paragraphs we have mostly treated so-called column transformations, that is, performing specific preprocessing or transformations that act on one column vector individually (single-variable transformations). Normalization is performed individually on the objects (samples), not on the variables (such as wavelength in spectroscopy). Each object vector is rescaled (normalized) into a common sum, for example, 1.00 or 100%. The row sum of all variable elements is computed for each object. Each variable element is then divided by this object sum. The result is that all objects now display a common size—they have become "normalized" to the same sum area in this case. Normalization is a row analogy to column scaling (1/standard deviation). Normalization is a common object transformation. For instance, in chromatography it is used to compensate for (smaller or larger) variations in the amount of analyte injected into the chromatograph. Clearly, it would be of considerable help in the analytical process if this particular measurement variance could be controlled by simple data analytic preprocessing like normalization, otherwise an entire extra PLS component would have to be included to model these input variations. There are several other data analysis problems where normalization can be used in a similar fashion.

6.4.6 ERROR MODELING

How well does a model fit to the X-data and to the Y-data? How small are the modeling residuals? One may perhaps feel that a good modeling fit implies good prediction ability, but this is generally not so, and it in fact happens only rarely. To minimize such chance, one should compare at least the standard error of calibration and validation (SECV), and biases, as discussed in preceding paragraphs.

Assessing Classifier Accuracy: Once a potentially useful classifier has been developed, its accuracy must be measured. Knowledge of the accuracy is necessary both in the application of the classifier and also in comparison of different classifiers.

Accuracy can be determined by applying the classifier to an independent training set of objects with known classifications. This is sometimes trickier than it seems. Since training sets are usually difficult to assemble, one rarely has the resources to construct yet another set of objects with known classifications purely for testing. One must avoid the temptation to train and test on the same set of objects. Once an object has been used for training, any test using it is necessarily biased.

We normally use fivefold cross-validation to measure the accuracy of our classifiers. The training set is divided into five randomly selected subsets with roughly equal numbers of objects. The classifier is then trained five times, excluding a single subset each time. The resulting classifier is tested on the excluded subset. Note that each training session must be completely independent of the excluded subset of objects; one cannot, for example, use the results of an earlier training session as a starting point.

The advantage of cross-validation is that all objects in the training set get used both as test objects and as training objects. This ensures that the classifier is tested on both rare and common types of objects. The cost of cross-validation is that the training process must be repeated many times, adding to the computational cost of the training. In most applications, though, the computer time necessary to repeat the training is more readily available than is the human expert time required to generate completely independent test and training sets.

6.4.7 VALIDATION OF THE SPECTRAL MODEL

Validation or performance tests of a model on an unknown set of samples that has not been used in calibration and whose actual results or specifications are known is essential in this type of data analysis. This new data set is called the "test set." It is used to test the model under realistic, future conditions specifically because it has been sampled to represent those future conditions. Indeed, if possible, one should even use several test sets! *Realistic* here means that the test set should be chosen from the same target population as the calibration set, and that the measuring conditions of both the training (calibration) set and the test set are as representative of the future use as possible. However, this does not mean that the test set should be too similar to the training set. For instance, the training set should not simply be divided into two halves, provided the original set is large enough, as has unfortunately sometimes been recommended in chemometrics. This would decidedly be wrong! The brief overview below is intended only to introduce important issues of validation that must be remembered when specifying a multivariate calibration. From a properly conducted validation, one gets some very important quantitative results, especially the "correct" number of components to use in the calibration model, as well as proper, statistically estimated assessments of future prediction error levels.

6.4.7.1 *Test Set Validation*

When a separate independent set other than the calibration set of data is used for testing the performance of a spectral model, it is called "test set validation." There is an important point here: One also has to know the pertinent Y-values for the test set, just as for the calibration set. The procedure involved in test set validation is to let the calibrated model predict the Y-values and then to compare these independently predicted values of the test set with the known, real Y-values, which have so far been kept out of the modeling and the prediction. An ideal test set situation is to have a sufficiently large number of training set measurements for both X and Y, appropriately sampled from the target population. This data set is then used for the calibration of the model. Now an independent, second sampling of the target population is carried out to produce a test set to be used exclusively for testing/validating of the model. The comparison results can be expressed as prediction errors, or residual variances, which now quantify both the accuracy and precision of the predicted *Y*-values, that is, the error levels that can be expected in future predictions.

6.4.7.2 Cross-validation

There is no better validation than test set validation—testing an entirely "new" data set. One should always strive to use validation by test set. *Test is the best!* There is, however, a price to pay. Test set validation entails taking twice as many samples as would be necessary with the training set alone. However desirable, there are admittedly situations in which this is manifestly not always possible, for example, when measuring the *Y*-values is (too) expensive, unacceptably dangerous, or the test set sampling is otherwise limited (e.g., for ethical reasons or when preparing samples is extremely difficult). For this situation, there is a viable alternative approach, called "cross-validation." Cross-validation can, in the most favorable of situations, be *almost* as good as test set validation, but it can never substitute for proper test set validation.

6.4.7.3 Leverage-Corrected Validation

Leverage-corrected validation is a "quick and dirty" validation method. It actually was being used initially because the concept of validation had not yet been introduced. This method uses the same calibration set, now "leverage-corrected," to also validate the model. It is obvious that this may be a questionable validation procedure; it all depends on the quality of the corrections used. Furthermore, leverage-corrected validation often gives results that are too optimistic. During initial modeling, however, when validation is not yet on the agenda, this method can be useful because it saves time.

6.5 PRACTICAL APPLICATIONS

Spectroscopy has been used for predicting the composition of biological materials for many commodities. Its commercial models using NIRS include, for example, measuring soluble solids content (Jha et al., 2001), firmness, and acidity of some fruits (Figure 6.36), and grain analyzers (Figure 6.37) for wheat, for example, are available. An official method to determine the protein content of wheat was

FIGURE 6.36

A near infrared—based fruit tester being used to check the maturity of apples and muskmelons.

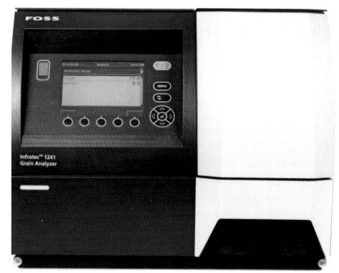

FIGURE 6.37

Grain analyzer.

established long ago using NIRS (AACC, 1983). Work on determining food safety using spectroscopy has recently gained momentum (Reid et al., 2006) and is being carried out worldwide; this is briefly presented below.

6.5.1 DETECTION OF LOW-VALUE FOOD IN HIGH-VALUE FOODS

With rising prices of food commodities, traders may resort to unscrupulous practices such as adulterating a high-value food commodity with a low-value commodity. Examples include orange juice and olive oil, which may contain some similar low-cost food additives or may be entirely replaced by a low-value food. Such practices may be referred to as an "economic adulteration." Table 6.7 provides some examples of high-value foods that may contain a low-value food. Such practices of economic

Table 6.7 Some Food Categories and Low-Value Foods That May Be Used to Adulterate Them		
S. No.	**Food Category**	**Lower-Value Food Additive/Alternative**
1	Extra virgin olive oil	Olive oil (an ordinary or less expensive variety, or mixed with some other nut or seed oil such as soybean and sunflower oil)
2	Fish and seafood (e.g., sushi)	Red snapper (actually tilefish); white tuna and butterfish
3	Milk and milk-based products	Buffalo milk, sheep milk, goat milk, etc.
4	Honey, maple syrup, and other natural sweeteners	Sugar syrup, corn syrup, high-fructose corn syrup
5	Fruit juice (orange, apple)	Lemon juice, grapefruit juice, pear juice, fig juice
6	Coffee and tea	Roasted corn, roasted barley
7	Spices (ground black pepper)	Papaya seeds, buckwheat
8	High-value food grains	Similar-looking low-value food grains or weed seeds

adulteration of food products may lead to illicit profits, unfair competition, and consumer fraud, and may damage the food industry in the long run. There is a strong need for stringent laws, but they can only be enacted effectively when rapid, scientifically accepted tests are available for the purpose. Numerous investigations to quantify added sugars and other low-value foods in different fruit juices (Tables 6.7 and 6.8) and honey (Tzayhri et al., 2009) have been conducted using NIRS and FT IRS

Table 6.8 Classification of Prepared and Commercial Mango Juice Samples into Adulterated and Unadulterated Groups Using Principle Component Analysis Projection Method at 5% Significance Level in the Wavenumber Range of 1476–912 cm^{-1} After Baseline Offset Correction (Jha and Gunasekaran, 2010)

Sample Type	Added Sucrose, %	Total Samples, n	Samples Classified, n		Correct Classification, %
			Unadulterated	**Adulterated**	
Sugar solution	0	5	5	5	0
	1	5	5	5	0
	5	5	0	5	100
	9	5	0	5	100
	13	5	0	5	100
	17	5	0	5	100
	21	5	0	5	100
	25	5	0	5	100
Prepared juice 1	0	5	5	5	0
	1	5	5	5	0
	5	5	5	5	0
	9	5	0	5	100
	13	5	0	5	100
	17	5	0	5	100
	21	5	0	5	100
	25	5	0	5	100
Prepared juice 2	0	5	5	5	0
	3	5	5	5	0
	7	5	0	5	100
	11	5	0	5	100
	15	5	0	5	100
	19	5	0	5	100
	23	5	0	5	100
	27	5	0	5	100
Commercial juice 1	3.6	5	4	1	20
	7.1	5	0	5	100
	10.7	5	0	5	100
Commercial juice 2	12.8	30	0	30	100

(Vardin et al., 2008; Luis et al., 2001; Lijuan et al., 2009; Bureau et al., 2009; Jha and Gunasekaran, 2010) and can be used for practical purposes. It is beyond the scope of this book to describe all reported work here; however, a few studies based on the detection of low-value foods and toxins, microbes, and so on, are discussed in the subsequent sections.

6.5.1.1 Detection of Soymilk in Bovine Milk

Various new and increasingly sophisticated techniques have been tried and developed worldwide to detect the presence of nondairy ingredients in dairy products (Abernethy and Higgs, 2013; Mannso, 2002; Gutierrez et al., 2009; Ntakatsane et al., 2013). Gutierrez et al. (2009) reported detection of nonmilk fat in milk fat by gas chromatography in combination with chemometric analyses. ATR-FTIR was used in combination with chemometric analysis of spectral data to predict the presence of soymilk (SM) in milk. A clear difference in the absorption values of SM and milk was observed (Figure 6.38). The models developed without any data treatment could correctly (93%) classify the test samples into their respective class using SIMCA except for 6% of SM in the spectral range of 1639–1635 cm^{-1} (Table 6.9). The coefficient of determination for quantitative prediction of SM was found to be 0.99 and 0.92 for calibration and validation, respectively, in the wavenumber range of 1472–1241 cm^{-1} using MLR. The detection limit at a 5% significance level was as low as 2%, which indicated that ATR-FTIR had the potential to detect SM in milk without much sample preparation and data treatments (Jaiswal et al., 2015).

6.5.1.2 Detection of Sunflower Oil in Olive Oil

Downey (2006) reported use of NIRS for the detection and quantification of sunflower oil in olive oils. A mathematical model was developed to describe pure olive oils and then was applied to the spectra of pure and adulterated samples. Visible and NIR transflectance spectra (400–2498 nm) were recorded on a NIR Systems 6500 scanning monochromator (FOSS NIR Systems, Silver Spring, MD, USA) fitted with a sample transport accessory. SIMCA was used to study a number of different spectral ranges and pretreatments. The most accurate model was produced using a first derivative of spectral data from 400 to 2498 nm; in this case, 100% of pure olive oils were correctly identified, as were the adulterated samples. No adulterated samples were wrongly identified as authentic. Quantification of the adulterant sunflower oil was done with PLS regression using raw spectral data in the range of 400–2498 nm. It was possible to predict sunflower oil content with a SEP equal to 0.8% and a minimum detection limit of approximately 3% w/w.

6.5.1.3 Detection of Horse Meat in Beef

Kamruzzaman et al. (2015) investigated adulteration of minced beef meat with horse meat using a visible–NIR hyperspectral imaging system (400–1000 nm). A calibration model was developed and optimized using partial least squares regression (PLSR) with full internal cross-validation; it then was validated by external validation using an independent validation set. The established PLSR models based on raw spectra had coefficients of determination (R^2) of 0.99, 0.99, and 0.98, and standard errors of 1.14, 1.56, and 2.23%, for calibration, cross-validation, and prediction, respectively. Four important wavelengths (515, 595, 650, and 880 nm) were selected using regression coefficients resulting from the best PLSR model. Using these important wavelengths, an image-processing algorithm was developed to predict the adulteration level in each pixel across the whole surface of the samples.

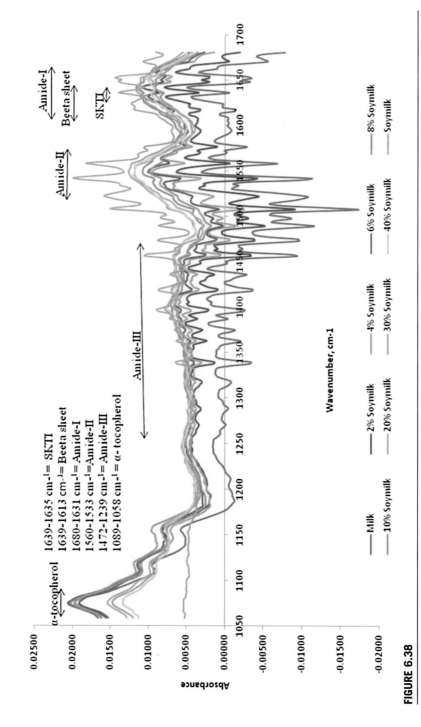

FIGURE 6.38

Spectra of pure and different percentages of soymilk-adulterated cow/buffalo milk.

Table 6.9 Soft, Independent Modeling by Class Analogy Classification Result of Milk Adulterated with Different Concentrations of Soymilk at 20 °C

Wavenumber Range, cm^{-1}	Soymilk, %	Samples Classified, n[a]			Classification, Efficiency, %
		Milk	Adulterated Milk (2–40%)	Soymilk	
1472–1239	0	15	0	0	100
	2	0	15	0	100
	4	0	14	0	93
	6	0	15	0	100
	8	0	15	0	100
	10	0	14	0	93
	20	0	15	0	100
	30	0	14	0	93
	40	0	15	0	100
	100	0	0	15	100
1560–1533	0	15	0	0	100
	2	0	15	0	100
	4	0	14	0	93
	6	0	14	0	93
	8	0	15	0	100
	10	0	15	0	100
	20	0	15	0	100
	30	0	15	0	100
	40	0	15	0	100
	100	0	0	15	100
1680–1631	0	15	0	0	100
	2	0	15	0	100
	4	0	14	0	93
	6	0	15	0	100
	8	0	15	0	100
	10	0	14	0	93
	20	0	15	0	100
	30	0	15	0	100
	40	0	15	0	100
	100	0	0	14	93
1639–1613	0	15	0	0	100
	2	0	14	0	93
	4	0	14	0	93
	6	0	15	0	100
	8	0	15	0	100
	10	0	15	0	100
	20	0	15	0	100

Continued

Table 6.9 Soft, Independent Modeling by Class Analogy Classification Result of Milk Adulterated with Different Concentrations of Soymilk at 20 °C—cont'd

Wavenumber Range, cm^{-1}	Soymilk, %	Samples Classified, n[a]			Classification, Efficiency, %
		Milk	Adulterated Milk (2–40%)	Soymilk	
	30	0	15	0	100
	40	0	15	0	100
	100	0	0	14	93
1639–1635	0	15	0	0	100
	2	0	15	0	100
	4	0	14	0	93
	6	0	13	0	86
	8	0	15	0	100
	10	0	14	0	93
	20	0	15	0	100
	30	0	15	0	100
	40	0	15	0	100
	100	0	0	14	93
1089–1058	0	15	0	0	100
	2	0	15	0	100
	4	0	15	0	100
	6	0	15	0	100
	8	0	15	0	100
	10	0	15	0	100
	20	0	15	0	100
	30	0	14	0	93
	40	0	15	0	100
	100	0	0	15	100

[a]*Total number of samples = 150 (15 for each level of soymilk).*

6.5.2 DETECTION OF ADULTERANTS AND CONTAMINANTS

Food adulteration (e.g., spice adulteration) may be categorized into two separate groups, namely, incidental and intentional adulteration. Incidental adulteration occurs when foreign substances are added to a food as a result of ignorance, negligence, or improper facilities. Intentional adulteration involves the deliberate addition of inferior materials to a food to heighten appearance qualities and to gain greater profits. These inferior substances include ground material (e.g., sawdust), leaves, powdered products (e.g., starches), and other spice species (ASTA, 2004). Such substances may cause serious damage to human health. An example in this regard is the adulteration of milk with synthetic milk that contains harmful substances such as urea, caustic soda, or vegetable oil. Although 180–400 mg/L urea is present naturally in milk (Jonker et al., 1998), the cutoff limit is a concentration of 700 mg/L (FSSAI, 2012). The concentration of urea beyond the cutoff limit may cause

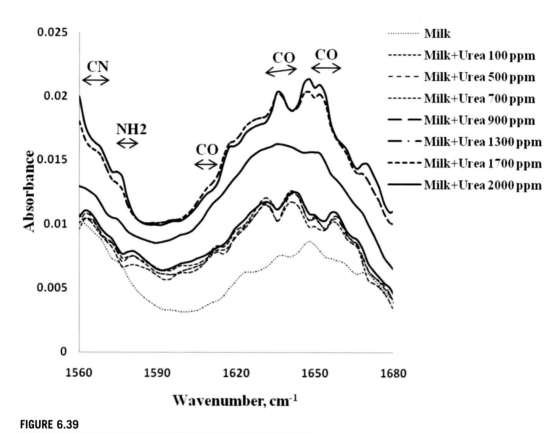

FIGURE 6.39

Typical spectra of milk and urea-mixed milk in the spectral range of $1680-1560$ cm^{-1}.

diseases and disorders such as indigestion, acidity, ulcers, cancer, and kidney malfunction (Trivedi et al., 2009).

Jha (2007) explained adulterations in juices, milk, and dairy products and developed a technique (Jha et al., 2015) for the detection and quantification of urea in milk using ATR-FTIR. A clear difference in absorption values of milk adulterated with urea compared with those of pure milk was observed in the spectral region of $1670-1564$ cm^{-1} (Figure 6.39), and PCA indicated sharp clustering of samples adulterated with different concentrations of commercial urea (Figure 6.40). The models developed without any preprocessing could correctly classify more than 86% of the test samples into their respective class using SIMCA (Table 6.10). The coefficients of determination using PLS for quantitative prediction of urea were found to be 0.906 for calibration and 0.879 for validation in the wavenumber ranges of $1649-1621$ and $1611-1580$ cm^{-1}, respectively. The root mean standard error (RMSE) of calibration and the RMSE of prediction for the corresponding range were found to be 214.53 and 244.55 for the calibration and validation sets of samples, respectively. Similarly, Santos et al. (2013) also detected and quantified various adulterants in milk using IR microspectroscopy and chemometrics.

FIGURE 6.40

Principal component scores plot depicting clusters of milk and urea-mixed milk in the wavenumber range of 1670–1564 cm^{-1}.

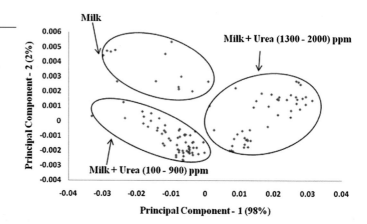

NIR hyperspectral imaging in conjunction with multivariate image analysis was used for the detection of millet and buckwheat flour in ground black pepper. Mid-IRS was also used for the quantification of millet and buckwheat flour in ground black pepper. Preprocessed NIR spectra revealed adulterant-specific absorption bands (1743, 2112, and 2167 nm), whereas preprocessed mid-IR spectra revealed a buckwheat-specific signal at 1574 cm^{-1}. The model created with millet-adulterated black pepper samples had a classification accuracy of 77%; a classification accuracy of 70% was obtained for the buckwheat-adulterated black pepper samples (September, 2011).

6.5.3 DETECTION OF MICROBES AND TOXINS

6.5.3.1 Detection of Microbes

Viability assessment of bacterial cells is vital in monitoring the quality of food samples. My team has worked on spectroscopic detection of live and dead *Escherichia coli* using chemometrics and found potential in addition to saving time, cost, and labor when compared with the traditional method. The method was based on the use of a UV–visual spectrometer for detection and quantitative prediction of dead and live cells of *E. coli* (ATCC 8739). A live bacterial suspension showed an absorption peak at 260 nm. A clear decrease in amplitude with an increase in the proportion of dead cells was observed. PCA of spectral data showed clear clustering of samples with a significance level of 5%. The models developed without any data treatment could correctly (100%) classify the test samples into their respective class using SIMCA and PLS discriminant analysis approach. The MLR model successfully predicted the percentage of live and dead cells in a small wavelength range (259–261 nm); the multiple correlation coefficient (R) values for calibration and validation for quantitative prediction of live and dead cells were 0.98 and 0.977, respectively.

Grewal et al. (2015) successfully used FT IRS in combination with chemometrics to differentiate bacteria directly on the surface of poultry meat. Based on the analysis of the obtained FTIR spectral profile with chemometric tools, a spectral window of 1800–1200 cm^{-1} was identified as the region with the potential for classification into different groups. However, 100% correct classification was achieved using PLS discriminant analysis in the wavenumber ranges of 3000–2500 and 4000–575 cm^{-1}.

Table 6.10 Soft, Independent Modeling by Class Analogy Classification of Unadulterated and Adulterated Milk with Different Concentrations of Urea (Jha et al., 2015)

Wavenumber Range, cm^{-1}	Urea, ppm	Total Samples, n	Selected Classes, n			Misclassified	Classification Efficiency, %
			0	100–900	1300–2000		
1670–1653 [CO]	0	15	13	2		2	86.67
	100–900	60		58	7	7	96.66
	1300–2000	45	4	4	44	8	97.78
1649–1621 (+) 1611–1580 [CO & NH_2]	0	15	13			0	86.67
	100–900	60		58	1	1	96.66
	1300–2000	45	1	2	43	3	95.56
1643–1635 (+) 1611–1580 [CO & NH_2]	0	15	12			0	80.00
	100–900	60		55		0	91.67
	1300–2000	45			43	0	95.56
1643–1635 [CO]	0	15	13	1	1	2	86.67
	100–900	60		58		0	96.66
	1300–2000	45	2		42	2	93.33
1615–1607 [CO]	0	15	13	1		1	86.67
	100–900	60		56	1	1	93.33
	1300–2000	45		1	43	1	95.56
1594–1572 [NH_2]	0	15	13			0	86.67
	100–900	60		58		0	96.66
	1300–2000	45		9	43	9	95.56
1594–1564 [CN]	0	15	12	1		1	80.00
	100–900	60		57		0	95.00
	1300–2000	45	1	12	44	13	97.78

6.5.3.2 Detection of Toxins

Aflatoxins are a group of toxic and carcinogenic secondary metabolites produced by *Aspergillus flavus* and *Aspergillus parasiticus* (Rustom, 1997). These pose major health and economic problems worldwide (Hussain et al., 2010; Iqbal et al., 2011) and account for losses of millions of dollars annually in human health, animal health, and condemned agricultural products, disposal of contaminated foods and feeds, and investment in research and applications to reduce the severity of the mycotoxin problem worldwide (Vasanthi and Bhat, 1998). Aflatoxin (AF) B1, the most common toxin, has been classified as a group 1 carcinogen by the International Agency for Research on Cancer (IARC) (IARC, 2002). AFB1 is metabolized to AFM1 in mammals after consumption of AFB1-contaminated feed, which is then secreted in their milk (Ruangwises and Ruangwises, 2010). AFM1 is as acutely hepatotoxic, like AFB1, but its carcinogenicity is approximately 2–10% more than that of AFB1 (Asi et al., 2012). It was initially classified by IARC as an agent in group 2B, with a possible carcinogenic effect on humans (IARC, 1993), but it was later reclassified as a group 1 carcinogenic agent because of its high toxicity (IARC, 2002). Moreover, AFM1 is a very stable aflatoxin—it is not destroyed even by heat treatment such as pasteurization, autoclaving, or other methods used in the production of fluid milk (Tajkarimi et al., 2008). Since consumption of milk and milk products is high, particularly among infants and young children, the risk of exposure to AFM1 is of great concern (Baskaya et al., 2006). Researchers have shown the hazardous effects of AFM1 through dairy products using diverse methods (Pathirana et al., 2010).

Fernández-Ibañez et al. (2009) used NIRS for rapid detection of mycotoxigenic fungi such as AFB1. The best predictive model to detect AFB1 in maize was obtained using standard normal variate and detrending as scatter correction ($r^2 = 0.80$ and 0.82; SECV = 0.211 and 0.200 for grating and FT NIRS instruments, respectively). In the case of barley, the best predictive model was developed using standard normal variate and detrending on the dispersive NIRS instrument ($r^2 = 0.85$; SECV = 0.176) and using spectral data as log 1/R for FT NIRS ($r^2 = 0.84$; SECV = 0.183).

Tripathi and Mishra (2009) used FT NIRS in diffuse reflectance mode combined with appropriate chemometric techniques for measuring AFB1 in red chili powder. Different spectral preprocessing methods were investigated and optimized based on the lowest values of RMSE of cross-validation. Spectral wavenumber ranges of 6900.3–4998.8 and 4902.3–3999.8 cm^{-1} and a straight-line subtraction preprocessing technique predicted AFB1 content with the best accuracy with the lowest RMSE of cross-validation value (0.654%) and the maximum correlation coefficient for validation plots ($R^2 = 96.7$). Similar work is in progress for the detection of other toxins, such as patulin in apple juice. All these reports confirm that spectroscopy is a potential tool for the detection and quantification of adulterants and contaminants, including microbes, in food for future commercial use.

REFERENCES

AACC, March 1983. Approved Methods of the American Association of Cereal Chemists, eighth ed. St. Paul Minn.

Abernethy, G., Higgs, K., 2013. Rapid detection of economic adulterants in fresh milk by liquid chromatography-tandem mass spectrometry. J. Chromatogr. A 1288, 10–20.

Anon, 2010. Acousto-optical Tunable Filter. http://www.sciner.com/Acousto-Optics/acoustooptical_tunable_filters.htm (accessed on 20–21 July).

Asi, M.R., Iqbal, S.Z., Arino, A., Hussain, A., 2012. Effect of seasonal variations and lactation times on aflatoxin M1 contamination in milk of different species from Punjab, Pakistan. Food Control 45 (1), 34−38.

ASTA, 2004. Spice Adulteration, White Paper. American Spice Trade Association ASTA, New York.

Baskaya, R., Aydin, A., Yildiz, A., Bostan, K., 2006. Aflatoxin M1 levels of some cheese varieties in Turkey. Med. Weter. 62, 778−780. Galvano et al. (1996).

Bellamy, L.J., 1975. The Infra-red Spectra of Complex Molecules, third ed. Chapman and Hall.

Born, M., Wolf, E., 1977. Principles of Optics. Pergamon Press.

Bureau, S., Ruiz, D., Reich, M., et al., 2009. Application of ATR-FTIR for a rapid and simultaneous determination of sugars and organic acids in apricot fruit. Food Chem. 115, 1133−1140.

Clancy, P.J., 2002. Transfer of calibration between on-farm whole grain analysers. In: Cho, R.K., Davies, A.M.C. (Eds.), Near Infrared Spectroscopy: Proceedings of the 10th Int. Conference on Near Infrared Spectroscopy, Kuonjgu, Korea. NIR Publications, Chichester, UK.

Curcio, J.A., Petty, C.C., 1951. The near infrared absorption spectrum of liquid water. J. Opt. Soc. Am. 41 (5), 302−304.

Downey, G. (Ed.), 2006. Food Authentication Using Infrared Spectroscopic Methods. Agriculture and Food Development Authority, Teagasc Oak Park Carlow Co., Carlow, Dublin.

Elliott, A., Hanby, W.E., Malcolm, B.R., 1954. The near infrared absorption spectra of natural and synthetic fibres. Br. J. Appl. Phys. 5, 377−381.

Fearn, F.R.B., 1982. Near infrared reflectance as an analytical technique, part 3. New advances. Lab. Pract. 31 (7), 658−660.

Fernández-Ibañez, V., Soldado, A., Martínez-Fernández, A., de la Roza-Delgado, B., 2009. Application of near infrared spectroscopy for rapid detection of aflatoxin B1 in maize and barley as analytical quality assessment. Food Chem. 113, 629−634.

Foster, G.N., Row, S.B., Griskey, R.G., 1964. Infrared spectrometry of polymers in the overtone and combination regions. J. Appl. Polym. Sci. 8, 1357−1361.

FSSAI, 2012. Manual of Analysis of Methods of Foods, Milk and Milk Products. Food Safety and Standards Authority of India, p. 12.

Glatt, L., Ellis, J.W., 1951. Near infrared pleochroism II. The 0.8−2.5μ region of some linear polymers. J. Chem. Phy. 19, 449−457.

Goddu, R.F., 1960. Near-infrared spectrophotometry. Adv. Anal. Chem. Instrum. 1, 347−417.

Greensill, C.V., Walsh, K.V., 2002. Standardization of near infrared spectra across miniature photodiode array-based spectrometers in the near infrared assessment of citrus soluble solids content. In: Cho, R.K., Davies, A.M.C. (Eds.), Near Infrared Spectroscopy: Proceedings of the 10th Int. Conference on Near Infrared Spectroscopy, Kuonjgu, Korea. NIR Publications, Chichester, UK.

Grewal, M.K., Jaiswal, P., Jha, S.N., 2015. Detection of poultry meat specific bacteria using FTIR spectroscopy and chemometrics. J. Food Sci. Technol. 52 http://dx.doi.org/10.1007/s13197-014-1457-9.

Gutierrez, R., Vega, S., Diaz, G., Sanchez, J., Coronado, M., Ramirez, A., Perez, J., Gonzalez, M., Schettino, B., 2009. Detection of non-milk fat in milk by gas chromatography and linear discriminant analysis. J. Dairy Sci. 92, 1846−1855.

Hammaker, R.M., Graham, J.A., Tilotta, D.C., et al., 1986. What is Hammard transform spectroscopy. In: Durig, J.R. (Ed.), Vibrational Spectra and Structure, vol. 15. Elsevier, Amsterdam, pp. 401−485.

Hecht, K.T., Wood, D.L., 1956. The near infra-red spectrum of the peptide group. Proc. Royal Soc. 235, 174−188.

Holland, J.K., Newnham, D.A., Mills, I.M., 1990. Vibrational overtone spectra of monofluoroacetylene: a preliminary report. Mol. Phy. 70, 319−330.

Holman, R.T., Edmondson, P.R., 1956. Near infrared spectra of fatty acids and related substances. Anal. Chem. 28, 1533−1538.

Hussain, I., Anwar, J., Asi, M.R., Munawar, M.A., Kashif, M., 2010. Aflatoxin M1 contamination in milk from five dairy species in Pakistan. Food Control 21, 122−124.

IARC, 1993. Some Naturally Occurring Substances: Food İtems and Constituents, Heterocyclic Aromatic Amines and Mycotoxins. International Agency for Research on Cancer, Lyon, France, p. 599.

IARC, 2002. Some traditional herbal medicines, some mycotoxins, naphthalene and styrene. In: IARC Monographs on the Evaluation of Carcinogenic Risks to Humans, vol. 82, pp. 171−300. Tajkarimi et al. (2008).

Iqbal, S.Z., Paterson, R.R.M., Bhatti, I.A., Asi, M.R., 2011. Comparing aflatoxins contamination in chillies from Punjab, Pakistan, produced in summer and winter. Mycotoxin Res. 27, 75−80.

Iwamoto, M., Uozumi, J., Nishinari, K., 1987. Preliminary investigation of the state of water in foods by near infrared spectroscopy. In: Hollo, J., Kaffka, K.J., Gonczy, J.L. (Eds.), Near Infrared Diffuse Reflectance/Transmittance Spectroscopy. Akademiai Kiado, Budapest, pp. 3−12.

Jaiswal, P., Jha, S.N., Borah, A., Gautam, A., Grewal, M.K., Jindal, G., 2015. Detection and quantification of soymilk in cow-buffalo milk using attenuated total reflectance fourier transform infrared spectroscopy. Food Chem. 168. http://dx.doi.org/10.1016/j.foodchem.2014.07.010.

Jaquinot, P., 1958. J. de Physique Radium 19, 223.

Jha, S.N., Matsuoka, T., 2000. Non-destructive techniques for quality evaluation of intact fruits and vegetables. Food Sci. Technol. Res. 6 (4), 248−251.

Jha, S.N., Gunasekaran, S., 2010. Authentication of sweetness of mango juice using Fourier transform infrared − attenuated total reflection spectroscopy. J. Food Eng. 101 (3), 337−342. http://dx.doi.org/10.1016/j/jfoodeng2010.07.19.

Jha, S.N. (Ed.), 2010. Nondestructive Evaluation of Food Quality: Theory and Practice. Springer−Verlag GmbH Berlin Heidelberg, Germany, p. 288.

Jha, S.N., Matsuoka, T., 2004a. Detection of adulterants in milk using near infrared spectroscopy. J. Food Sci. Technol. 41 (3), 313−316.

Jha, S.N., Matsuoka, T., 2004b. Nondestructive determination of acid brix ratio (ABR) of tomato juice using near infrared (NIR) spectroscopy. Int. J. Food Sci. Technol. 39 (4), 425−430.

Jha, S.N., Garg, R., 2010. Nondestructive prediction of quality of intact apple using near infrared spectroscopy. J. Food Sci. Technol. 47 (2), 207−213.

Jha, S.N., Jaiswal, P., Borah, A., Gautam, A.K., Srivastava, N., 2015. Detection and quantification of urea in milk using attenuated total reflectance-Fourier Transform Infrared Spectroscopy. Food Bioprocess Technol. 8 (4), 926−933.

Jha, S.N., Matsuoka, T., Kawano, S., 2001. A simple NIR instrument for liquid type samples. In: Proceedings of Annual Meeting of Japanese Society of Agric Structures, pp. 146−147. Paper No. C-20.

Jha, S.N., 2007. Nondestructive methods for quality evaluation of dairy and food products. Beverage Food World 34 (1), 80−83.

Jonker, J.S., Kohn, A., Erdman, R.A., 1998. Using milk urea nitrogen to predict nitrogen excretion and utilization efficiency in lactating dairy cows. J. Dairy Sci. 81 (10), 2681−2692.

Kamruzzaman, M., Makino, Y., Oshita, S., Liu, S., January 2015. Assessment of visible near-infrared hyperspectral imaging as a tool for detection of horsemeat adulteration in minced beef. Food Bioprocess Technol. 8.

Kawano, S., Watanabe, H., Iwamoto, M., 1992. Determination of sugar content in intact peaches by near infrared spectroscopy with fibre optics in interactance mode. J. Jpn. Soc. Hort. Sci. 61, 445−451.

Kaye, W., 1954. Near infrared spectroscopy: I. Spectral identification and analytical applications. Spectrochim. Acta 6, 257−287.

Kaye, W., Canon, C., Devaney, R.G., 1951. Modification of a Beckman model DU spectrophotometer for automatic operation at 210−2700 μm. J. Opt. Soc. Am. 41 (10), 658−664.

Krikorian, S.E., Mahpour, M., 1973. The identification and origin of N−H overtone and combination bands in the near-infrared spectra of simple primary and secondary amides. Spectrochim. Acta 29A, 1233−1246.

Lauer, J.L., Rosenbaum, E.J., 1952. Near infrared absorption spectrophotometry. App. Spec. 6 (5), 29–46.

Law, D.P., Tkachuk, R., 1977. Near infrared diffuse reflectance spectra of wheat and wheat components. Cereal Chem. 54 (2), 256–265.

Liddel, U., Kasper, C., 1933. Spectral differentiation of pure hydrocarbons: a near infrared absorption study. J. Res. Natl. Bur. Stand. 11, 599–618.

Lijuan, X., Ye, X., Liu, D., Ying, Y., 2009. Quantification of glucose, fructose and sucrose in bayberry juice by NIR and PLS. Food Chem. 114, 1135–1140.

Luis, E.R.-S., Fedrick, S.F., Michael, A.M., 2001. Rapid analysis of sugars in fruit juices by FT-NIR spectroscopy. Carbohydr. Res. 336, 63–74.

Mannso, M.A., 2002. Determination of vegetal proteins in milk powder by Sodium Dodecyl Sulfate–Capillary gel Electrophoresis: Inter laboratory study. J. AOAC Int. 85 (5), 1090–1095.

Meurens, M., 1984. Analysis of aqueous solutions by NIR reflectance on glass fibre. In: Proceedings of the Third Annual Users Conference for NIR Researchers Pacific Scientific: MD USA.

Morimoto, S., 2002. A nondestructive NIR spectrometer: development of a portable fruit quality meter. In: Cho, R.K., Davies, A.M.C. (Eds.), Near Infrared Spectroscopy: Proceedings of the 10th Int. Conference on Near Infrared Spectroscopy, Kuonjgu, Korea. NIR Publications, Chichester, UK.

Morimoto, S., McClure, W.F., Stanfield, D.L., 2001. Handheld NIR spectrometry: Part I. An instrument based upon gap-second derivative theory. Appl. Spectrsc. 55 (1), 182–189.

Murray, I., 1988. Aspects of interpretations of near infrared spectra. Food Sci. Technol. Today 2, 135–140.

Murray, M., 1987. In: Hollo, J., Kaffka, K.J., Gonczy, J.L. (Eds.), The NIR Spectra of Homologous Series of Organic Compounds in Near Infrared Diffuse Reflectance/Transmittance Spectroscopy. Akademia Kiado, Budapest, pp. 13–28.

Norris, K.H., 1984. Multivariate analysis of raw materials. In: Shemilt, L.W. (Ed.), Chemistry and *World Food Supplies: The New Frontiers*, Chemrawn II. Pergamon Press, pp. 527–535.

Ntakatsane, M.P., Liu, X.M., Zhou, P., 2013. Short communication: rapid detection of milk fat adulteration with vegetable oil by fluorescence spectroscopy. J. Dairy Sci. 96 (4), 2130–2136.

Osborne, B.G., Douglas, S., 1981. Measurement of degree of starch damage in flour by near infrared reflectance analysis. J. Sci. Food Agric. 32, 328–332.

Osborne, B.G., Fearn, T., Hindle, P.H., 1983. Practical NIR Spectroscopy in Food and Beverage Analysis. Longman Scientific and Technical, UK.

Pathirana, U.P.D., Wimalasiri, K.M.S., Silva, K.F.S.T., Gunarathne, S.P., 2010. Investigation of farm gate cow milk for aflatoxin M1. Trop. Agric. Res. 21 (2), 119–125.

Reid, L.M., O'Donnell, C.P., Downey, G., 2006. Recent technological advances for the determination of food authenticity. Trends Food Sci. Technol. 17, 344–353.

Rose Jr., F.W., 1938. Quantitative analysis with respect to the component structural groups of the infrared (1 to 2 μ) molar absorptive indices of 55 hydrocarbons. J. Res. Natl. Bur. Stand. 20, 129–157.

Ruangwises, N., Ruangwises, S., 2010. Aflatoxin M1 contamination in raw milk within the central region of Thailand. Bull. Environ. Contam. Toxicol. 85 (2), 195–198.

Rustom, I.S., 1997. Aflatoxin in food and feed: occurrence, legislation and inactivation by physical methods. Food Chem. 59, 57–67.

Santos, P.M., Pereira-Filho, E.R., Rodriguez-Saona, L.E., 2013. Rapid detection and quantification of milk adulteration using infrared microspectroscopy and chemometric analysis. Food Chem. 138, 19–24.

September, D.J.F., 2011. Detection and Quantification of Spice Adulteration by Near Infrared Hyperspectral Imaging (Master's Thesis). University of Stellenbosch.

Tajkarimi, M., Aliabadi-Sh, F., Salah Nejad, A., Poursoltani, H., Motallebi, A.A., Mahdavi, H., 2008. Aflatoxin M1contamination in winter and summer milk in 14 states in Iran. Food Control 19 (11), 1033–1036.

Tilotta, D.C., Hammaker, R.M., Fateley, W.G., 1987. A visible-near-infrared Hadamard transform spectrometer based on a liquid crystal spatial light modulator array: a new approach in spectroscopy. Appl. Spectrosc. 41 (6), 727–734.

Tosi, C., Pinto, A., 1972. Near-infrared spectroscopy of hydrocarbon functional groups. Spectrochim. Acta 28A, 585–597.

Tripathi, S., Mishra, H.N., 2009. A rapid FT-NIR method for estimation of aflatoxin B1 in red chili powder. Food Control 20 (9), 840–846.

Trivedi, U.B., Lakshminarayana, D., Kothari, I.L., Patel, N.G., Kapse, H.N., Makhija, K.K., et al., 2009. Potentiometric biosensor for urea determination in milk. Sens. Actuators, B 140 (1), 260–266.

Trott, G.F., Woodside, E.E., Taylor, K.G., et al., 1973. Physicochemical characterization of carbohydrate-solvent interactions by near-infrared spectroscopy. Carbohydr. Res. 27 (2), 415–435.

Tzayhri, G.V., Guillermo, O.S., Marlene, Z.L., et al., 2009. Application of FTIR-HATR spectroscopy and multivariate analysis to the quantification of adulterants in Mexican honeys. Food Res. Int. 42, 313–318.

Vardin, V., Tay, A., Ozen, B., et al., 2008. Authentication of pomegranate juice concentrate using FTIR spectroscopy and chemometrics. Food Chem. 108, 742–748.

Vasanthi, S., Bhat, R.V., 1998. Mycotoxins in foods occurrence, health and economic significance and food control measures. Indian J. Med. Res. 108, 212–222. IARC (2002).

Wheeler, O.H., 1959. Near infrared spectra of organic compounds. Chem. Rev. 59, 629–666.

Williams, P., Norris, K., 1987. Near Infrared Technology in the Agricultural and Food Industries. American Association of Cereal Chemists Inc., St. Paul, pp. 247–290.

IMAGING METHODS

There are basic techniques, biosensors, and spectroscopy combined with chemometrics for the detection and quantification of adulterants and contaminants. The first method usually destroys the sample, whereas the second and third ones give surface or subsurface information only (if the samples are solid). It is therefore difficult to know the infected, rotten areas (microbes, etc.) if they are inside a solid food (e.g., a fruit fly or stone weevil in mango or guava) and any other internal undesirable artifacts in any solid food. These internal infected areas and/or bacteria or fruit flies can be only seen using suitable imaging techniques. The basic theory of X-ray, computed tomography, magnetic resonance imaging, and ultrasonic and hyperspectral imaging, and their applications for the detection of these internal defects in foods, are described briefly in this chapter so that one can have a complete spectrum of knowledge to test food materials with regard to internal artifacts.

7.1 X-RAY IMAGING

X-ray imaging is one of the most successful techniques for the nondestructive detection of the internal aspects of any body. X-rays, because of their high energy, can penetrate through many objects. There are, however, differences in penetration through different materials because of the differences in the material properties. Photons in an X-ray beam are either transmitted, scattered, or absorbed when passed through a body. Radiography intends to capture the difference in the transmitted X-ray beam that is caused by material differences in the form of a visual contrast in the image. This contrast can be a measure of the spatial and quantitative distribution of a certain material(s) within a composite. Radiography can not only determine the extent of internal damage but also estimate the volume of different internal features. One of the major problems associated with use of X-rays is that high-energy electromagnetic radiation, like X-rays, can ionize and kill biological cells. It is therefore mandatory to provide a shield between the radiation source and people working in the vicinity. Equipment designed for radiography therefore needs to fulfill functional as well as radiation safety requirements (Kotwaliwale, 2010).

An X-ray, also known as a roentgen ray, is produced when a fast-moving electron, emanated from a heated cathode, impinges on a heavy-metal anode. There are generally two types of X-ray emission (Figure 7.1): (1) Bremsstrahlung, in which a free electron is attracted to the nucleus, and to conserve momentum a photon is created with an energy dependent on the change in the electron's trajectory; and (2) K-shell or characteristic X-rays (a), in which an electron from the cathode dislodges orbital electrons in the target material and produces excited atoms, which are stabilized by X-ray emissions (b). Generally less than 1% of the energy received by the anode is converted into X-rays. The X-rays

Rapid Detection of Food Adulterants and Contaminants. http://dx.doi.org/10.1016/B978-0-12-420084-5.00007-X
Copyright © 2016 Elsevier Inc. All rights reserved.

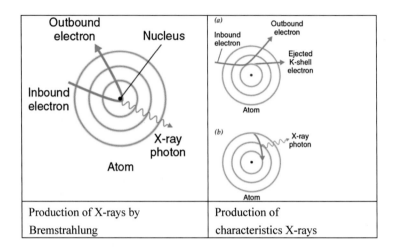

FIGURE 7.1

Types of X-rays based on their production.

emitted from the focal spot travel in a conical beam through a window in the X-ray tube. The efficiency, quantity, quality, and energy of produced X-rays generally depend on the anode material. A material with a higher atomic number produces more X-rays. X-rays (or roentgen rays) are a form of electromagnetic radiation with a wavelength in the range of 10 to 0.01 nm, corresponding to frequencies in the range 30 to 30,000 PHz (1 PHz = 10^{15} Hz). As a subsurface technique, X-ray radiation can give three-dimensional information about the spatial density differences and changes in the atomic number within a sample (Zwiggelaar et al., 1997). X-ray imaging techniques have mainly been developed for medical, particle, and material physics; its usage for food materials is scant.

7.1.1 X-RAY PROPERTIES

An X-ray is a short electromagnetic wave that behaves like a particle as well as a wave when interacting with matter. X-ray particles are discrete bundles of energy and are called "photons" or "quanta." If a photon has 15 eV or more energy, it is capable of ionizing atoms and molecules, and is called "ionizing radiation." An atom is ionized when it loses an electron. The energy of a photon is given by $E = h\nu$, where h is Plank's constant (4.13×10^{-18} keV s or 6.62×10^{-34} J s), and ν is the frequency of the photon wave ($\nu = \frac{c}{\lambda}$, where c is the speed of light [3×10^{8} m/s], and λ is the wavelength). Rearranging and substituting values of h and c, we get $E = \frac{12.4}{\lambda}$.

7.1.2 ATTENUATION COEFFICIENT

The photons in a soft X-ray beam are either transmitted, scattered (Compton scattering), or absorbed (photoelectric collision) when passed through an object. As a result, the energy of incident photons is reduced exponentially (Curry et al., 1990) and is represented by Eqn (7.1) (Kotwaliwale, 2010):

$$I = I_0 e^{-\mu_m z \rho} \tag{7.1}$$

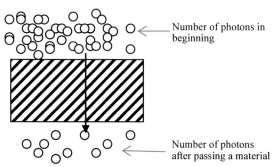

Number of photons in beginning

Number of photons after passing a material

FIGURE 7.2

Attenuation of X-ray beam when passing through a matter.

where I is the energy (in kiloelectron volts) of photons exiting through a body; μ_m is mass attenuation coefficient in millimeter square per gram; ρ is the material density in gram per cubic millimeter; and z is the thickness (in millimeters) through which the X-rays pass. The mass attenuation coefficient for a material is a function of the atomic number of the absorbing material and the incident photon energy. The exiting photon energy depends on material properties, including thickness. If the absorbing material consists of more than one element, the mass attenuation coefficient of the composite material is a function of the mass attenuation coefficients of the individual elements and their mass fraction in the path of the photon beam (Hubbell and Seltzer, 1995). The attenuation coefficient of a material changes with thickness when measured under polychromatic X-rays (Paiva et al., 1998; Kotwaliwale et al., 2006). Buzzell and Pintauro (2003) defined the "R value" of a material as the ratio of the attenuation coefficient at a low energy level to that at a high energy level. Figure 7.2 represents the attenuation of X-ray beam when passing through a material.

7.1.3 COMPONENTS USED IN X-RAY IMAGING

There are four basic components in any type of X-ray imaging: (1) the X-ray source; (2) an X-ray converter; (3) the imaging medium; and (4) casing for the imaging medium. The X-ray converter (e.g., a phosphor screen) stops X-rays from reaching the imaging medium and produces a visible output proportional to the number of incident X-ray photons. The imaging medium (e.g., a photographic medium) captures the image while the casing protects the imaging medium from visible radiation surrounding it.

7.1.3.1 X-ray Tube

In X-ray tubes, X-rays are generated by interactions between the energetic electrons and atoms of the target. Radioactive substances and X-ray tubes are the two sources of X-rays. Radioactive substances may generate monochromatic X-rays (almost all the photons have the same energy level), whereas X-ray tubes generate a polychromatic beam. Bombardment must take place in a vacuum to prevent the ionization of air. The X-ray source is one of the most important system components in determining the overall image quality. The literature shows that different types of X-ray tubes have been used as X-ray sources in radiography of agricultural produce. Maximum variations in tubes are in their voltage, current, focal spot size, window material, electrode material, and tube cooling system, among others. The major parts of an X-ray tube (Figure 7.3) are the evacuated envelop (special glass or metal tube),

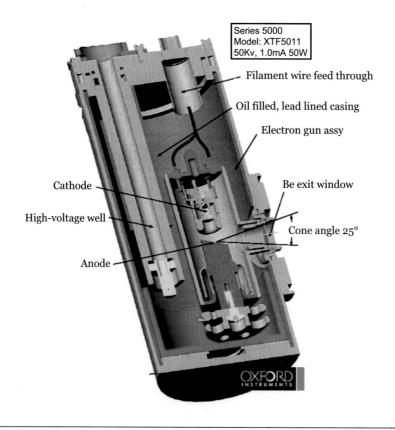

FIGURE 7.3

Major parts of an X-ray tube in a cut-away view.

Courtesy: Oxford Instruments, USA.

a heated tungsten filament (cathode), an anode, a tungsten target attached to the anode, an electron focusing cup, and a high-voltage power supply.

7.1.3.2 Imaging Medium

In general, the acquisition of X-ray images can be either film-based or digital. In film-based X-ray imaging, which is similar to conventional photography, the X-ray is transmitted through the inspected object and a sensing film is exposed to form the object's image. After developing the film, a high-resolution X-ray image can be obtained. Digital images created by scanning film radiographs have been reported, which helped in the digital image processing and electronic storage of radiographs. Different resolutions and bit depths have been reported for different products.

7.1.3.3 Photographic Plate

The detection of X-rays is based on various methods. The most commonly known methods use a photographic plate, an X-ray film in a cassette, and rare earth screens. In earlier periods a photographic plate was used to produce radiographic images. The images were produced directly on the glass plates.

Film replaces these plates. Since photographic plates are sensitive to X-rays, they provide a means of recording the image, but require a lot of exposure (of the patient) to the X-rays, so intensifying screens were devised. They allow a lower dose to the patient by taking the X-ray information and intensify it so that it can be recorded on film positioned next to the intensifying screen. The object to be radiographed is placed between the X-ray source and the image receptor to produce a shadow of the internal structure of that particular body. X-rays are partially blocked ("attenuated") by dense tissues and pass more easily through soft tissues. The X-rays strike darkens in solid areas when developed, causing bones to appear lighter than the surrounding soft tissue.

7.1.3.4 Photostimulable Phosphors

An increasingly common method uses photostimulated luminescence wherein a photostimulable phosphor plate is used in place of a photographic plate. After the plate is X-rayed, excited electrons in the phosphor material remain "trapped" in "color centers" in the crystal lattice until stimulated by a laser beam passed over the plate's surface. The light given off during laser stimulation is collected by a photomultiplier tube, and the resulting signal is converted into a digital image by computer technology, which gives this process its common name: computed radiography (also referred to as digital radiography). The photostimulable phosphor plate can be reused, and existing X-ray equipment requires no modification for the purpose.

7.1.3.5 Scintillators

Some materials such as sodium iodide can "convert" an X-ray photon to a visible photon; an electronic detector can be built by adding a photomultiplier. These detectors are called "scintillators," "film screens," or "scintillation counters." The main advantage of using these is to get a maximum image while subjecting the patient to a much lower dose of X-rays.

7.1.3.6 Scintillators Plus Semiconductor Detectors (Indirect Detection)

With the availability of large semiconductor array detectors, it has become possible to design detector systems using a scintillator screen to convert X-rays to visible light, which is then converted to electrical signals in an array detector. Indirect flat-panel detectors are widely used today in medical, dental, veterinary, and industrial applications.

7.1.3.7 Direct Semiconductor Detectors

Since the 1970s, new semiconductor detectors have been developed (silicon or germanium doped with lithium). X-ray photons are converted to electron–hole pairs in the semiconductor and are collected to detect the X-rays. When the temperature is low enough, it is possible to directly determine the X-ray energy spectrum; this method is called "energy-dispersive X-ray spectroscopy"; it is often used in small X-ray fluorescence spectrometers. These detectors are sometimes called "solid-state detectors." Detectors based on cadmium telluride and its alloy with zinc, cadmium zinc telluride, which allows lower doses of X-rays to be used, have an increased sensitivity.

7.1.3.8 X-ray Camera

A digitized X-ray image can be acquired and analyzed in real time using digital X-ray scanning sensors for the online inspection of materials. Applications of digital X-ray imaging in industries have increased significantly in recent years. Two types of digital cameras are typically used: line-scan cameras and two-dimensional (2D) cameras. In line-scan cameras, the relative movement of the sample and the camera is required to acquire a digital radiograph. Two types of arrangements exist for 2D radiography: a digital plate comprising an X-ray converter, a charge-coupled device or complementary metal-oxide semiconductor array, and a casing that replaces the conventional "film cassette," which is then read through an image reader. Alternatively, digital X-ray cameras giving instantaneous or "online" readouts are also available. An X-ray digital camera typically has a 2D photodiode array or complementary metal-oxide semiconductor array having pixels spaced apart in a sensing area. A scintillator screen, placed in direct contact with the photodiode array, converts incident X-ray photons to light, which in turn is detected by the photodiodes. To protect the sensitive electronics from accidental damage caused by ambient light, graphite window (Figure 7.4) shields are usually used. Resolution of the digitized analog signal, the number of A/D channels, the dynamic range (defined as the maximum signal divided by the read noise), the type of frame grabber, and the frame rate are some of the features that determine the quality of the camera (Kotwaliwale et al., 2007a).

7.1.3.9 Shield

X-rays in the range of 10–50 keV are about 10,000 times more energetic than visible light. Because of their high energy content, the photons can penetrate almost all materials. These electromagnetic radiations also have ionization properties that can kill biological cells; hence proper shielding is required when dealing with X-rays. The maximum dose is generated when the X-ray tube operates at its peak voltage and current. In thicker shields, a phenomenon of buildup from scattering must be accounted for. The thicker and taller the shield, the larger is the buildup of scatter component. Also, the energy of

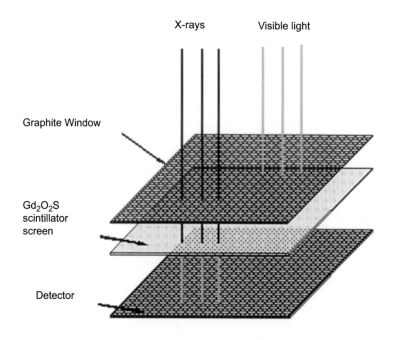

FIGURE 7.4

Arrangement of a detector and graphite window screen.

the source affects the contribution of the scatter factor to the exposure range. For primary X-rays, the buildup caused by shielding transmission is expressed by Eqn (7.2):

$$B_x = 1.67 \times 10^{-5} \frac{Hd^2}{DT} \tag{7.2}$$

where B_x is the shielding transmission, H is the maximum permissible dose equivalent (mrem/h), d is the distance (in meters) between the X-ray source and the reference point, D is the absolute dose index rate (rad m^2/min), and T is the area occupancy factor.

The effect of material thickness on the penetration of neutrons, as a modified form of Eqn (7.1), is given as Eqn (7.3):

$$I(z) = BI_0 e^{-\Sigma_t z} \tag{7.3}$$

where Σ_t is the macroscopic total cross section for neutrons and B is the buildup factor. The buildup factor could be expressed in the form of $B (\mu \cdot z) = 1 + (\mu \cdot z)$, and thus:

$$I(z) = (1 + \mu z)I_0 e^{-\Sigma_t z} \tag{7.4}$$

Approximating Σ_t by μ for X-rays, we get

$$I(z) = (1 + \mu z)I_0 e^{-\mu \cdot z} \tag{7.5}$$

This equation could be solved for z to get the thickness of shield required to reduce the radiation dose by a desired factor.

Shield thickness to protect against this dose is determined using two approaches. A safe reduction of photon energy through shielding is considered as 1 million times reduction, that is, $\frac{I}{I_0} = 10^{-6}$.

Approach I: Incorporating buildup caused by the shield, Eqn (7.1) thus becomes

$$I = I_0(1 + B_x)e^{-\mu_m z \rho} \tag{7.6}$$

A conservative estimate of B_x is 1.148×10^{-9}, calculated at a distance of 0.25 m away from the source (a 3-MeV, 2-mA, 10-mm-diameter electron beam) and an area occupancy factor of 1. Eqn (7.6) is then solved for shield thickness z.

Approach II: Using Eqn (7.5) and solving for shield thickness z.

The mass attenuation coefficient of P_b for 50 keV, $\mu_m = 6.74$ cm^2/g. Linear attenuation coefficient $\mu = \mu_m^*$. Density $\rightarrow 6.74$ cm^2/g $\times$ 11.35 g/cm$^3 = 76.49$ cm^{-1}.

7.1.3.10 Working Principle

The principle of soft X-ray inspection is based on the density of the product and the contaminant. As an X-ray penetrates a food product, it loses some of its energy. A dense area, such as contaminant, reduces the energy even further. As the X-ray exits the product, it reaches a sensor. The sensor then converts the energy signal into an image of the interior of the food product. Foreign matter appears as a darker shade of gray that helps to identify foreign contaminants. An X-ray imaging system can also produce three-dimensional (3D) information that can be manipulated numerically for rapid and noninvasive assessment of the quality of food.

The object for which a radiograph is generated is kept between the X-ray tube and the imaging medium, as shown in Figure 7.5(a). This is in contrast to photography, where light source and the camera are normally facing the same direction compared with the object being photographed. This

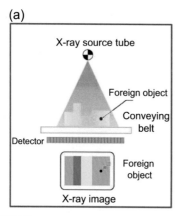 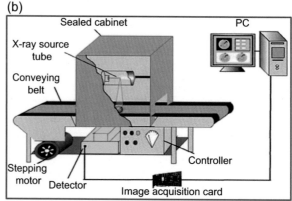

FIGURE 7.5

Soft X-ray imaging system: (a) working principle, (b) schematic equipment setup (Chen et al., 2013).

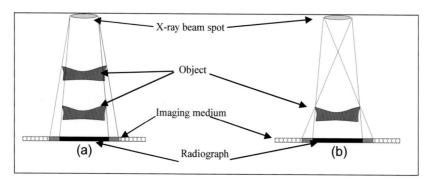

FIGURE 7.6

Image size changes because of the distance of the subject from (a) the imaging medium and (b) geometric unsharpness as a result of the size of the beam spot.

situation poses a limitation that the imaging medium (film or camera's sensing area) should be larger than the object. Line-scan detectors help to alleviate this problem to some extent. Further, the distance between the object and the imaging medium is also vital if the dimensions of features in the radiographs have any significance. The spot size of the X-ray beam at the source should be as small as possible. Possible artifacts caused by the distance between the object and the imaging medium and by larger spot size are shown in Figure 7.6.

7.1.4 X-RAY DIGITAL IMAGING EQUIPMENT

The soft X-ray inspection system, shown in Figure 7.5(b), mainly comprises a computer-controlled X-ray generator (i.e., the X-ray source tube), a line-scanning sensor for X-ray detection, a

FIGURE 7.7

Mango X-ray imaging—based sorting system.

conveying belt, a stepping motor, an image-acquisition card, and a computer. A mango X-ray scanning system is shown in Figure 7.7 (Jha et al., 2012).

7.2 COMPUTED TOMOGRAPHY

Computed tomography (CT) is another powerful technique for imaging the internal parts of an object and is commonly used for medical diagnostics. It is a proven method for evaluating a cross section of an object using a movable X-ray source and detector assembly to accumulate data from thin projected sample slices. The basic principle of CT is that the internal structure of an object can be reconstructed from multiple X-ray projections of an object using tomography. Figure 7.8 depicts an illustration of a CT scanner. CT thus is also a type of X-ray technique that enables the acquisition of 2D X-ray images of thin "slices" of an object. It was previously known as computed axial tomography and body section roentenography (Kotwaliwale, 2010).

The first commercially available CT scanner using X-rays was developed by Godfrey Newbold Hounsfield at Thorn EMI Central Research Laboratories in the United Kingdom. Hounsfield conceived his idea in 1967, and it was publicly announced in 1972. Allan McLeod Cormack of Tufts University (Massachusetts, USA) independently constructed a similar process and shared a Nobel Prize in medicine in 1979. The first prototype took 160 parallel readings through 180 angles, each 1° apart; each scan took a little over 5 min. A large computer took almost 2.5 h to process the images from these scans using algebraic reconstruction methods. The first practical CT instrument was developed in 1971 by Hounsfield in England and was used to image the brain. The projection data were acquired in approximately 5 min, and the tomographic image was reconstructed in approximately 20 min. Since then, CT technology has developed dramatically, and the technique has become a standard imaging procedure for virtually all parts of the body.

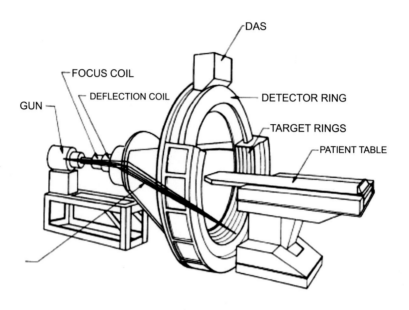

FIGURE 7.8

Schematic of the internal part of a computed tomography scan machine.

Cunningham and Judy (2000).

7.2.1 CT GENERATIONS

Five generations of scanners with differences in tube-detector configuration and scanning speed have been developed to date (Table 7.1). The X-ray beam was not wide enough to cover the entire width of the slice of interest or across the field of view in the first and second generations. A mechanical arrangement was required to move the X-ray source and detector horizontally. In the third- and fourth-generation designs, the X-ray beam was able to cover the entire field of view of the scanner. The first- and fourth-generation CT are compared graphically in Figure 7.9. This avoids the need for any horizontal motion. An entire "line" can be captured almost instantly. This allowed for the simplification of motion to rotation of the X-ray source. Third- and fourth-generation designs differ in the

Table 7.1 Different Generations of Computed Tomography Scanners				
Generation	**Configuration**	**Detectors, n**	**Beam**	**Minimum Scan Time**
First	Translate-rotate	1–2	Pencil thin	2.5 min
Second	Translate-rotate	3–52	Narrow fan	10 s
Third	Rotate-rotate	256–1000	Wide fan	0.5 s
Fourth	Rotate-fixed	600–4800	Wide fan	1 s
Fifth	Electron beam	1284	Wide fan electron beam	33 ms

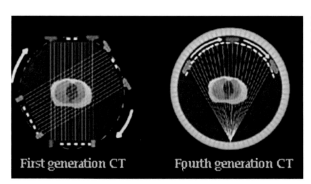

FIGURE 7.9

Configuration of two generations of computed tomography (CT).

First generation CT   Fourth generation CT

arrangement of the detectors. In the third generation, the detector array is as wide as the beam and must therefore rotate as the source rotates. The object is placed in the center of the gantry, while the gantry rotates around it. In the fourth generation, an entire ring of stationary detectors is used. A modern CT scanner without a cover is shown in Figure 7.10. The X-ray tube (T) and the detectors (D) are mounted on a ring-shaped gantry. Instead of rotating a conventional X-ray tube around the object, the electron beam CT machine houses a huge vacuum tube in which an electron beam is electromagnetically steered toward an array of tungsten X-ray anodes arranged circularly around the object. Each anode is hit in turn by the electron beam and emits X-rays that are collimated and detected as in conventional CT. The lack of moving parts allows very quick scanning, with single-slice acquisition in 50–100 ms.

Helical or spiral CT was introduced in the early 1990s. In this version, X-ray sources (and detectors in 3rd generation designs) are attached to a freely rotating gantry. During a scan, the table moves the

FIGURE 7.10

Computed tomography scanner without a cover to illustrate the principle of operation.

FIGURE 7.11

Depiction of the arrangement in helical computed tomography.

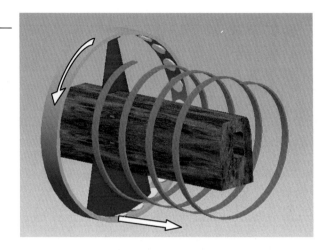

object smoothly through the scanner; the name derives from the helical path traced by the X-ray beam (Figure 7.11). It was the development of two technologies that made spiral CT practical: slip rings to transfer power and data on and off the rotating gantry, and the switched mode power supply, which is powerful enough to supply the X-ray tube but small enough to be installed on the gantry. The major advantage of spiral scanning compared with the traditional shoot-and-step approach is speed; a large volume can be covered in 20–60 s. This is advantageous for a number of reasons: (1) often the patient can hold their breath for the entire study, reducing motion artifacts; (2) it allows for more optimal use of intravenous contrast enhancement; and (3) the study is quicker than the equivalent conventional CT, permitting the use of higher-resolution acquisitions in comparatively the same amount of time. Data obtained from helical CT are often well-suited for 3D imaging because of the absence of motion caused by breath in the body. In addition, spiral CT has also slightly lower z-axis resolution than step-and-shoot.

A multislice CT scanner is similar in concept to helical or spiral CT, having 64 or more detector rings. Recent models have up to three rotations per second and an isotropic resolution of 0.35-mm voxels, with a z-axis scan speed of up to 18 cm/s. The major benefit of multislice CT is the increased speed of volume coverage, which allows large volumes to be scanned. Inverse geometry CT (IGCT) is another novel concept being investigated to refine the third-generation design. IGCT reverses the shapes of the detector and X-ray source. Conventional third-generation CT geometry uses a point source of X-rays, which diverge in a fan-like beam to act on a linear array of detectors. In multidetector computed tomography, this is extended in three dimensions to a conical beam acting on a 2D array of detectors. The IGCT concept, conversely, uses an array of highly collimated X-ray sources that act on a point detector. The individual sources can be activated in turn by steering an electron beam onto each source target using a principle similar to that in electron beam tomography. The rationale behind IGCT is that it avoids the disadvantages of the cone-beam geometry of third-generation multidetector computed tomography. As the z-axis width of the cone beam increases, the quantity of scattered radiation reaching the detector increases, and the z-axis resolution is degraded because of the increasing z-axis distance that each ray must traverse. This reversal of roles has extremely high intrinsic resistance to scatter and a reduced number of detectors per slice.

7.2.2 **CT NUMBER**

An X-ray source illuminates an object from a particular direction to create a CT image. The intensities of X-rays going through the object are measured by detectors, digitized, and used in the reconstruction of image of the object using the CT number, which is based on linear X-ray absorption coefficients and, in general, expressed by brightness data in an image. A CT number is therefore defined using Eqn (7.7). If $k = 1000$, the CT number is also called a Hounsfield unit.

$$CT \ number = (\mu - \mu_w) \times \frac{k}{\mu \cdot w} \tag{7.7}$$

where μ is the object's linear X-ray absorption coefficient (per meter), μ_w is the linear X-ray absorption coefficient of water (per meter), and k is a constant (1000). Ogawa et al. (1998) computed CT numbers in the range of -1000 to $+4000$, using 1000 for air and 0 for water.

7.2.3 **TERMINOLOGY IN CT IMAGING**

In CT imaging, pixels are displayed in terms of relative radiodensity based on the mean attenuation of the tissue(s) on a scale of -1024 to $+3071$ Hounsfield units (HU). A pixel is a 2D unit based on the matrix size and the field of view. When 3D CT slice thickness is used, the unit is known as a voxel. The phenomenon that one part of the detector cannot differ between different tissues is called the "partial volume effect." Water has an attenuation of 0 HU, whereas air is -1000 HU, and cancellous bone is typically $+400$ HU. The attenuation of metallic insertions/impurities depends on the atomic number of the element used. Titanium usually has an amount of $+1000$ HU, whereas iron steel can completely extinguish the X-ray.

Windowing is the process of making an image using the calculated HU. The various radiodensity amplitudes are mapped to 256 shades of gray. These shades can be distributed over a wide range of HU values to get an overview of structures that attenuate the beam to widely varying degrees. Alternatively, these shades of gray can be distributed over a narrow range of HU values (called a "narrow window") centered over the average HU value of the particular structure to be evaluated. In this way, subtle variations in the internal makeup of the structure can be discerned; this is called "contrast compression" in image processing.

Contemporary CT scanners do not restrict the display of images as in conventional axial images, because CT offers isotropic, or near isotropic, resolution. Instead, it is possible for a software program to build a volume by "stacking" the individual slices one on top of the other. The program may then display the volume in an alternative manner; this process is called "3D image reconstruction." In multiplanar reconstruction, a volume is built by stacking axial slices. The software cuts slices through the volume in a different plane, usually orthogonal. Another option a special projection method such as maximum-intensity projection or minimum-intensity projection can be used to build the reconstructed slices.

The term *ionizing radiation* refers to those subatomic particles and photons whose energy is sufficient to cause ionization in the matter with which they interact. The ionization process consists of removing an electron from an initially neutral atom or molecule. Ionizing radiation, such as X-rays, alpha rays, beta rays, and gamma rays, remains undetectable by the human senses, and the damage it causes to the body is cumulative, related to the total dose received. Since the photons are uncharged, they do not interact through the coulomb force and therefore can pass over large distances through matter without significant interaction. The average distance traveled between interactions is called the "mean free path," and in solid materials this ranges from a few millimeters for low-energy X-rays to tens of centimeters for high-energy gamma rays. When an interaction does occur, however, it is catastrophic in the sense that a single interaction can profoundly affect the energy and direction of the photon or can make it disappear entirely. In such an interaction, all or part of the photon energy is transferred to one or more electrons in the absorber material. Because the secondary electrons thus produced are energetic and charged, they interact in much the same way as described earlier for primary fast electrons. The fact that the presence of an original X-ray or gamma ray is indicated by the appearance of secondary electrons means that information on the energy carried by the incident photons can be inferred by measuring the energy of these electrons. The three major types of such interactions are photoelectric absorption, Compton scattering, and pair production.

7.2.3.1 Photoelectric Absorption

In photoelectric absorption, the incident X-ray or gamma-ray photon interacts with an atom of the absorbing material, and the photon completely disappears. Its energy is transferred to one of the orbital electrons of the atom. Because this energy in general far exceeds the binding energy of the electron in the host atom, the electron is ejected at a high velocity. The kinetic energy of this secondary electron is equal to the incoming energy of the photon minus the binding energy of the electron in the original atomic shell. The process leaves the atom with a vacancy in one of the normally filled electron shells, which is then refilled after a short time by a nearby free electron. This filling process again liberates the binding energy in the form of a characteristic X-ray photon, which then typically interacts with electrons from less tightly bound shells in nearby atoms, producing additional fast electrons. The overall effect is the complete conversion of the photon energy into the energy carried by fast electrons. The fast electrons, detectable through their coulomb interactions, indicate the presence of the original gamma-ray or X-ray photon, and a measurement of their energy is tantamount to measuring the energy of the incoming photon.

7.2.3.2 Compton Scattering

An incoming gamma-ray photon can interact with a single free electron in the absorber through the process of Compton scattering. In this process, the photon abruptly changes direction and transfers a portion of its original energy to the electron from which it scattered, producing an energetic recoil electron.

7.2.3.3 Pair Production

This third type of interaction is possible when the incoming photon energy is above 1.02 MeV. In the field of a nucleus of the absorber material, the photon may disappear and be replaced by the formation of an electron–positron pair.

Monitoring radiation inside a facility and monitoring the dose received by an individual are the two important aspects of radiation control. Installed or portable instruments are available to monitor radiation in an area. The detectors used in portable instruments are ionizing chambers, Geiger-Müller counters, and scintillation detectors. Most common detection methods were initially based on the ionization of gases, as in the Geiger-Müller counter: a sealed volume, usually a cylinder, with a mica, polymer, or thin metal window contains a gas, a cylindrical cathode, and a wire anode; a high voltage is applied between the cathode and the anode. When an X-ray photon enters the cylinder, it ionizes the gas and forms ions and electrons. Electrons accelerate toward the anode, causing further ionization along their trajectory in the process. Figure 7.12 shows one portable instrument that can measure instantaneous radiation or dose accumulated over a period of time. Common types of wearable dosimeters for ionizing radiation include a quartz fiber dosimeter, film badge dosimeter, thermoluminescent dosimeter, and solid-state (silicon diode) dosimeter.

FIGURE 7.12

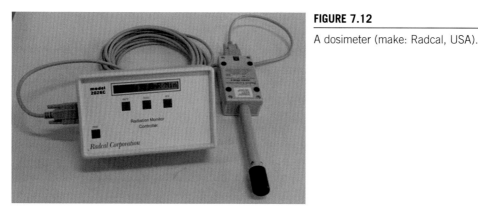

A dosimeter (make: Radcal, USA).

A real dose (dose-unit as centigray or absorbed radiation dose) reflects something objective: the energy deposited by X-rays (g^{-1} of irradiated body tissue). By contrast, an "effective" dose is a calculation that estimates what dose, if given to the entire body, might produce approximately the same amount of risk as would the real dose actually received by the irradiated sections. Effective doses (dose-unit as cSv [centiSievert, 1 Sv = 10 rem] or rem) incorporate a crude adjustment for the different types of ionizing radiation, plus "tissue weighting factors" that attempt (despite woefully inadequate evidence) to assess the attributable probability of fatal cancer in different organs, the additional detriment from nonfatal cancer and hereditary disorders, and the different latency periods for cancers in various tissues.

7.2.4 UNITS USED IN RADIATION

The unit used to measure the energy of photons is electron volts (eV). An electron volt is the amount of energy that an electron gains as it is accelerated by a potential difference of 1 V. In the case of X-rays, it is generally expressed as energy from 0.12 to 120 keV. The higher the energy, the stronger the transmittance.

The unit for radioactivity is becquerel (Bq), which is s^{-1} in SI units, measured as decay of one nucleus per second (dps). Curie was the unit previously used for radioactivity, which corresponds to 3.7×10^{10} dps. The unit for absorbed dose is roentgen (R or A·s/kg or coulomb/kg), the quantity of radiation that produces one electrostatic unit of positive or negative electricity per cubic centimeter of air at standard temperature and pressure, or the quantity of radiation that will produce 2.083×10^{09} ion pairs per cubic centimeter of dry air, or radiation received in 1 h from 1 g source of radiation at a distance of 1 m; other known units for dose are rad (J/kg) and the SI unit gray (Gy); $1 \text{ R} = 8.72 \times 10^{-03}$ Gy, which is m^2/s^2 in SI units. Equal absorbed doses from different radiations do not necessarily have equal biological effects. Therefore, the term *dose equivalent* (DE), which is equal to the absorbed dose multiplied by a factor for the energy distribution in tissues. The unit for DE is sievert (Sv), which is equal to Gy × F. The value of F is taken as 1 for gamma rays, X-rays, and beta particles. The unit "rem" was previously used for DE, where 1 Sv = 100 rem.

The X-ray tube current, measured in milliamperes, refers to the number of electrons flowing per second from the filament to the target. Where electron volt or kiloelectron volt is the unit to represent photon energy in X-rays, kilovolt (peak) is the unit commonly used to represent the peak energy of photons in the polychromatic X-ray beam generated by an X-ray tube.

7.3 MAGNETIC RESONANCE IMAGING

Nuclear magnetic resonance (NMR), or magnetic resonance imaging (MRI), is a unique technology that measures the magnetic properties of spins that can then be related to the physical or chemical properties of subjects. NMR is the physical process whereby the nucleus, whose magnetic moment is not zero, resonantly absorbs radiation of a certain frequency under an external magnetic field. Detectors detect and receive NMR signals released as electromagnetic radiation; these signals can then be sent to a computer and be converted into an image through data processing. MRI machines make use of the fact that food tissue contains an enormous amount of water, which is aligned in a large magnetic field. Each water molecule has two hydrogen nuclei or protons. When food is put in a powerful magnetic field, the average magnetic moment of many protons becomes aligned with the direction of the field. A radiofrequency (RF) transmitter is briefly turned on, producing a varying electromagnetic field. This electromagnetic field has just the right frequency, known as the resonance frequency, to be absorbed and flip the spin of the protons in the magnetic field. After the electromagnetic field is turned off, the spins of the protons return to thermodynamic equilibrium, and the bulk magnetization becomes realigned with the static magnetic field. During this relaxation, an RF signal is generated and can be measured with receiver coils. Information about the origin of the signal in 3D space can be learned by applying additional magnetic fields during the scan. A 3D image is compiled from multiple 2D images, which are produced from any plane of view. The image can be rotated and manipulated to be better able to detect tiny changes of structures within the food object. These fields, generated by passing electric currents through gradient coils, make the magnetic field strength vary depending on the position within the magnet. Because this makes the frequency of the released radio signal also depend on its origin in a predictable manner, the distribution of protons in the food can be mathematically recovered from the signal, typically using the inverse Fourier transform. In the images, each pixel value reflects the NMR signal intensity of a voxel in the measured material, which relates to the resonance density and the two main parameters (i.e., relaxation times T_1 and T_2). An MRI shows an image of the object structure, making visible its physical and chemical information. In brief, an MRI system includes a magnet and power supply that can produce a wide range of uniform, stable, and constant magnetic fields; a set of gradient magnetic field coils, a controller and power-driven equipment; an RF system; a computer system with large storage capacity for data collection and processing; and some auxiliary equipment.

In brief, the MRI technique works in five steps: put the subject in a big magnetic field; transmit radio waves into the subject (2−10 ms); turn off the radio wave transmitter; receive radio waves retransmitted by the subject; and convert measured RF data to an image.

7.3.1 PHYSICS IN MRI

Spin is like electrical charge and mass: It is a fundamental property of nature. It is in multiples of 1/2 and can be positive or negative. Protons, electrons, and neutrons possess spin. Individual unpaired electrons, protons, and neutrons each possess a spin of 1/2. To understand how particles with spin behave in a magnetic field, consider a proton with spin as a magnetic moment vector, causing the proton to behave like a tiny magnet with a north and south pole. In the absence of any external magnetic field, the nuclei's spin angular momentum has random orientations (Figure 7.13) in their atomic or molecular environment.

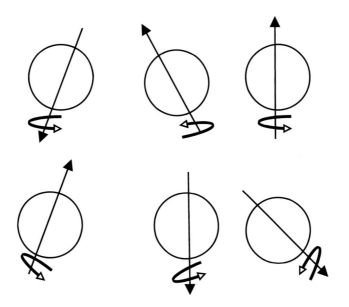

FIGURE 7.13

Random orientation of a nucleus.

When a proton is placed in an external magnetic field, the spin vector of the particle aligns itself with the external field (Figure 7.14), just like a magnet does. There is a low energy configuration or state where the poles are aligned in a high-energy state. This particle then undergoes a transition between the two energy states by absorbing a photon. A particle in the lower-energy state absorbs a photon and ends up in the higher-energy state. The energy of this photon exactly matches the energy difference between the two states. The energy E of a photon is related to its frequency γ by Planck's constant ($h = 6.626 \times 10^{-34}$ J s), as given in Eqn (7.8):

$$E = h\gamma \tag{7.8}$$

When the energy of the photon matches the energy difference between the two spin states, absorption of energy occurs. In NMR spectroscopy, γ is between 60 and 800 MHz for hydrogen nuclei. It is typically between 15 and 80 MHz for hydrogen in clinical MRI.

When certain types of nuclei, such as protons, cabon-13, and fluorine-19, are placed in a strong static magnetic field (B_0), they absorb electromagnetic radiation in the RF range. Such nuclei are said to be "spin-active" and to resonate. The precise frequencies at which spin-active nuclei resonate can be picked up and displayed by instruments. Imposing a linear magnetic gradient on the external magnetic field causes the proton resonance frequency to vary; thus the position of resonating nuclei can be determined and represented as an image (Chen et al., 1989; Clark et al., 1997). At equilibrium, the net magnetization vector lies along the direction of the applied magnetic field B_0 and is called the "equilibrium magnetization" (M_0). In this configuration, the Z component of magnetization M_z equals M_0. M_z is referred to as the longitudinal magnetization. There is no transverse magnetization (M_x or M_y) here. The time constant, which describes how M_z returns to its equilibrium value, is called the

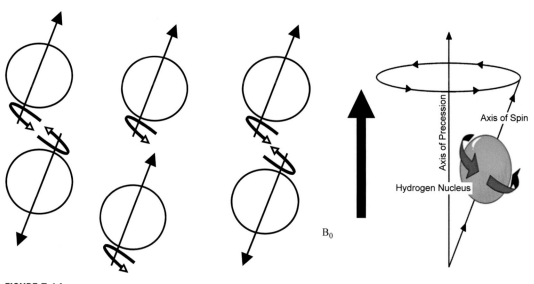

FIGURE 7.14

A nucleus oriented after the application of an external magnetic field (B_0).

"spin lattice relaxation time" (T_1). The equation governing this behavior as a function of the time t after its displacement is shown in Eqn (7.9):

$$M_z = M_0\left(1 - e^{-t/T_1}\right) \tag{7.9}$$

T_1 is the time to reduce the difference between the longitudinal magnetization (M_z) and its equilibrium value by a factor e. If the net magnetization is placed along the $-z$-axis, it gradually returns to equilibrium position along the $+z$-axis at a rate governed by T_1. The time constant that describes the return to equilibrium of the transverse magnetization, MXY, is called the "spin–spin relaxation time" (T_2), the time to reduce the transverse magnetization by a factor e, represented in Eqn (7.10):

$$MXY = MXY_o e^{-t/T_2} \tag{7.10}$$

Both spin lattice relaxation and spin–spin relaxation processes occur simultaneously; the only restriction is that T_2 is less than or equal to T_1.

The signal intensity on the magnetic resonance image is determined by four basic parameters: proton density, relaxation time T_1, relaxation time T_2, and flow. Proton density is the concentration of protons in the tissue in the form of water and macromolecules (proteins, fat, etc.). The T_1 and T_2 relaxation times define the way that the protons revert back to their resting states after the initial RF pulse.

7.3.2 IMAGE FORMATION

The main principle behind all MRI is the resonance equation, which states that the resonance frequency ν (also known as the Larmor frequency) of a spin is proportional to the magnetic field B_0 (Eqn (7.11)):

$$\nu = \gamma B_0 \tag{7.11}$$

where γ is the gyromagnetic ratio. For hydrogen, $\gamma = 42.58$ MHz/T.

The magnetic field gradient is a variation in magnetic field with respect to position, which allows imaging of regions of spin. It could be one-dimensional (1D) or multidimensional. The most useful type of gradient in MRI is a 1D linear magnetic field gradient. A 1D magnetic field gradient along the x-axis in a magnetic field B_0 indicates that the magnetic field is increasing in the x-direction. Here, the length of the vectors represents the magnitude of the magnetic field. G_x, G_y, and G_z are used as symbols for a magnetic field gradient in the x-, y-, and z-directions, respectively.

The isocenter of a magnet is the center point of the magnet where $(x, y, z) = 0, 0, 0$. The magnetic field at the isocenter is B_0 and the resonant frequency is v_0. When a linear magnetic field gradient is applied, different regions in the subject experience different magnetic fields. The result is an NMR spectrum with more than one signal. The amplitude of the signal is proportional to the number of spins in a plane perpendicular to the gradient. This procedure is known as frequency encoding and causes the resonance frequency to be proportional to the position of the spin (Eqns (7.12) and (7.13)).

$$v = \gamma(B_0 + xG_x) = v_0 + \gamma x G_x \tag{7.12}$$

$$x = \frac{v - v_0}{\gamma G_x} \tag{7.13}$$

Among the first forms of MRI is back projection, in which an object is first placed in a magnetic field. A 1D field gradient is applied at several angles (between $0°$ and $359°$), and the NMR spectrum for each gradient is recorded. After recording these data, the same are back-projected using a computer. After suppressing the background intensity, an image is seen. The back-projection technique is also known as the inverse radon transform.

7.3.2.1 Imaging Coordinates

Clinical imagers do not use the *XYZ* magnetic resonance coordinate system for collection and presentation of images. Instead, the anatomic coordinate system is used. In this system the axes are referenced to the body. The three axes are left-right (*L/R*), superior-inferior (*S/I*), and anterior-posterior (*A/P*) (Figure 7.15). Similarly, on clinical imagers the terms *XY, XZ,* and *YZ* are not used to indicate the

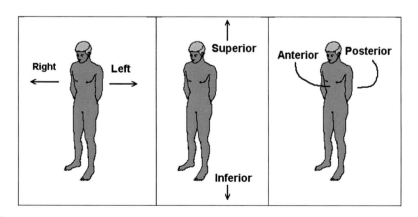

FIGURE 7.15

Depiction of three imaging coordinates.

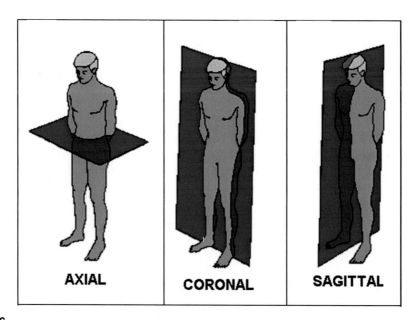

FIGURE 7.16

Depiction of magnetic resonance imaging in three planes.

imaged planes. An imaged plane perpendicular to the long axis of the body is called an "axial plane." The sides of this plane are *L/R* and *A/P*. A plane bisecting the front of the body from the back is referred to as a "coronal plane." The sides of this plane are *L/R* and *S/I*. A plane bisecting the left and right sides of the body is called a "sagittal plane." The sides of this plane are *S/I* and *A/P* (Figure 7.16).

7.4 ULTRASOUND

Ultrasound is mechanical waves at frequencies above 20 kHz, beyond the upper limit of the human auditory acoustic frequency range (namely, 20–20,000 Hz). These waves are propagated by vibration of particles in the medium and may be reflected and transmitted, passing from one medium to another (Cho, 2010). Detailed information about the physical properties of materials can be acquired through the amount of energy reflected or transmitted through them, depending on their relative acoustic impedances. In addition, the time of flight and velocity could also indicate a material property or changes in material characteristics, since ultrasound velocity depends on the density and the elastic property of the medium. Like light waves, incident ultrasound captures objects, and ultrasound energy attenuation differs based on the internal structure of an object, which produces an echo that leads to a series of points of light displayed on the screen, that is, the ultrasound image. The image contrast primarily depends on differences in densities and speeds of sound because these properties determine the scattering and the reflectivity of tissue. Since the 1960s, ultrasound imaging has undergone considerable development as a result of rapid development of modern electronic and computer technologies and signal-processing techniques. There are several different modes of ultrasound, including A-mode (amplitude mode), B-mode (brightness mode), C-mode, motion mode, Doppler mode, pulse-inversion mode, and harmonic mode. Of these, the B-mode instrument has become the most commonly used in diagnostics. Ultrasound imaging, in comparison to other internal imaging techniques, is cheap, easy, and without complicated image-processing procedures. It is now a mature tool for medical diagnostics, and more than a quarter of all clinical imaging procedures in the United Kingdom is ultrasound imaging. In recent years, ultrasonic imaging proved its merit as one of the most promising techniques for food quality and safety assessment because of its nondestructive nature, rapidity, and online applicability (Chen et al., 2013).

Two parameters mostly used in ultrasonic measurements are ultrasonic velocity and the attenuation coefficient. Velocity is the most widely used parameter because of its simplicity. However, a limitation of this method is that the distance traveled by the wave must be known. Ultrasonic velocity and the attenuation coefficient could serve as good indicators of a material's property or a change in material characteristics such as density, elastic modulus, and viscosity. The ultrasonic parameters therefore can provide detailed information about physicochemical properties of media. In addition, foreign objects such as bone, glass, or metal fragment residue in or on food products can precisely be seen using ultrasonic imaging as a result of the strong reflection and refraction of ultrasonic waves at the interfaces of the host tissue and the foreign object(s).

The traditional contact ultrasonic method uses a coupling medium between the transducer and the test specimen to overcome the high attenuation resulting from the large acoustic impedance mismatch between air and the material. The use of a couplant, however, might change or destroy liquid-sensitive, porous, or continuously formed food materials. In addition, the use of a couplant makes rapid measurement and process control cumbersome. Several novel methods of noncontact ultrasonic imaging have therefore been developed. The advantage of noncontact measurement is its ability to measure the ultrasonic velocity, attenuation coefficient, and thickness without contacting the sample. Noncontact ultrasound has become an exciting alternative to the traditional ultrasonic method, especially for quality evaluation of certain foods (Cho, 2010).

7.4.1 THEORY

The mechanism of ultrasonic wave propagation can be described using stress and strain. When stress is applied to a material, it produces elastic waves carrying changes in stress and strain. Wave propagation is created by a balance between the stress and strain when fractions of a medium are distorted by traction forces, as shown in Figure 7.17. From the definition of stress and strain (Eqns (7.14) and (7.15)), the wave equation can be derived as Eqn (7.22) (Cho, 2010):

$$T = \frac{F}{A} \tag{7.14}$$

where T is the stress, F is the longitudinal force, and A is the cross-sectional area.

If dz is assumed to be small,

$$\xi + d\xi = \xi + \left(\frac{\partial \xi}{\partial z}\right) dz \tag{7.15}$$

$$d\xi = \left(\frac{\partial \xi}{\partial z}\right) dz \tag{7.16}$$

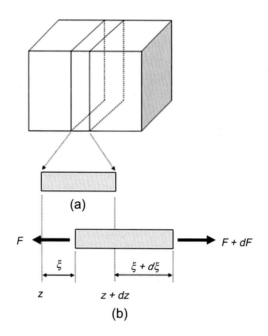

FIGURE 7.17

Depiction of an (a) undistorted and (b) distorted medium.

$$S = \frac{d\xi}{dz} = \frac{\left(\frac{\partial \xi}{\partial z}\right) dz}{dz} = \frac{\partial \xi}{\partial z} \tag{7.17}$$

where ξ is the longitudinal displacement of a particle, S is the strain, and z is the displacement in the longitudinal direction.

$$dF = (F + dF) - F = \left(F + \frac{\partial F}{\partial z} dz\right) - F = \frac{\partial F}{\partial z} dz = A \frac{\partial T}{\partial z} dz \tag{7.18}$$

From the Newton's law,

$$F = ma = \rho A dz \frac{\partial^2 \xi}{\partial t^2} = TA \tag{7.19}$$

$$\frac{\partial T}{\partial z} = \rho \frac{\partial^2 \xi}{\partial t^2} \tag{7.20}$$

where ρ is the density of a medium.

Particle velocity v is obtained from the time derivative of particle displacement:

$$v = \frac{\partial \xi}{\partial t} \tag{7.21}$$

Hence, the wave equation becomes

$$\frac{\partial T}{\partial z} = \rho \frac{\partial v}{\partial t} \tag{7.22}$$

Equation (7.22) is the 1D wave equation. This partial differential equation can be used to analyze ultrasonic wave propagation in an elastic medium.

7.4.2 ULTRASONIC PARAMETERS

As mentioned earlier, the two parameters that are most widely used in ultrasonic measurements are ultrasonic velocity and the attenuation coefficient. These parameters are related to the physical properties of media, such as the structure, texture, and physical state of the components.

7.4.2.1 Velocity

Ultrasonic velocity is a constant quantity for a material in a given state and depends on its physical properties. Velocity is a vector quantity that describes both the magnitude and the direction, whereas speed is a scalar quantity that provides only the magnitude of velocity. In fact, "ultrasonic speed" is a more accurate expression; however, both terms have been used equally in engineering since the direction of ultrasonic velocity is always considered as the same direction as the ultrasonic propagation. In the case of solid material, the ultrasonic velocity is related to the modulus of elasticity E and the density ρ of the solid material. The theoretical background of acoustic wave velocity in viscoelastic materials follows (Cho, 2010).

Stress (T) in a viscoelastic material is related to the elastic modulus and viscosity by the relationship

$$T = ES + \eta \frac{dS}{dt} \tag{7.23}$$

where E is the elastic modulus and η is the viscosity.

$$\frac{\partial S}{\partial t} = \frac{\partial}{\partial t}\left(\frac{\partial \xi}{\partial z}\right) = \frac{\partial v}{\partial z} \tag{7.24}$$

Substituting Eqn (7.23) into Eqn (7.22),

$$\frac{\partial}{\partial z}\left(ES + \eta \frac{\partial S}{\partial t}\right) = \rho \frac{\partial v}{\partial t} \tag{7.25}$$

$$\frac{\partial^2}{\partial t \partial z}\left(ES + \eta \frac{\partial S}{\partial t}\right) = \rho \frac{\partial^2 v}{\partial t^2} \tag{7.26}$$

From Eqn (7.24),

$$\frac{\partial^2 S}{\partial z \partial t} = \frac{\partial^2 v}{\partial z^2} \tag{7.27}$$

Substituting Eqn (7.24) and Eqn (7.27) into Eqn (7.26),

$$\rho \frac{\partial^2 v}{\partial t^2} = E \frac{\partial^2 v}{\partial z^2} + \eta \frac{\partial^3 v}{\partial t \partial z^2} \tag{7.28}$$

If the solution of the above equation is a plane wave, the velocity can be defined as

$$v = Ce^{j(wt - \widehat{k}z)} \tag{7.29}$$

where $\widehat{k}$ is the wavenumber, w is the angular velocity (equal to πf), and C is a constant.

If attenuation by absorption is included, the only effect is that the wavenumber becomes complex. It can be assumed as

$$\widehat{k} = \beta - j\alpha \tag{7.30}$$

where α is the attenuation coefficient by absorption and β is the wave propagation constant defined as 2π divided by the wavelength $\left(= \frac{2\pi}{\lambda}\right)$.

From Eqn (7.29),

$$\frac{\partial^2 v}{\partial t^2} = -w^2 v \tag{7.31}$$

$$\frac{\partial^2 v}{\partial z^2} = -\widehat{k}^2 v \tag{7.32}$$

Substituting Eqn (7.31) and Eqn (7.32) into Eqn (7.28),

$$-\rho w^2 = -E\widehat{k}^2 - j\eta \widehat{k}^2 w \tag{7.33}$$

where

$$\widehat{k}^2 = \beta^2 - \alpha^2 - j2\beta\alpha \tag{7.34}$$

Hence, Eqn (7.23) becomes

$$-\rho w^2 = -E\beta^2 + E\alpha^2 - 2\eta\beta\alpha w + j\left(2E\beta\alpha - \eta\beta^2 w + \eta\alpha^2 w\right) \tag{7.35}$$

Using the real part of Eqn (7.35),

$$-\rho w^2 = -E\beta^2 + E\alpha^2 - 2\eta\beta\alpha w \tag{7.36}$$

If η, $\alpha << \beta$, w, the acoustic wave velocity is obtained as

$$v_a = \frac{w}{\beta} = \sqrt{\frac{E}{\rho}} \tag{7.37}$$

The wave propagation in an extended solid medium implies that the lateral dimension of the material is more than five times greater than the wavelength of ultrasonic waves in the material and is denoted as bulk propagation. In this case, the modulus of elasticity should be described by the combination of the bulk modulus and the shear modulus, as in Eqn (7.38) (Kinsler et al., 2000):

$$E = K + \frac{4}{3}G \tag{7.38}$$

where K and G are the bulk and shear moduli, respectively.

Hence the longitudinal ultrasonic velocity in solids is defined as

$$v = \sqrt{\frac{\left(K + \frac{4}{3}G\right)}{\rho}} \tag{7.39}$$

Equation (7.39) holds for plane longitudinal waves traveling in homogenous and isotropic solids.

Ultrasonic waves are different in the case of a liquid medium than those in solids because liquids in equilibrium are always homogeneous, isotropic, and compressible. In addition, the pressure in liquids is scalar and uniform on a volume element. Hence the shear modulus need not be considered; only the bulk modulus is appropriate for longitudinal wave propagation in a liquid medium. The ultrasonic velocity in liquid is simplified as (Cho, 2010)

$$v = \sqrt{\frac{B}{\rho}} = \sqrt{\frac{\gamma B_T}{\rho}} \tag{7.40}$$

where B is the adiabatic bulk modulus, B_T is the isothermal bulk modulus, and γ is the ratio of the specific heat.

7.4.2.2 Measurement of Ultrasonic Wave Velocity

A simple experimental arrangement, as shown in Figure 7.18, can be developed for measuring ultrasonic wave velocity (Cho, 2010). The system consists of an ultrasonic transducer, a pulser-receiver, and a position control system linked to a personal computer. The measurement technique uses a pulse-echo method. The sample is placed in the bottom and the wave propagation velocity through the sample is computed. An electrical pulse generated from the pulser-receiver is converted into an ultrasonic pulse by the transducer. The ultrasonic pulse travels through the sample and reaches the bottom, where it is echoed back to the transducer. The ultrasonic pulse is then converted back into an electrical pulse by the transducer, and the electrical pulse is sent to the oscilloscope for display and further processing. Velocity is calculated by dividing the total distance traveled by the time of flight,

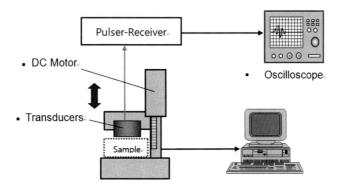

FIGURE 7.18

Schematic of an ultrasonic system for measuring velocity (Cho et al., 2001).

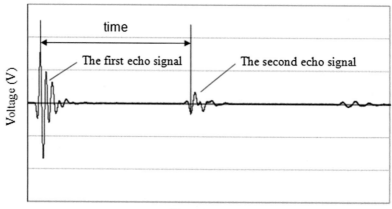

FIGURE 7.19

Ultrasonic pulse-echo signal for a sample (Cho et al., 2001).

which is the time interval between the peak of the first pulse through the sample and the peak of the second pulse reflected from the bottom. Figure 7.19 shows a typical ultrasonic pulse-echo signal for measuring wave propagation velocity.

7.4.2.3 Attenuation

Attenuation is the amount of decreasing power of a wave while traveling through a material. The major reasons for attenuation are absorption and scattering as the wave propagates through the medium. In the absorption process, ultrasonic energy is permanently converted into heat energy, which may cause a temperature increase in the material. Absorption is caused by a variety of mechanisms, such as internal friction caused by viscosity, thermal conduction, and molecular relaxation. When an ultrasonic wave travels through a nonuniform medium, scattering occurs, in which part of the wave changes its initial direction and propagates separately from the original incident wave, distorting and interfering with the initial wave. Ultrasonic scattering does not reduce the mechanical energy, however; because of the change of wave direction it is difficult to detect. Discontinuities within a medium, such as cracks, holes, and foam, play a role in scattering. The effect of scattering is less than that of absorption in homogeneous media. Attenuation is also affected by the frequency of the ultrasonic wave. In general, attenuation of ultrasonic signals increases as a function of frequency. Theoretically, attenuation can be derived from Eqn (7.35). From the imaginary part of the equation we can write it as follows (Cho, 2010):

$$2E\beta\alpha - \eta\beta^2 w + \eta\alpha^2 w = 0 \tag{7.41}$$

If $\eta, \alpha << \beta, w$, the attenuation can be defined as

$$\alpha = \frac{\eta w \beta}{2E} = \frac{\eta w^2}{2v_a^3 \rho} \tag{7.42}$$

As shown in Eqn (7.42), attenuation is related to the wave velocity, density, viscosity, and frequency. When an ultrasonic wave is propagated through a material, the total loss of ultrasonic energy can be described as in Eqn (7.43):

$$W = 20 \log \left(\frac{A_0}{A} \right) \tag{7.43}$$

where W is the total loss (in decibels), A_0 is the amplitude of the transmitted intensity without a test material between the transducers, and A is the amplitude of the transmitted intensity through the test material.

The attenuation is linearly dependent on the thickness of a material, where other factors affecting the attenuation, such as reflection and coupling loss, are not related to thickness. The reflection and coupling loss for a material and transducer are constant. Hence the total loss W can be expressed as:

$$W = \alpha x + \beta \tag{7.44}$$

where α is the attenuation coefficient (decibels per millimeter), x is the thickness of the test material (millimeters), and β is the reflection and coupling loss (decibels).

Attenuations of two materials with different thickness are defined in Eqns (7.45) and (7.46):

$$W_1 = 20 \log \left(\frac{A_0}{A_1} \right) = \alpha x_1 + \beta \tag{7.45}$$

and

$$W_2 = 20 \log \left(\frac{A_0}{A_2} \right) = \alpha x_2 + \beta \tag{7.46}$$

where W_1 is the total loss after traveling the distance x_1, W_2 is the total loss after traveling the distance x_2, A_1 is the amplitude of the received signal after traveling the distance x_1, and A_2 is the amplitude of the received signal after traveling a distance x_2.

The attenuation coefficient can therefore be calculated by subtracting Eqn (7.45) from Eqn (7.46) using Eqn (7.47):

$$20 \log \left(\frac{A_0}{A_2} \right) - 20 \log \left(\frac{A_0}{A_1} \right) = \alpha(x_2 - x_1) \tag{7.47}$$

$$\alpha = \frac{20 \log(A_1/A_2)}{x_2 - x_1} \tag{7.48}$$

There are two types of attenuation coefficient: an apparent attenuation coefficient, measured in the time domain, and a frequency-dependent attenuation coefficient. The apparent attenuation coefficient is determined by measuring the peak amplitudes of transmitted ultrasonic pulses in the time domain at two different sample thicknesses. If the frequency of the transducer has a narrow band, the apparent attenuation is acceptable; in case of a broad band, however, the frequency-dependent attenuation needs to be used.

7.4.3 NONCONTACT ULTRASONIC MEASUREMENT

As mentioned earlier, the traditional contact ultrasonic method uses a couplant, such as a gel, between the transducer and the test sample to overcome the high attenuation caused by the large acoustic impedance mismatch between air and the material. The limitations of conventional contact ultrasonic techniques can be overcome using noncontact (or air-coupled) ultrasonic transducers. The most important element in developing an air-coupled transducer is the matching layer, which determines the efficiency of ultrasonic transmission from piezoelectric material to the medium. For perfect transmission of ultrasound, a specific matching layer should be developed. The thickness of the matching layer should be a quarter of a wavelength and a specific acoustic impedance of 0.1 MRayl (Hayward, 1997).

To overcome the high acoustic impedance mismatch between air and the test material, highly sensitive noncontact transducers need to be developed. Fox et al. (1983) developed noncontact ultrasonic transducers with a 1- and 2-MHz central frequency using silicon rubber as the matching layer. They demonstrated the ability of the transducers to measure the distance in air from 20 to 400 mm, with an accuracy of 0.5 mm. Haller and Khuri-Yakub (1992) enhanced the transmission efficiency of noncontact ultrasonic transducers using a specially designed matching layer with tiny glass spheres in the matrix of silicone rubber. Bhardwaj (1997, 1998) demonstrated a highly efficient acoustic matching layer for noncontact ultrasonic transducers using soft polymers.

7.4.3.1 Noncontact Ultrasonic Velocity Measurement

A noncontact ultrasonic system can be set up using two transducers, which can operate both as a transmitter and a receiver, as shown in Figure 7.20 (Cho and Irudayaraj, 2003c). Since the system has

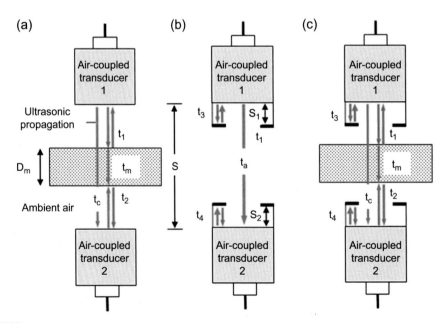

FIGURE 7.20

Schematic of (a) the noncontact ultrasonic measurement with a sample material, and (b) the noncontact air instability compensation ultrasonic measurement (b) without and (c) with a sample material (Cho and Irudayaraj, 2003c).

two channels for data acquisition (with two amplifiers and two analog I/O boards), it can provide four operation modes: two reflection (one for each of the two transducers) and two transmission modes (one used as a transmitter and the other as a receiver, and vice versa). After a simple calibration for the air velocity (Va) using a calibration material with a known thickness, the thickness and ultrasonic velocity of a sample can be calculated directly using Eqns (7.49) and (7.50) (Bhardwaj, 2000):

$$D_m = V_a \left(t_a - \frac{t_1 + t_2}{2} \right) = S - V_a \frac{t_1 + t_2}{2} \tag{7.49}$$

$$V_m = \frac{D_m}{t_m} = \frac{D_m}{t_c - \frac{t_1 + t_2}{2}} \tag{7.50}$$

where D_m is the sample thickness, V_a and V_m are the respective velocities of ultrasound in air and through the sample, S is the distance between transducer 1 and transducer 2, t_m is the time of flight in the test material, t_a is the time of flight between transducer 1 and transducer 2 in air, t_c is the time of flight between transducer 1 and transducer 2 with sample, t_1 is the round-trip time of flight between transducer 1 and the sample, and t_2 is the round-trip time of flight between the sample and transducer 2.

In the case of food sample measurements, the signal of noncontact ultrasound is weak and mixed with random noise; hence it is difficult to identify the original signal from the mixed signal when the noise level is higher than the signal level. An original signal is shown in Figure 7.21(a), and the simulated real signal is displayed in Figure 7.21(b), which is the combination of the random noise and the original signal with a time delay (Cho and Irudayaraj, 2003a). The original signal is hidden in the random noise, of which the level is twice that of the original signal amplitude. To eliminate the random noise from the mixed signal, a cross-correlation method can be used. The correlated result is shown in Figure 7.21(c), in which the noise is much reduced. The advantages of the cross-correlation method are that it not only reduces noise but also estimates the time shift between the signals. Since the position of the maximum output correlation indicates the shifted time between the original signal and the transmitted signal at a time delay, the time of flight of the ultrasonic wave through the sample can be determined by reading the location in time of the maximum peak. For precise identification, the maximum peaks from the correlated signal, the signal envelope, was made using Hilbert transformation (Oppenheim et al., 1999). As shown in Figure 7.21(d), the waveform clearly shows where the main peak is located. In addition, the area underneath the most significant peak above −6 dB from the transmitted signal (in decibels), denoted as integrated response, can be depicted as ultrasonic energy. The integrated response provides information about the energy attenuation during transmission of the ultrasonic wave through air and/or a specimen (Bhardwaj, 2000).

7.4.3.2 Noncontact Ultrasonic Attenuation Measurement

A typical noncontact ultrasound through-transmission signal of a solid material is shown in Figure 7.22 (in right side of figure). The left side figure show in graphical form that the first peak is the signal transmitted directly through air and sample material; the second peak is caused by internal reflection of the transmitted signal. Other periodic peaks are multireflected by a sample material and are hardly observed in materials with high attenuation rate. The attenuation coefficient can be calculated by dividing the difference between the integrated response of the first and the second peaks by the sample thickness.

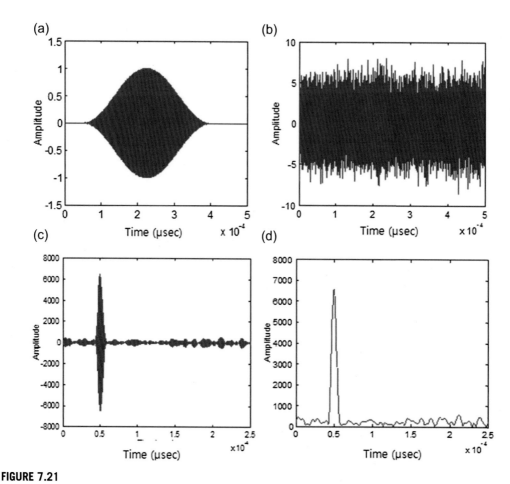

FIGURE 7.21

Transmitted signal (a) that was embedded in noise (b); (c) pulse-compressed signal of (b); (d) enveloped signal of (c) (Cho and Irudayaraj, 2003a).

7.4.3.3 Calibration of Ultrasonic Measurement

A noncontact ultrasonic system should be calibrated before measurement using a known property material, such as a polystyrene block or a Dow Corning silicon fluid. After aligning the transducers parallel to each other and verifying the shape of the received signal, the polystyrene block is placed between the two transducers. The measured thickness is compared with the actual thickness. If the difference is not acceptable, the system parameters, such as the computer-generated chirp used for a transmitted signal, and the parallelism of the transducers need to be adjusted within 0.1% error (Cho, 2010).

7.4.3.4 Noncontact Ultrasonic Image Measurement

To obtain a noncontact ultrasonic image of samples, an X-Y positioning system can be made (Cho and Irudayaraj, 2003b) (Figure 7.23). All scanning and data acquisition are controlled in real time using a

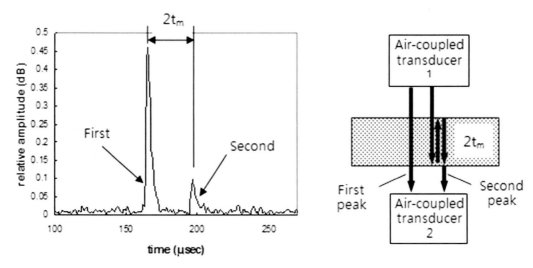

FIGURE 7.22

A typical noncontact ultrasonic signal transmitted through a solid sample (Cho and Irudayaraj, 2003a).

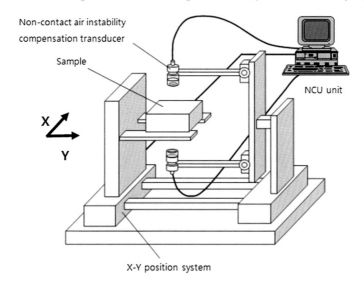

FIGURE 7.23

Schematic of a noncontact ultrasonic (NCU) imaging system (Cho and Irudayaraj, 2003b).

computer. The velocity and attenuation coefficient through the sample can be measured easily. Figure 7.24 shows ultrasonic attenuation images of a metal fragment in a poultry breast and cracks in cheese. Since ultrasonic attenuation is more sensitive to the difference in product characteristics than velocity, the attenuation provides better images than those of velocity in general. The results of

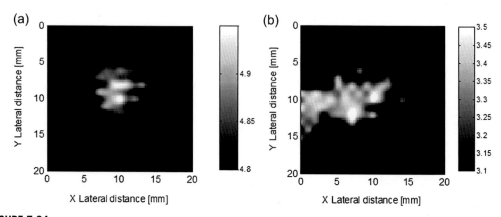

FIGURE 7.24

Modified noncontact ultrasonic attenuation images of a metal fragment (5×3 mm^2) in a poultry breast (a) and a crack in extra sharp cheddar cheese (b) (Cho and Irudayaraj, 2003b).

noncontact ultrasonic images demonstrate its potential to detect the presence of foreign objects and defects inside food materials.

7.4.3.5 Ultrasonic Images

Ultrasonic images are 2D ultrasonic representations of the internal structure of materials. Of the ultrasonic imaging methods, a C-scan is the most common; this is a sequence of waveforms taken at points on a grid overlaid on the surface of a material, as illustrated in Figure 7.25 (Cho, 2010). The waveform to be imaged is defined through the use of an electronic gate. Along with the C-scan there are two other types of ultrasonic data display: A- and B-scans. An A-scan (Figure 7.26) is a displayed waveform of an ultrasonic signal for a point on a sample material, in which one axis represents time

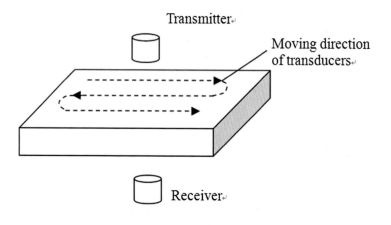

FIGURE 7.25

Illustration of test arrangement for a C-scan.

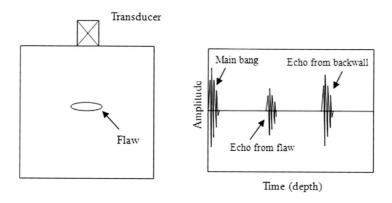

FIGURE 7.26

Illustration of an A-scan in an ultrasonic system.

and the other corresponds to the intensity of the signal. The A-scan provides 1D depth information along the line of the beam's propagation. A B-mode measures the "brightness" modulation of the displayed signal. A B-scan is a sequence of A-scans taken at points along a line on the surface of a material and displayed side by side to represent a cross section of the material, as shown in Figure 7.27 (Cho, 2010). The scans can be made using either pulse-echo or through-transmission inspection techniques.

Another widely used imaging type is a steered image, which is made by steered sweeping beams of a linear array of transducers. The measurement is performed in the pulse-echo mode, so there is no separate moving receiver. The beam is controlled by increasing time delays across the transducer array,

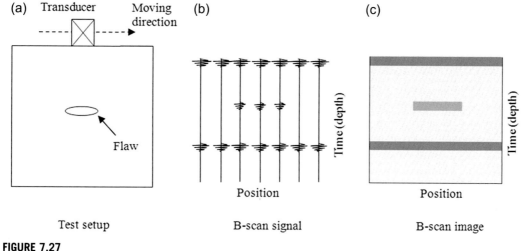

FIGURE 7.27

Illustration of a B-scan in an ultrasonic system.

FIGURE 7.28

Steered wave form of an ultrasonic system.

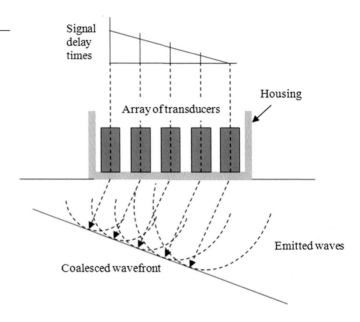

as shown in Figure 7.28. The orientation of the wave front is steered by changing the excitation times across the array. The technique is usually operated in the pulse-echo mode and is used extensively in medical imaging.

Ultrasonic imaging techniques have been used to provide information about meat quality, such as the ratio of fat to lean tissue, both in live animals and in carcasses. Commercialized ultrasonic inspection systems for meat quality are widely used. This technique has also been applied to monitor creaming and sedimentation processes in emulsions and suspensions, as well as to detect foreign objects in some foods.

7.5 HYPERSPECTRAL IMAGING

The hyperspectral imaging (HSI) technique is a relatively recent chemical or spectroscopic imaging analytical tool that is gaining the interest of researchers analyzing food materials. It is an emerging technique that integrates both conventional imaging and spectroscopy to attain both spatial and spectral information about an object. The images captured, usually known as hypercubes, are constructed as 3D data cubes using hundreds of contiguous wavebands for each spatial position of a target (Figure 7.29). The spectra of each pixel can therefore be used to characterize the composition of that specific position, and surface-feature information can be obtained according to the spatial images. There are two conventional methods for HSI acquisition: the "staring imager" configuration and "push-broom" acquisition (Gowen et al., 2007). Push-broom acquisition involves the acquisition of simultaneous spectral measurements from a series of adjacent spatial positions; this requires relative movement between the object and the detector. Some instruments produce HSIs based on a point step and acquiring mode: Spectra are obtained at single points on a sample, and then the sample

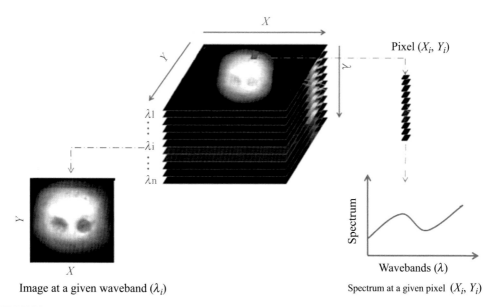

Pixel (X_i, Y_i)

Image at a given waveband (λ_i)

Spectrum at a given pixel (X_i, Y_i)

FIGURE 7.29

Relationship between spectral and spatial dimensions of hypercubes.

is moved and another spectrum is measured. Hypercubes obtained using this configuration are stored as the band interleaved by pixel format. Advances in detector technology have reduced the time required to acquire hypercubes. Line-mapping instruments record the spectrum of each pixel in a line of a sample that is simultaneously recorded by an array detector; the resultant hypercube is stored in the band interleaved by line format (Gowen et al., 2007). HSI systems based on push-broom acquisition typically contain components such as an objective lens, spectrograph, camera, acquisition system, translation stage, illumination source, and computer. Figure 7.30(a) shows an HSI system based on push-broom acquisition, and Figure 7.30(b) shows a hypercube obtained using this system.

The hypercube obtained through the push-broom acquisition system consumes too much time and acquires a huge amount of data for processing. This system is relatively costly, too. Push-broom acquisition generally requires spectral preprocessing, image processing, and model recognition to reach the intended goals in a laboratory; some optimum images at specific wavebands are found from the hypercube. Thus, the right filters corresponding to some specific wavebands are selected to develop a multispectral imaging system for real-time usage.

HSI acquisition using a multispectral imaging system is also called the "staring imager" configuration; this involves keeping the image field of view fixed and obtaining images one wavelength after another. Hypercubes obtained using this configuration thus comprise a 3D stack of images, stored in what is known as the band sequential format, shown in Figure 7.30(d). Wavelength in the staring imager configuration is typically moderated using a tunable filter (Gowen et al., 2007). The simplest method to obtain images at a discrete spectral region is by positioning a bandpass filter (or interference filter) in front of a monochrome camera lens. Hypercubes can be obtained by capturing a series of

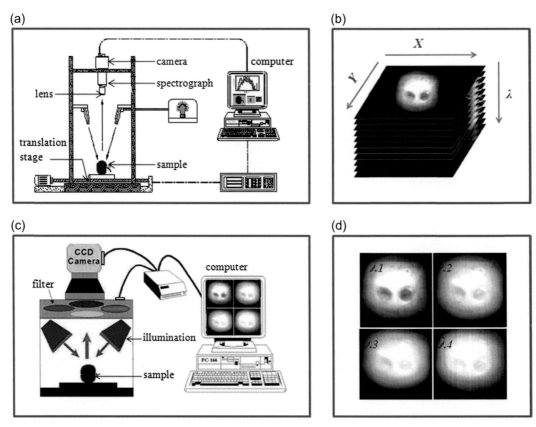

FIGURE 7.30

A hyperspectral imaging (HSI) system and three-dimensional hypercube. (a) An HSI system based on "push-broom" acquisition. (b) A hypercube created by push-broom acquisition. (c) An HSI system based on "staring imager" acquisition. (d) A hypercube obtained using staring imager acquisition (Chen et al., 2013).

spectral images by sequentially changing filters in front of the camera. Figure 7.30(c) shows a multispectral imaging system based on a rotating filter wheel. By integrating spectroscopy with image analysis, the HSI technique has shown its superiority over commonly used imaging methods. HSI may be carried out in reflectance, transmission, or fluorescence modes; the majority of the current published literature, however, uses reflectance mode only. This technique has recently emerged as a powerful analytical tool and a rapid, noncontact, and nondestructive technique for food quality and safety evaluation.

7.5.1 THEORY

In the case of HSI, basic theory about the interaction of incident light or electromagnetic waves with molecules of object is almost the same as that of spectroscopy described in Chapter 6.

7.5.2 ELECTROMAGNETIC PENETRATION DEPTH

Electromagnetic or light penetration depth is defined as the depth at which the incident light is reduced by 99% (Huang et al., 2014). It can vary according to the status, the type of sample, and the detection waveband. Optical features of the light penetration depths are mainly determined by strong absorbing constituents in the sample. Research regarding the penetration depth of light in the visible and near infrared (NIR) range is very limited. Lammertyn et al. (2000) proved that light penetration depth in apples was dependent on the detection wavelength by putting forward a nonlinear model describing the correlation between the reflectance and thickness of apple slices. The penetration of apple is up to 4 mm in the 700–900 nm range and between 2 and 3 mm in the 900–1900 nm range. In research by Qin and Lu (2008), the light penetration depth in tissues of apple, peach, pear, kiwifruit, tomato, zucchini, cucumber, and plum was calculated according to the absorption and reduced scattering spectra of the test samples at different wavelengths. The minimum light penetration depths ranged from 7.1 mm at 535 nm for the plum to 13.8 mm at 720 nm for the zucchini. The wavelengths were correlated to the absorption peaks of the major pigments in the fruits and vegetables. The maximum penetration depths ranged from 18.3 mm for the apple to 65.2 mm for the zucchini. This study highlighted that penetration depth varies to a great extent depending on the type of object being studied and the applied depth. The penetration depth could have effects on the HSI detection. Most of the studies about penetration depth were conducted using fruits. Further research concerning penetration depth would prove beneficial by providing references for thickness determination, and could be valuable for designing an appropriate and accurate sensing configuration in both spectroscopy and HSI techniques for food analysis.

In addition to the above imaging methods, some research on thermal imaging, fluorescence imaging, and odor imaging are also being used to inspect food quality and safety, but they are in the early stages.

7.6 PRACTICAL APPLICATIONS OF IMAGING METHODS

There are numerous reported applications of different imaging methods for the detection of different adulterants and contaminants and the determination of food quality parameters, but the majority of them have been published in research journals, books, and reviews. It is difficult to describe all of them here. Therefore, only a few of them are discussed in the following sections. Readers are encouraged to scan the literature for further detailed applications of each imaging technique.

7.6.1 DETECTION OF LOW-VALUE FOOD IN HIGH-VALUE FOOD

Visual image analysis is a cost-effective and nondestructive digital technique that uses computer vision and can be used to detect the extent of adulteration. Tanck and Kaushal (2014) used image analysis to differentiate adulteration of genuine basmati rice with low-quality rice by analyzing the perimeter and aspect ratio of rice grain. This method was found to be fairly accurate and is an improvement over the traditional method of human inspection.

Determination or certification of honey of unifloral origin is an important issue because low-grade honey may be used to adulterate highly priced variants with the motive of earning higher profits. Common techniques to determine or certify the unifloral origin of honeys are either time-consuming or costly. NMR spectroscopy has been used to assess the botanical origin of several food products and quantify some major compounds in them. Ohmenhaeuser et al. (2013) used a combination of MRI and chemometrics to provide qualitative information about botanical origins, identifying marker compounds for the specific honey types. Principal component analysis helped identify clusters of honeys from the same botanical origin based on the glucose and fructose present in them. Soft independent modeling of class analogy and partial least squares discriminant analysis was used to automatically classify honeys according to their botanical origin with 95–100% accuracy. Further, 13 compounds (carbohydrates, aldehydes, aliphatic and aromatic acids) also were quantified using external calibration curves.

^{1}H NMR fingerprint technologies were used to identify 20 different chemical compounds found in tea from different origins (Chen et al., 2006). Use of MRI, principal component analysis, and cluster analysis helped to distinguish different types of tea from different origins. Similarly, beers of different origin were identified using proton (^{1}H) NMR spectra fingerprinting (Zhao et al., 2009). An NMR technique was also applied for detecting the incorporation of waste oil in fresh oil. In that study, the presence of solid fat content indicated the use of lower-grade oil. The method could detect even 1% adulteration of low-quality oil in edible oil, which was assumed to be zero in edible oil but very high in swill oil and waste oil (Wang et al., 2008).

Xu et al. (2014) developed a low-field unilateral NMR method for detecting the adulteration of virgin olive oil with sunflower oil and red palm oil. Using this technique, adulteration of virgin olive oil with 10% or more of sunflower or palm oil became feasible. 2D inverse Laplace transformation helped distinguish the transverse relaxation and self-diffusion behaviors of different oils. This study demonstrated the use of NMR as a nondestructive screening technique that can even be performed in sealed bottles.

7.6.2 DETECTION OF ADULTERANTS AND CONTAMINANTS

Adulterants and contaminants are rampant in food materials nowadays, and because of their wide variations it is a real challenge to monitor and control them. The soft X-ray technique is now successfully being used for detection of foreign matter, internal defects, and weevils inside grain and fruits (Diener et al., 1970; Han et al., 1992; Thomas et al., 1995; Keagy et al., 1996; Schatzki et al., 1997; Casasent et al., 1998, 2001; Abbott, 1999; Arslan et al., 2000; Kim and Schatzki, 2000, 2001; Haff and Slaughter, 2004; Karunakaran et al., 2004; Tollner et al., 2005; Neethirajan et al., 2006; Fornal et al., 2007; Kotwaliwale et al., 2007a, 2014; Narvankar et al., 2009). X-rays with energy ranging from 15 to 80 kVp at various currents have reportedly been beneficial.

Application of CT in the nondestructive evaluation of food safety and quality, however, is mostly in the research stage. Research is being carried out using medical CT scanners and software developed for analysis of human anatomy. Ogawa et al. (1998) used a medical CT scanner to detect selected nonmetallic materials embedded in various fluids and food materials; most recently, Jha et al. (2012) successfully detected undesirable soft-tissue and stone weevils in mango using CT, which no doubt gives better results than X-ray imaging. The biggest disadvantage of this technique, however, is the unavailability of small equipment, high cost, and requirement of trained manpower to operate the machines and analyze the data.

Hatsukade et al. (2013) developed an ultra-low frequency NMR and MRI system using a high-temperature superconducting–superconducting quantum interference device for the detection of contaminants in food and drink. The protons in a water sample were prepolarized using a permanent magnet, and NMR signals from water samples were measured with or without various contaminants, such as stainless steel, aluminum, and glass balls. Modified NMR signal intensities were recorded for both contaminated and pure samples and analyzed. NMR was also used to analyze the sugar adulteration in pure apple or orange juice using stable isotope ratio of ^{2}H to ^{1}H distribution of ethanol derived from the added sugars (Martin et al., 1996).

Milk is often adulterated with melamine to avoid the detection of added water. Milk contaminated with melamine may result in kidney disorders and may lead to death. Ultrasonic characterization of adulterated milk was used to detect gross levels of melamine contamination using sound speed and density in skim milk (Elvira et al., 2007; Elvira and Rodríguez, 2009). Nazário et al. (2007) also used ultrasonic and neural networks to detect adulteration of milk with water and serum using acoustic parameters, such as propagation velocity and attenuation coefficient under controlled temperature conditions, with an adulteration range of 1–10% added water/serum. Shah et al. (2001) developed a real-time approach for detecting seal defects in food packages using ultrasonic imaging. Ultrasonic imaging has also been applied to detect the fat content of animal products and to predict the lean content of carcasses (Youssao et al., 2002). Although the result of prediction was not satisfactory, real-time ultrasound could be a tool to predict the composition of pig carcasses.

Detection of defects in different kind of foods using HSI has also been investigated. Mehl et al. (2002) developed a multispectral system and detected bruises, fungi, scrabs, and soils in apples with considerable accuracy. Gowen et al. (2008) investigated damage on the caps of white mushrooms (*Agaricus bisporus*) using a push-broom line-scanning HSI instrument, and they illustrated the potential of the system for nondestructive monitoring of damaged mushrooms on the processing line. Lu et al. (2011) applied visible/NIR HSI combined with radial basis function support vector machine classification to inspect hidden bruises on kiwi fruits. Little research has been attempted for the application of HSI to food-quality classification. ElMasry et al. (2011) developed an NIR HSI system to assess the quality of cooked turkey with different ingredients and processing parameters. Maftoonazad et al. (2011) used an HSI technique in which multilayer artificial neural networks were used to model quality changes in avocados during storage at different temperatures. This inspection approach based on image analysis and processing also found satisfactory results in tea-quality assessment (Jiang et al., 2011; Li and He, 2009; Zhao et al., 2009; Wu et al., 2009; Li et al., 2008; Chen et al., 2008). Researchers have explored the applications of HSI in other food-quality analyses, such as pear (Zhao et al., 2010), citrus fruits (Gomez-Sanchis et al., 2008), tomato (Qin et al., 2011), and pork (Jun et al., 2007; Barbin et al., 2012) as well.

7.6.3 DETECTION OF MICROBES AND TOXINS

Rapid identification and quantification of toxic elements such as arsenic, lead, mercury, and selenium in cranberry juice, yogurt, and chocolate have been investigated in terms of selectivity, limits of detection, linear dynamic range, accuracy, and speed using portable X-ray fluorescence analyzers. Limits of detection for all four target elements were in the range of 5–10 ppm. The study revealed that although X-ray fluorescence was less sensitive than atomic spectrometry methods, it offered a number of significant advantages, including minimal sample preparation, rapid analysis, multielement detection, and true field use using hand-held analyzers. These capabilities make X-ray fluorescence a powerful tool for screening for toxic elements and rapidly responding to emergency situations that require the identification and quantification of toxic elements.

In food-quality inspection, the microbial integrity of flexible food packages usually depends on a zero-defect level in the fused-seam seal. Detection of microbes and toxins in food materials using MRI is not yet available; however, El-Boubbou et al. (2007) reported a nondestructive magnetic glyco-nanoparticle–based system for rapid detection of *Escherichia coli*. Based on the utility of nanoparticles as nanosensors, the identities of three different pathogenic *E. coli* strains were determined. Since bacteria use mammalian cell surface carbohydrates as anchors for attachment, a combination of magnetic nanocomposites and diverse carbohydrates were used to develop a biosensing system. This system could remove up to 88% of the target bacteria from the medium. The rapid detection and removal of pathogens provided an attractive possibility for pathogen decontamination and diagnostic applications.

Concerning food-safety control, HSI has made a great contribution. Fecal contamination has become an important aspect in various food materials. Yang et al. (2011) optimized and evaluated multispectral algorithms for detection of fecal contamination on apples. Yoon et al. (2007) differentiated fecal and nonfecal poultry carcasses using HSI combined with kernel density estimation. They also developed a prototype line-scan HSI system for inspecting poultry carcasses with fecal material and ingesta, and they provided a commercially viable imaging platform for fecal detection (Yoon et al., 2011). Research on determining the total viable count of microbes in chilled pork was carried out by Peng et al. (2010). They showed that HSI is a valid tool for assessing the quality and safety properties of chilled pork during storage. Tao et al. (2012) applied multiple linear regression models to predict pork tenderness and *E. coli* contamination based on hyperspectral scattering technique.

Nowadays, the problem of pesticide residues has seriously influenced food safety. Hu et al. (2006) carried out an initial experiment to detect pesticide residues on fruit surfaces using laser HSI, and they offered an encouraging method for pesticide residue detection. The HSI technique has so far been applied in the detection of defective fruits, damaged mushrooms, meat products, and tea. Among these, detection of contamination in meat products seems to attract the greatest attention. Image-processing methods or mathematical algorithms, such as artificial neural networks and multiple linear regression, have been attempted in these studies, leading to encouraging results. Detection of food safety aspects using imaging methods has gained momentum; however, much attention is needed for extensive research on their real field applications.

REFERENCES

Abbott, J., 1999. A quality measurement of fruits and vegetables. Postharvest Biol. Technol. 15, 207−225.

Arslan, S., Inanc, F., Gray, J.N., et al., 2000. Grain flow measurements with X-ray techniques. Comput. Electronics Agric. 26 (200), 65−80.

Barbin, D., Elmasry, G., Sun, D.-W., Allen, P., 2012. Near-infrared hyperspectral imaging for grading and classification of pork. Meat Sci. 90, 259−268.

Bhardwaj, M.C., 1997, 1998. Ultrasonic Transducer for High Transduction in Gases and Method for Ultrasonic Non-contact Transmission into Solid Materials. Int Patent.

Bhardwaj, M.C., 2000. High transduction piezoelectric transducers and introduction of non-contact analysis. e-J. Nondestr. Test. Ultrason. [Serial Online] 1, 1−21 available from NDT net. http://www.ndt.net (posted January 2000).

Buzzell, P., Pintauro, S., 2003. Dual Energy X-ray Absorptiometery. Department of Food Sciences and Nutrition. University of Vermont available at http://nutrition.uvm.edu/bodycomp/dexa/ (accessed 06.01.03. by using laser imaging, Acta Agri. Univ. Jiangxi 6, 013).

Casasent, D., Talukder, A., Keagy, P., et al., 2001. Detection and segmentation of items in X-ray imagery. Trans. ASAE 44 (2), 337–345.

Casasent, D.A., Sipe, M.A., Schatzki, T.F., et al., 1998. Neural net classification of X-ray pistachio nut data. Lebensm.-Wiss. Technol. 31 (2), 122–128.

Chen, P., McCarthy, M.J., Kauten, R., 1989. NMR for internal quality evaluation of fruits and vegetables. Trans. ASAE 32 (5), 1747–1753.

Chen, X.J., Wu, D., He, Y., 2008. Study on discrimination of tea based on color of multispectral image. Spectrosc. Spect. Anal. 28, 2527–2530.

Chen, B., Zhang, W., Kang, H., 2006. Fingerprinting tea by ^{1}H NMR. Chin. J. Magn. Reson. 23 (2), 169–180.

Chen, Q., Zhang, C., Zhao, J., Quyang, Q., 2013. Recent advances in imaging techniques for non-destructive detection of food quality and safety. Trends Anal. Chem. 52, 261–274.

Cho, B.K., 2010. Ultrasonic technology. In: Jha, S.N. (Ed.), Nondestructive Evaluation of Food Quality: Theory and Practice. Springer.

Cho, B.K., Irudayaraj, J.M.K., 2003a. A non-contact ultrasound approach for mechanical property determination of cheese. J. Food Sci. 68, 2243–2247.

Cho, B.K., Irudayaraj, J.M.K., 2003b. Foreign object and internal disorder detection in food materials using non-contact ultrasound imaging. J. Food Sci. 68, 967–974.

Cho, B.K., Irudayaraj, J.M.K., 2003c. Design and application of a non-contact ultrasound velocity measurement system with air instability compensation. Trans. ASAE 46, 901–909.

Cho, B., Irudayaraj, J.M.K., Omato, S., 2001. Acoustic sensor fusion approach for rapid measurement of modulus and hardness of cheddar cheese. Applied Engng. in Agric. 17, 827–832.

Clark, C.J., Hockings, P.D., Joyce, D.C., et al., 1997. Application of magnetic resonance imaging to pre- and post-harvest studies of fruits and vegetables. Postharvest Biol. Technol. 11, 1–21.

Cunningham, I.A., Judy, P.F., 2000. 'Computed Tomography.' In: Joseph, D. (Ed.), The Biomedical Engineering Handbook, second ed. Bronzino, Boca Raton: CRC Press LLC, available at http://www.kemt-old.fei.tuke.sk/. (accessed 30.10.15).

Curry, T.S., Dowdey, J.E., Murry, R.C., 1990. Christensen's Physics of Diagnostic Radiology, fourth ed. Williams and Wilkins, Baltimore.

Diener, R.G., Mitchell, J.P., Rhoten, M.L., 1970. Using an X-ray image scan to sort bruised apples. Agric. Eng. 4, 356–361.

El-Boubbou, K., Gruden, C., Huang, X., 2007. Magnetic glyco-nanoparticles: a unique tool for rapid pathogen detection, decontamination, and strain differentiation. J. Am. Chem. Soc. 129, 13392–13393.

ElMasry, G., Iqbal, A., Sun, D.-W., Allen, P., Ward, P., 2011. Quality classification of cooked, sliced turkey hams using NIR hyperspectral imaging system. J. Food Eng. 103, 333–344.

Elvira, L., Durán, C., Sierra, C., Resa, P., de Espinosa, F.M., 2007. Ultrasonic measurement device for the characterization of microbiological and biochemical processes in liquid media. Meas. Sci. Technol. 18, 2189–2196.

Elvira, L., Rodríguez, J., 2009. Sound speed and density characterization of milk adulterated with melamine. J. Acoust. Soc. Am. 125, 177–182.

Fornal, J., Jelinski, T., Sadowska, J., et al., 2007. Detection of granary weevil *Sitophilus granarius* (L.) eggs and internal stages in wheat grain using soft X-ray and image analysis. J. Stored Prod. Res. 43, 142–148.

Fox, J.D., Khuri-Yakub, B.T., Kino, G.S., 1983. High frequency wave measurements in air. IEEE Ultrason. Symp. 1, 581–592.

Gomez-Sanchis, J., Molto, E., Camps-Valls, G., Gomez-Chova, L., Aleixos, N., Blasco, J., 2008. Automatic correction of the effects of the light source on spherical objects. An application to the analysis of hyperspectral images of citrus fruits. J. Food Eng. 85, 191–200.

Gowen, A.A., O'Donnell, C.P., Cullen, P.J., Downey, G., Frias, J.M., 2007. Hyperspectral imaging — an emerging process analytical tool for food quality and safety control. Trends Food Sci. Technol. 18, 590—598.

Gowen, A., O'Donnell, C., Taghizadeh, M., Cullen, P., Frias, J., Downey, G., 2008. Hyperspectral imaging combined with principal component analysis for bruise damage detection on white mushrooms (*Agaricus bisporus*). J. Chemom 22, 259—267.

Haff, R.P., Slaughter, D.C., 2004. Real-time X-ray inspection of wheat for infestation by the granary weevil, *Sitophilus granarius* (L.). Trans. ASAE 47 (2), 531—537.

Haller, M.I., Khuri-Yakub, B.T., 1992. 1-3 Composites for ultrasonic air transducer. IEEE Ultrason. Symp. 2, 937—939.

Han, Y.J., Bowers, S.V., Dodd, R.B., 1992. Nondestructive detection of split-pit peaches. Trans. ASAE 35 (6), 2063—2067.

Hatsukade, Y., Tsunake, S., Yamamoto, M., Abe, T., 2013. Feasibility study of contaminants detection for food with ULF-NMR/MRI system using HTS-SQUID. Phys. C 11 (494), 199—2002. http://dx.doi.org/10.1016/j.physc.2013.04.004.

Hayward, G., 1997. Air coupled NDE — constraints and solutions for industrial implementation. IEEE Ultrasound Symp. 1, 665—673.

Hu, S., Liu, M., Liu, H., 2006. A study on detecting pesticide residuals on fruit surface by using laser imaging. Acta Agri. Univ. Jiangxi 6, 013.

Huang, H., Liu, L., Ngadi, M.O., 2014. Recent development in hyperspectral imaging for assessment of food quality and safety. Sensors, 14, 7248—7276. http://dx.doi.org/10.3390/s140407248.

Hubbell, J.H., Seltzer, S.M., 1995. Tables of X-ray Mass Attenuation Coefficients and Mass Energy Absorption Coefficients and Mass Energy Absorption Coefficients 1 keV to 20 MeV for Elements Z = 1 to 92 and 48 Additional Substances of Dosimetric Interest. NISTIR 5632. National Institute of Standards and Technology, US Department of Commerce, Gaithersburg, MD, USA.

Jha, S.N., Kar, A., Kotwaliwale, N., 2012. Development of Nondestructive Methods for Quality Evaluation of Mango. Final report submitted to NAIP. ICAR, New Delhi.

Jiang, F., Qiao, X., Zheng, H., Yang, Q., 2011. Grade discrimination of machine-fried Longjing tea based on hyperspectral technology. Trans. CSAE 27, 343—348.

Jun, Q., Ngadi, M., Wang, N., Gunenc, A., Monroy, M., Gariepy, C., Prasher, S., 2007. Pork quality classification using a hyperspectral imaging system and neural network. Int. J. Food Eng. 3, 1—12.

Karunakaran, C., Jayas, D.S., White, N.D.G., 2004. Identification of wheat kernels damaged by the red flour beetle using X-ray images. Biosystems Eng. 87 (3), 267—274.

Keagy, P.M., Parvin, B., Schatzki, T.F., 1996. Machine recognition of navel orange worm damage in X-ray images of pistachio nuts. Lebensm. Wiss. Technol. 29 (1&2), 140—145.

Kim, S., Schatzki, T.F., 2000. Apple water-core sorting system using X-ray imagery: I. Algorithm development. Trans. ASAE 43 (6), 1695—1702.

Kim, S., Schatzki, T.F., 2001. Detection of pinholes in almonds through X-ray imaging. Trans. ASAE 44 (4), 997—1003.

Kinsler, L.E., Frey, A.R., Coppens, A.B., et al., 2000. Fundamentals of Acoustics, fourth ed. John Wiley & Sons, New York.

Kotwaliwale, N., 2010. Radiography, CT and MRI. In: Jha, S.N. (Ed.), Nondestructive Evaluation of Food Quality: Theory and Practice. Springer.

Kotwaliwale, N., Weckler, P.R., Brusewitz, G.H., 2006. X-ray attenuation coefficients using polychromatic X-ray imaging of pecan components. Biosystems Eng. 94 (2), 199—206.

Kotwaliwale, N., Subbiah, J., Weckler, P.R., et al., 2007a. Calibration of a soft X-ray digital imaging system for biological materials. Trans. ASABE 50 (2), 661—666.

Kotwaliwale, N., Weckler, P.R., Brusewitz, G.H., et al., 2007b. Non-destructive quality determination of pecans using soft X-rays. Postharvest Biol. Technol. 45, 372—380.

Kotwaliwale, N., Singh, K., Kalne, A., Jha, S.N., Seth, N., Kar Abhijit, 2014. X-ray imaging methods for internal quality evaluation of agricultural produce. J. Food Sci. Technol. 51 (1), 1–15.

Lammertyn, J., Peirs, A., de Baerdemaeker, J., Nicolaı, B., 2000. Light penetration properties of NIR radiation in fruit with respect to non-destructive quality assessment. Postharv. Biol. Technol. 18, 121–132.

Li, X.L., He, Y., 2009. Classification of tea grades by multi-spectral images and combined features. Trans. CSAM 40, 113–118.

Li, X.L., He, Y., Qiu, Z.J., 2008. Textural feature extraction and optimization in wavelet sub-bands for discrimination of green tea brands. Proc. 17th Int. Conf. Mach. Learn. Cybern. 3, 1461–1466.

Lu, Q., TaNg, M., Cai, J., Zhao, J., Vittayapadu Ng, S., 2011. Vis/NIR hyperspectral imaging for detection of hidden bruises on kiwifruits. Czech. J. Food Sci. 29, 595–602.

Maftoonazad, N., Karimi, Y., Ramaswamy, H., Prasher, S., 2011. Artificial neural network modeling of hyperspectral radiometric data for quality changes associated with avocados during storage. J. Food Process. Preserv. 35, 432–446.

Martin, G.G., Hanote, V., Lees, M., Martin, Y.L., 1996. Interpretation of combined 2H SNIF/NMR and 13C SIRA/MS analyses of fruit juices to detect added sugar. J. Assoc. Off. Anal. Chemists 79 (1), 62–72.

Mehl, P., Chao, K., Kim, M., Chen, Y., 2002. Detection of defects on selected apple cultivars using hyperspectral and multispectral image analysis. Appl. Eng. Agric. 18, 219–226.

Narvankar, D.S., Singh, C.B., Jayas, D.S., et al., 2009. Assessment of soft X-ray imaging for detection of fungal infection in wheat. Biosys Eng. 103, 49–56.

Nazário, S.L.S., Kitano, C., Higuti, H.T., Isepon, J.S., Buiochi, F., 2007. Use of ultrasound and neural networks to detect adulteration in milk samples. In: Proceedings of the International Congress on Ultrasonics, Vienna, April 9–13. http://dx.doi.org/10.3728/ICUltrasonics.2007.Vienna.1398_nazario. Paper ID 1398, Session R03: Acoustic sensors.

Neethirajan, S., Karunakaran, C., Symonsc, S., et al., 2006. Classification of vitreousness in durum wheat using soft X-rays and transmitted light images. Comput. Electronics Agric. 53, 71–78.

Ogawa, Y., Morita, K., Tanaka, S., et al., 1998. Application of X-ray CT for detection of physical foreign materials in foods. Trans. ASAE 41 (1), 157–162.

Ohmenhaeuser, M., Monakhova, Y.B., Kuballa, T., Lachenmeier, D.W., 2013. Qualitative and quantitative control of honeys using NMR spectroscopy and chemometrics. ISRN Anal. Chem. 4. http://dx.doi.org/10.1155/2013/825318.

Oppenheim, A.V., Schafer, R.W., Buck, J.R., 1999. Discrete-time Signal Processing, second ed. Prentice Hall, New Jersey.

Paiva, R.F.D., Lynch, J., Rosenberg, E., et al., 1998. A beam hardening correction for X-ray microtomography. NDT E Int. 31 (1), 17–22.

Peng, Y., Tao, F., Li, Y., Wang, W., Chen, J., Wu, J., Dhakal, S., 2010. Rapid detection of total viable count of chilled pork using hyperspectral scattering technique. Int. Soc. Opt. Photonics 7676, 76760K1–76760K8.

Qin, J.W., Chao, K.L., Kim, M.S., 2011. Evaluating carotenoid changes in tomatoes during postharvest ripening using Raman chemical imaging. Sens. Agric. Food Qual. Saf. 8027, 03–11.

Qin, J., Lu, R., 2008. Measurement of the optical properties of fruits and vegetables using spatially resolved hyperspectral diffuse reflectance imaging technique. Postharv. Biol. Technol. 49, 355–365.

Schatzki, T.F., Haff, R.P., Young, R., et al., 1997. Defect detection in apples by means of X-ray imaging. Trans. ASAE 40 (5), 1407–1415.

Shah, N.N., Rooney, P.K., Ozguler, A., Morris, S.A., Obrien, W.D., 2001. A real-time approach to detect seal defects in food packages using ultrasonic imaging. J. Food Prot. 64, 1392–1398.

Tanck, P., Kaushal, B., 2014. Adulteration detection in basmati rice when mixed with low premium rice brand. In: Proceedings of 4th SARC International Conference, 30th March-2014, Nagpur, India, pp. 53–56.

Tao, F., Peng, Y., Li, Y., Chao, K., Dhakal, S., 2012. Simultaneous determination of tenderness and *Escherichia coli* contamination of pork using hyperspectral scattering technique. Meat Sci. 90 (2012), 851–857.

Thomas, P., Kannan, A., Degwekar, V.H., et al., 1995. Non-destructive detection of seed weevil-infested mango fruits by X-ray imaging. Postharvest Bio Technol. 5 (1–2), 161–165.

Tollner, E.W., Gitaitis, R.D., Seebold, K.W., et al., 2005. Experiences with a food product X-ray inspection system for classifying onions. Trans. ASAE 21 (5), 907–912.

Wang, L., Li, Y., Hu, G.-H., 2008. Discrimination of edible vegetable oil adulterated waste cooking oil by nuclearmagnetic resonance. China Oils and Fats, 33 (10), 75–77.

Wu, D., Chen, X.J., He, Y., 2009. Application of multispectral image texture to discriminating tea categories based on DCT and LS-SVM. Spectrosc. Spect. Anal. 29, 1382–1385.

Xu, Z., Morris, R.H., Bencsik, M., Newton, M.I., 2014. Detection of virgin olive oil adulteration using low field unilateral NMR. Sensors 14, 2028–2035. http://dx.doi.org/10.3390/s140202028.

Yang, C.C., Kim, M.S., Kang, S., Tao, T., Chao, K., Lefcourt, A.M., 2011. The development of a simple multi-spectral algorithm for detection of fecal contamination on apples using a hyperspectral line-scan imaging system. Sensing Instrumentation Food Qual. Saf. 5 (1), 10–18. http://dx.doi.org/10.1007/s11694-010-9105-1.

Yoon, S., Lawrence, K., Park, B., Windham, W., 2007. Optimization of fecal detection using hyperspectral imaging and kernel density estimation. Trans. ASABE 50, 1063–1071.

Yoon, S.C., Park, B., Lawrence, K.C., Windham, W.R., Heitschmidt, G.W., 2011. Line-scan hyperspectral imaging system for real-time inspection of poultry carcasses with fecal material and ingesta. Comput. Electron. Agric. 79, 159–168.

Youssao, I., Verleyen, V., Leroy, P., 2002. Prediction of carcass lean content by real time ultrasound in Pietrain and negative stress Pietrain. Anim. Sci. 75, 25–32.

Zhao, J.W., Chen, Q.S., Cai, J.R., Ouyang, Q., 2009. Automated tea quality classification by hyperspectral imaging. Appl. Opt. 48, 3557–3564.

Zhao, J., Ouyang, Q., Chen, Q., Wang, J., 2010. Detection of bruise on pear by hyperspectral imaging sensor with different classification algorithms. Sens. Lett. 8, 570–576.

Zwiggelaar, R., Bull, C.R., Mooney, M.J., 1997. The detection of "soft" materials by selective energy X-ray transmission imaging. J. Agric. Engng Res. 66 (3), 203–212.

Index

Note: Page numbers followed by "f" and "t" indicates figures and tables respectively.

Printed in the United States
By Bookmasters